Contemporary Ring Theory 2011

Proceedings of the Sixth China–Japan–Korea
International Conference on Ring Theory

THE SIXTH CHINA–JAPAN–KOREA INTERNATIONAL CONFERENCE ON RING AND MODULE THEORY June 27–July 2, 2011

Contemporary Ring Theory 2011

Proceedings of the Sixth China–Japan–Korea
International Conference on Ring Theory

editors

Jin Yong Kim
Kyung Hee University, Korea

Chan Huh
Pusan National University, Korea

Yang Lee
Pusan National University, Korea

Tai Keun Kwak
Daejin University, Korea

 **World Scientific**

NEW JERSEY · LONDON · SINGAPORE · BEIJING · SHANGHAI · HONG KONG · TAIPEI · CHENNAI

Published by

World Scientific Publishing Co. Pte. Ltd.

5 Toh Tuck Link, Singapore 596224

USA office: 27 Warren Street, Suite 401-402, Hackensack, NJ 07601

UK office: 57 Shelton Street, Covent Garden, London WC2H 9HE

British Library Cataloguing-in-Publication Data
A catalogue record for this book is available from the British Library.

CONTEMPORARY RING THEORY 2011
Proceedings of the Sixth China–Japan–Korea International Conference on Ring Theory

Copyright © 2012 by World Scientific Publishing Co. Pte. Ltd.

ISBN-13 978-981-4397-67-4
ISBN-10 981-4397-67-9

Printed in Singapore by World Scientific Printers.

PREFACE

The Sixth China–Japan–Korea International Conference on Ring Theory was held at Kyung Hee University, Suwon, Korea during June 27–July 2, 2011. The present volume contains the texts of selected talks delivered at the conference. We would like to express our hearty thanks to all of the referees for their helpful suggestions and advices.

The place of the meeting was located at Suwon and the environment was calm enough for participants to develop active discussion and investigation. We would like to express our appreciation to all the participants who have taken time to join us in this conference. We were quite happy to realize the conference which devotes extremely to important topics. It is very delightful for us that more than 107 people (including family) from 13 countries and regions came to Suwon to attend this conference.

In 1991, this conference was started by the efforts of Prof. Shae Xue Liu, Prof. Hiroyuki Tachikawa and Prof. Manabu Harada. We believe the presentations and discussions in these conferences contributed to create collaborative works in this field of mathematics. The past conferences were held at Guilin (China), Okayama (Japan), Kyungju (Korea), Nanjing (China) and Tokyo (Japan). The main aim of this conference is to foster an exchange of investigation among ring and module theorists in this area. The number of participants for this conference has increased tremendously. The next China–Japan–Korea International Conference on Ring Theory will be held in China.

January 2012

The Main Organizers
Jin Yong Kim
Chan Huh
Yang Lee
Tai Kuen Kwak

ORGANIZING COMMITTEES

EDITORIAL BOARD

Jin Yong Kim, Kyung Hee University, Korea
Chan Huh, Pusan National University, Korea
Yang Lee, Pusan National University, Korea
Tai Keun Kwak, Daejin University, Korea

ORGANIZING COMMITTEE

- Korea

Jin Yong Kim, Kyung Hee University
Chan Huh, Pusan National University
Yang Lee, Pusan National University
Juncheol Han, Pusan National University
Tai Keun Kwak, Daejin University

- China

Jianlong Chen, Southeast University
Nanqing Ding, Nanjing University
Quanshui Wu, Fudan University
Yingbo Zhang, Beijing Normal University

- Japan

Masahisa Sato, Yamanashi University
Hideto Asashiba, Shizouka University
Yoshitomo Baba, Osaka Kyoiku University
Hiroshi Yoshimura, Yamaguchi University

ASSOCIATE ORGANIZERS

Chan Yong Hong, Kyung Hee University
Hong Kee Kim, Gyeongsang National University
Sangwon Park, Dong-A University
Sang Bok Nam, Kyung Dong University
Nam Kyun Kim, Hanbat National University

PARTICIPANTS LIST

Australia	Nazer Halimi	The University of Queensland
		`n.halimi@uq.edu.au`
Brazil	Claus Haetinger	Univates University Center
		`chaet@univates.br`
Brazil	Miguel Ferrero	UFRGS
		`mferrero@mat.ufrgs.br`
Brunei	John Kauta	University Brunei Darussalam
		`john.kauta@ubd.edu.bn`
Canada	W. Keith Nicholson	University of Calgary
		`wknichol@ucalgary.ca`
Canada	Yiqiang Zhou	Memorial Univ. of Newfoundland
		`zhou@mun.ca`
China	Jianlong Chen	Southeast University
		`jlchen@seu.edu.cn`
China	Jian Cui	Southeast University
		`seujcui@126.com`
China	Nanqing Ding	Nanjing University
		`nqding@nju.edu.cn`
China	Ruizhu Han	Southeast University
		`daodao-777@163.com`
China	Jiafeng Lu	Zhejiang Normal University
		`jiafenglv@zjnu.edu.cn`
China	Lixin Mao	Nanjing Institute of Technology
		`maolx2@hotmail.com`
China	Liang Shen	Southeast University
		`lshen@seu.edu.cn`
China	Yanhua Wang	Shanghai University of Finance and Eonomic
		`yhw@mail.shufe.edu.cn`
China	Changchang Xi	Beijing Normal University
		`xicc@bnu.edu.cn`
China	Wang Xian	China University of Mining and Technology
		`wx2008117@163.com`
China	Hong You	Soochow University
		`youhong@suda.edu.cn`

India	Mohammad Yahya Abbasi	Jamia Millia Islamia `yahya_jmi@rediffmail.com`
India	Asma Ali	Aligarh Muslim University `asma.ali345@gmail.com`
India	Shakir Ali	Aligarh Muslim University `shakir50@rediffmail.com`
India	Mohammad Ashraf	Aligarh Muslim University `mashraf80@hotmail.com`
India	Nadeem ur Rehman	Aligarh Muslim University `rehman100@gmail.com`
India	Fahad Sikander	Aligarh Muslim University `fahadsikander@gmail.com`
Iran	Mohammad Javad	Islamic Azad University `mjnematollahi@gmail.com`
Iran	Yahya Talebi	University of Mazandaran `talebi@umz.ac.ir`
Iran	Fatemeh Dehghani-Zadeh	Islamic Azad University `fdzadeh@gmail.com`
Japan	Hideto Asashiba	Shizuoka University `shasash@ipc.shizuoka.ac.jp`
Japan	Takahiko Furuya	Tokyo University of Science `furuya@ma.kagu.tus.ac.jp`
Japan	Takao Hayami	Hokkai-Gakuen University `hayami@ma.kagu.tus.ac.jp`
Japan	Yasuyuki Hirano	Naruto University of Education `yahirano@naruto-u.ac.jp`
Japan	Song Huiling	Hiroshima University `huiling1978@hotmail.com`
Japan	Hirotaka KOGA	University of Tsukuba `koga@math.tsukuba.ac.jp`
Japan	Fujio Kubo	Hiroshima University `remakubo@hiroshima-u.ac.jp`
Japan	Yosuke Kuratomi	Kitakyushu N.C.T `kuratomi@kct.ac.jp`
Japan	Mamoru Kutami	Yamaguchi University `kutami@yamaguchi-u.ac.jp`
Japan	Hidetoshi Marubayashi	Tokushima Bunri University `marubaya@kagawa.bunriu.ac.jp`
Japan	Sato Masahisa	Yamanashi University `msato@yamanashi.ac.jp`
Japan	Hiroki Miyahara	Yamanashi University `hmiyahara@yamanashi.ac.jp`

Japan	Izuru Mori	Shizuoka University
		simouri@ipc.shizuoka.ac.jp
Japan	Kazunori Nakamoto	University of Yamanashi
		nakamoto@yamanashi.ac.jp
Japan	Nakashima Norihiro	Hokkaido University
		naka_n@math.sci.hokudai.ac.jp
Japan	Kiyoichi Oshiro	Yamaguchi University
		oshiro@yamaguchi-u.ac.jp
Japan	Kazuho Ozeki	Meiji University
		kozeki@math.meiji.ac.jp
Japan	Fumiya Suenobu	Hiroshima University
		suenobu@amath.hiroshimau.ac.jp
Japan	Kenta Ueyama	Shizuoka University
		speedmathter@gmail.com
Japan	Kunio Yamagata	TUAT
		yamagata@cc.tuat.ac.jp
Japan	Takehana Yasuhiko	Hakodate National College
		takehana@hakodate-ct.ac.jp
Korea	Jonguk Baik	Kyung Hee University
		jubaik@khu.ac.kr
Korea	Wooyoung Chin	Korea Science Academy of KAIST
		ycjeon@kaist.ac.kr
Korea	Yong Uk Cho	Silla University
		yucho@silla.ac.kr
Korea	Juncheol Han	Pusan National University
		jchan@pusan.ac.kr
Korea	Chan Yong Hong	Kyung Hee University
		hcy@khu.ac.kr
Korea	Chan Huh	Pusan National University
		chuh@pusan.ac.kr
Korea	Da Woon Jung	Pusan National University
		jungdw@pusan.ac.kr
Korea	Hong Kee Kim	Gyeongsang National University
		hkkim@gnu.ac.kr
Korea	Jin Yong Kim	Kyung Hee University
		jykim@khu.ac.kr
Korea	Nam Kyun Kim	Hanbat National University
		nkkim@hanbat.ac.kr
Korea	Tai Keun Kwak	Daejin University
		tkkwak@daejin.ac.kr
Korea	Chang Ik Lee	Pusan National University
		cilee@pusan.ac.kr

Korea	Sang Cheol Lee	Chonbuk National University
		scl@chonbuk.ac.kr
Korea	Yang Lee	Pusan National University
		ylee@pusan.ac.kr
Korea	Jung Wook Lim	POSTECH
		lovemath@postech.ac.kr
Korea	Sang Bok Nam	Kyungdong University
		sbnam@k1.ac.kr
Korea	Sei-Qwon Oh	Chungnam National University
		sqoh@cnu.ac.kr
Korea	Jae Keol Park	Pusan National University
		jkpark@pusan.ac.kr
Korea	Ji Yeon Park	Kyung Hee University
		jypark0902@khu.ac.kr
Korea	Sangwon Park	Dong-A University
		swpark@donga.ac.kr
Romania	Toma Albu	Romanian Academy
		Toma.Albu@imar.ro
Slovenia	Ajda Fošner	University of Primorska
		ajda.fosner@fm-kp.si
Slovenia	Janez Ster	University of Ljubljana
		janez.ster@imfm.si
Taiwan	Pjek-Hwee Lee	National Taiwan University
		phlee@math.ntu.edu.tw
Taiwan	Tsiu-Kwen Lee	National Taiwan University
		tklee@math.ntu.edu.tw
Thailand	Sarapee Chairat	Thaksin University
		c_sarapee@hotmail.com
Thailand	Van Sanh Nguyen	Mahidol University
		nguyend2@ohio.edu
Thailand	Chitlada Somsup	Kasetsart University
		s_chitlada@yahoo.com
Turkey	Tugba Guroglu	Celal Bayar University
		tugba.guroglu@deu.edu.tr
U.S.A.	Alexander Diesl	Wellesley College
		adiesl@wellesley.edu
U.S.A.	Thomas Dorsey	CCR-La Jolla
		thomasjdorsey@gmail.com
U.S.A.	Nguyen Viet Dung	Ohio University
		nguyend2@ohio.edu
U.S.A.	Gangyong Lee	The Ohio State University
		lgy999@hanmail.net

U.S.A.	Pace Nielsen	Brigham Young University
		pace@math.byu.edu
U.S.A.	Syed Tariq Rizvi	The Ohio State University
		rizvi.1@osu.edu
U.S.A.	Cosmin Roman	The Ohio State University
		cosmin@math.ohio-state
U.S.A.	Chris Ryan	University of Louisiana
		cryan@louisiana.edu
U.S.A.	Hisaya Tsutsui	Embry-Riddle University
		Hisaya.Tsutsui@erau.edu
U.S.A.	Larry Xue	Bradley University
		lxue@bradley.edu
U.S.A.	Mohamed Yousif	The Ohio State University
		yousif.1@osu.edu

PROGRAM

Tuesday, June 28

09:00 - 09:15   Opening Ceremony Room: 211-1
                 Jin Yong Kim, Kyung Hee University, Korea

Plenary Session

09:20 - 09:50   Derived equivalences and Grothendieck constructions of
                 lax functors
                 Hideto Asashiba, Shizuoka University, Japan
09:55 - 10:25   On quasipolar rings
                 Jianlong Chen, Southeast University, China
10:30 - 11:00   Radicals of skew polynomial rings and skew Laurent
                 polynomial rings
                 Yang Lee, Pusan National University, Korea
11:10 - 11:40   Trace of symmetric bi-additive mapping in rings
                 M. Ashraf, Aligarh Muslim University, India
11:45 - 12:15   The Osofsky-Smith Theorem for modular lattices, and
                 applications
                 *Toma Albu, Simion Stoilow Institute of Mathematics of
                 the Romanian Academy, Romania*

Branch Session I

14:00 - 14:20   Some decomposition theorems for rings
                 Rekha Rani, N.R.E.C., College, India
14:25 - 14:45   On a class of hereditary crossed-product orders
                 John S. Kauta, University Brunei Darussalam, Brunei
14:50 - 15:10   On derivations in *-rings and H*-algebras
                 Shakir Ali, Aligarh Muslim University, India
15:20 - 15:40   Von Neumann regular rings with generalized almost
                 comparability
                 Mamoru Kutami, Yamaguchi University, Japan
15:45 - 16:05   On unitification problem of weakly rickart *-rings
                 B. N. Waphare, University of Pune, India

16:15 - 16:35   The McCoy condition on modules
Jian Cui*, Southeast University, China
Jianlong Chen, Southeast University, China

16:40 - 17:00   Generalized commuting idempotents in rings
Juncheol Han*, Pusan National University, Korea
Sangwon Park, Dong-A University, Korea

Branch Session II

14:00 - 14:20   A new pseudorandom number generator using an Artin-Schreier tower
Huiling Song*, Hiroshima University and Harbin Finance University, Japan
Hiroyuki Ito, Tokyo University of Science, Japan

14:25 - 14:45   On countably Σ-C2 rings
Liang Shen*, Southeast University, China
Jianlong Chen, Southeast University, China

14:50 - 15:10   A generalization of hereditary torsion theory and their dualization
Yasuhiko Takehana, Hakodate national college, Japan

15:20 - 15:40   The Galois map and its induced maps
Larry Xue*, Bradley University, U.S.A.
George Szeto, Bradley University, U.S.A.

15:45 - 16:05   The Quillen splitting lemma
Sang Cheol Lee*, Chonbuk National Univ., Korea
Yeong Moo Song, Sunchon National Univ., Korea

16:15 - 16:35   Generalization of basic and large submodules of QTAG-modules
Fahad Sikander*, Aligarh Muslim University, India
Sabah A R K Naji, Aligarh Muslim University, India

Branch Session III

14:00 - 14:20   On Lie structure of prime rings with generalized (α,β)-derivations
Claus Haetinger*, Univates University Center, Brazil
Nadeem ur Rehman, Aligarh Muslim University, India
Radwan Alomary, Aligarh Muslim University, India

14:25 - 14:45   A class of non-Hopf bi-Frobenius algebras
Wang Yanhua, Shanghai University of Finance and Economics, China

Wednesday, June 29

Plenary Session

09:00 - 09:30   Tilting modules and stratification of derived module
                categories
                *Changchang Xi**, Beijing Normal University, China
                Hongxing Chen, Beijing Normal University, China
09:35 - 10:05   McKay type correspondence for AS-regular algebras
                Izuru Mori, Shizuoka University, Japan
10:10 - 10:40   Strong lifting splits
                *W. Keith Nicholson**, University of Calgary, Canada
                M. Alkan, Akdeniz University, Turkey
                A. Cigdem Ozcan, Hacettepe University, Turkey*
10:50 - 11:20   Herstein's questions on simple rings revisited
                Pjek-Hwee Lee, National Taiwan University, Taiwan
11:25 - 11:55   Direct sum problem for Baer and Rickart modules
                S. Tariq Rizvi, The Ohio State University, U.S.A.
12:00 - 12:30   Dedekind-finite strongly clean rings
                Pace P. Nielsen, Brigham Young University, U.S.A.

Branch Session I

14:00 - 14:20   Derivations modulo elementary operators
                *Tsiu-Kwen Lee**, National Taiwan Univ., Taiwan
                Chen-Lian Chuang, National Taiwan Univ., Taiwan*
14:25 - 14:45   Simple-Baer rings and minannihilator modules
                Lixin Mao, Nanjing Institute of Technology, China
14:50 - 15:10   On fully bounded Noetherian modules
                Nguyen Van Sanh, Mahidol University, Thailand
15:20 - 15:40   Decomposition theory of modules and applications to
                quasi-Baer rings
                *Chris Ryan**, University of Louisiana at Lafayette,
                U.S.A.*
                *Gary Birkenmeier, University of Louisiana at
                Lafayette, U.S.A.*
15:45 - 16:05   On fully prime rings
                Hisaya Tsutsui, Embry-Riddle University, U.S.A.
16:15 - 16:35   Homogeneous functions on rings
                *Yasuyuki Hirano, Naruto University of Education,
                Japan*

16:40 - 17:00   A characterization of some integral domains of the form
                 $A + B[\Gamma^*]$
                 Jung Wook Lim*, POSTECH, Korea
                 Byung Gyun Kang, POSTECH, Korea

Branch Session II

14:00 - 14:20   Local derivations of a matrix algebra over a commutative ring
                 Xian Wang, China University of Mining and Technology, China

14:25 - 14:45   Relative mono-injective modules and relative mono-ojective modules
                 Yosuke Kuratomi*, Kitakyushu National College of Technology, Japan
                 Derya Keskin Tutuncu, University of Hacettepe, Turkey

14:50 - 15:10   HF-modules and isomorphic high submodules of QTAG-modules
                 Mohd. Yahya Abbasi, Jamia Millia Islamia, India

15:20 - 15:40   Rings whose left modules are direct sums of finitely generated modules
                 Nguyen Viet Dung, Ohio University, U.S.A.

15:45 - 16:05   Finiteness properties generalized local cohomology with respect to an ideal containing the irrelevant ideal
                 Fatemeh Dehghani-Zadeh, Islamic Azad University (Yazd Branch), Iran

16:15 - 16:35   On endoregular modules
                 Gangyong Lee*, The Ohio State University, U.S.A.
                 S. Tariq Rizvi, The Ohio State University, U.S.A.
                 Cosmin Roman, The Ohio State University, U.S.A.

16:40 - 17:00   Modules whose endomorphism rings are von Neumann regular
                 Cosmin Roman*, The Ohio State University, U.S.A.
                 Gangyong Lee, The Ohio State University, U.S.A.
                 S. Tariq Rizvi, The Ohio State University, U.S.A.

Branch Session III

14:00 - 14:20   On n-commuting and n-skew-commuting maps with generalized derivations in rings
                 Nadeem ur Rehman, Aligarh Muslim Univ., India

14:25 - 14:45   Some results on s.g. near-rings and $\langle R, S \rangle$-groups
                 Yong Uk Cho, Silla University, Korea

14:50 - 15:10   On decomposition theorems for near rings
                M. Shadab Khan, Aligarh Muslim Univ., India
15:20 - 15:40   On mutation of tilting modules over noetherian algebras
                Hirotaka Koga, University of Tsukuba, Japan
15:45 - 16:05   Modules of differential operators of a generic hyperplane
                arrangement
                Norihiro Nakashima*, Hokkaido University, Japan
                Go Okuyama, Hokkaido Institute of Technology, Japan
                Mutsumi Saito, Hokkaido University, Japan
16:15 - 16:35   Topics on the moduli of representations of degree 2
                Kazunori Nakamoto, University of Yamanashi, Japan

Friday, July 1

Plenary Session

09:00 - 09:30   On the Faith conjecture
                Kiyoichi Oshiro, Yamaguchi University, Japan
09:35 - 10:05   The minimal prime spectrum of rings with annihilator
                conditions and property (A)
                Chan Yong Hong, Kyung Hee University, Korea
10:10 - 10:40   On Gorenstein modules
                Nanqing Ding, Nanjing University, China
10:50 - 11:20   Partial actions of groups on semiprime rings
                *Miguel Ferrero, Universidade Federal do Rio Grande de
                Sul, Brazil*
11:25 - 11:55   Recent developments on projective and injective
                modules
                Mohamed Yousif, The Ohio State University, U.S.A.

Branch Session I

13:30 - 13:50   A class of clean rings
                *Yiqiang Zhou, Memorial University of Newfoundland,
                Canada*
13:55 - 14:15   Strongly clean matrix rings
                Thomas Dorsey*, CCR-La Jolla, U.S.A.
                Alexander Diesl, Wellesley College, U.S.A.
14:20 - 14:40   Star operation on orders in simple artinian rings
                Nazer Halimi, The University of Queensland, Australia

14:50 - 15:10   Generalized derivations on prime rings
                 Muzibur Rahman Mozumder*, National Taiwan University, Taiwan
                 Tsiu-Kwen Lee, National Taiwan University, Taiwan
15:15 - 15:35   Some generalization of IFP rings and McCoy rings
                 Chang Ik Lee*, Pusan National University, Korea
                 Yang Lee, Pusan National University, Korea
15:40 - 16:00   Some results and new questions about clean rings
                 Alexander Diesl, Wellesley College, U.S.A.
16:10 - 16:30   Nil-Armendariz rings and upper nilradicals
                 Da Woon Jung*, Pusan National University, Korea
                 Yang Lee, Pusan National University, Korea
                 Sung Pil Yang, Pusan National University, Korea
                 Nam Kyun Kim, Hanbat National University, Korea
16:35 - 16:55   Insertion-of-factors-property on nilpotent elements
                 Wooyoung Chin*, Korea Science Academy of KAIST, Korea
                 Jineon Baek, Korea Science Academy of KAIST, Korea
                 Jiwoong Choi, Korea Science Academy of KAIST, Korea
                 Taehyun Eom, Korea Science Academy of KAIST, Korea
                 Young Cheol Jeon, Korea Science Academy of KAIST, Korea

Branch Session II

13:30 - 13:50   On rings over which the injective hull of each cyclic module is Sigma-extending
                 Sarapee Chairat*, Thaksin University, Thailand
                 Chitlada Somsup, Thaksin University, Thailand
                 Maliwan Tunapan, Thaksin University, Thailand
                 Dinh Van Huynh, Ohio University, U.S.A.
13:55 - 14:15   Structure of augmentation quotients for integral group rings
                 Hong You*, Soochow University, China
                 Qingxia Zhou, Harbin Institute of Technology, China
14:20 - 14:40   Hilbert coefficients of parameter ideals
                 Kazuho Ozeki, Meiji University, Japan
14:50 - 15:10   Derivations on near rings
                 Asma Ali, Aligarh Muslim University, India

15:15 - 15:35 H_δ-supplemented modules
                  Mohammad Javad Nematollahi, Islamic Azad University (Arsanjan Branch), Iran
15:40 - 16:00   Modules whose non-cosingular submodules are direct summand
                  Yahya Talebi*, University of Mazandaran, Iran
                  M. Hosseinpour, University of Mazandaran, Iran
                  A. R. Moniri Hamzekolaei, University of Mazandaran, Iran
16:10 - 16:30   A note on variation of supplemented modules
                  Tugba Guroglu*, Celal Bayar University, Turkey
                  Gokhan Bilhan, Dokuz Eylul University, Turkey

Branch Session III

13:30 - 13:50   Some results on AS-Gorenstein algebras
                  Kenta Ueyama, Shizuoka University, Japan
13:55 - 14:15   Hochschild cohomology ring of the integral group ring of the semidihedral group
                  Takao Hayami, Hokkai-Gakuen University, Japan
14:20 - 14:40   Study on the algebraic structures in terms of geometry and deformation theory
                  Fumiya Suenobu*, Hiroshima University, Japan
                  Fujio Kubo, Hiroshima University, Japan
14:50 - 15:10   Support varieties for modules over stacked monomial algebras
                  Takahiko Furuya*, Tokyo University of Science, Japan
                  Nicole Snashall, University of Leicester, United Kingdom
15:15 - 15:35   Maps preserving matrix pairs with zero Lie or Jordan product
                  Ajda Fošner, University of Primorska, Slovenia
15:40 - 16:00   Introduction to piecewise-Koszul algebras
                  Jia-Feng Lu, Zhejiang Normal University, China
16:10 - 16:30   On near modules over skew polynomials
                  Ebrahim Hashemi, Shahrood University of Technology, Iran

CONTENTS

Proceedings of the Sixth China-Japan-Korea
International Conference on Ring Theory
June 27-July 2, 2011 Suwon, Korea

RINGS OVER WHICH POLYNOMIAL RINGS ARE NI

JUNCHEOL HAN, YANG LEE, AND SUNG PIL YANG*

ABSTRACT. Marks called a ring *NI* if the upper nilradical contains
all nilpotent elements. Smoktunowicz constructed an NI ring over
which the polynomial ring is not NI. We study some conditions
under which polynomial rings are NI. In the process we call a ring
R *polynomial-NI* if $R[Y]$ is NI for any finite set Y of commuting
indeterminates over R. We show that R is polynomial-NI if and
only if $R[X]$ is polynomial-NI for any (possibly infinite) set X
of commuting indeterminates over R. It is also shown that R is
polynomial-NI if and only if $R[x]$ is NI when R is a ring such that
every finite subset of R generates a subring of bounded index of
nilpotency, where x is an indeterminate over R.

1. Ring Theory

Throughout every ring is associative with identity unless otherwise
stated. Let R be a ring and X be a (possibly infinite) set of commuting
indeterminates over R. The polynomial ring over R with X is denoted
by $R[X]$, and if $X = \{x\}$ then we write $R[x]$ in place of $R[\{x\}]$. The n
by n full (resp. upper triangular) matrix ring over a ring R is denoted
by $Mat_n(R)$ (resp. $U_n(R)$), and e_{ij} denotes the n by n matrix with
(i,j)-entry 1 and zero elsewhere.

A subset S of a ring R is called *locally nilpotent* if for any finite subset
T of S there exists a positive integer n such that any product of n
elements from T is zero. Equivalently, S is locally nilpotent if and only
if any subring without identity generated by a finite number of elements
in S in nilpotent. Given a ring R, $N_*(R)$, L-rad(R), $N^*(R)$, and $N(R)$
denote the prime radical, the Levitzki radical (i.e., the sum of all locally
nilpotent ideals), the upper nilradical (i.e., sum of nil ideals), and the set
of all nilpotent elements in R, respectively. Note $N_*(R) \subseteq$ L-rad$(R) \subseteq$
$N^*(R) \subseteq N(R)$.

A prime ideal P of a ring R is called *completely prime* if R/P is a
domain. According to Kim et al. [8], a ring is called *nil-semisimple* if

---

2010 Mathematics Subject Classification : 16D25, 16S36.
Keywords : Polynomial-NI ring, NI ring, nilradical, polynomial ring.
*Corresponding author.

it has no nonzero nil ideals. Nil-semisimple rings are clearly semiprime, but semiprime rings need not be nil-semisimple as can be seen by [7, Example 1.2 and Proposition 1.3]. Due to Rowen [13, Definition 2.6.5], an ideal P of a ring R is called *strongly prime* if P is prime and R/P is nil-semisimple. Maximal ideals and completely prime ideals are clearly strongly prime. Nil-semisimple rings need not be prime as can be seen by direct products of reduced rings; and prime rings also need not be nil-semisimple as can be seen by [7, Example 1.2 and Proposition 1.3]. Note that any strongly prime ideal contains a minimal strongly prime ideal. $N^*(R)$ of a ring R is the unique maximal nil ideal of R by [13, Proposition 2.6.2], and moreover $N_*(R)$ is the intersection of all minimal strongly prime ideals of R by [13, Proposition 2.6.7].

A ring R is called *reduced* if $N(R) = 0$. Due to Marks [11], a ring R is called *NI* if $N^*(R) = N(R)$. Reduced rings are clearly NI, but $Mat_n(A)$ $(n \geq 2)$ cannot be NI for any ring A as we see in the computation that $e_{12}, e_{21} \in N(Mat_n(A))$ and $e_{12} + e_{21} \notin N(Mat_n(A))$. Note that R is NI if and only if $N(R)$ forms an ideal if and only if $R/N^*(R)$ is reduced. Hong et al. [5, Corollary 13] showed that a ring R is NI if and only if every minimal strongly prime ideal of R is completely prime. Following to Birkenmeier et al. [2], a ring R is called *2-primal* if $N_*(R) = N(R)$. Note that R is 2-primal if and only if $R/N_*(R)$ is reduced. Shin [15, Proposition 1.11] showed that a ring R is 2-primal if and only if every minimal prime ideal of R is completely prime.

It was shown in [6, Lemma 1.1] that a ring R is 2-primal if and only if every proper subring (possibly without identity) of R is 2-primal and $N(R) = N^*(R)$. Based on this equivalence, a ring is called *locally 2-primal* if each finite subset generates a 2-primal ring in [6]. 2-primal rings are locally 2-primal by [6, Lemma 1.1], but not conversely by [6, Theorem 1.2]. Locally 2-primal rings are NI by [6, Proposition 1.4]. Local 2-primal property can go up to polynomial rings by [6, Theorem 3.2], but the polynomial ring extension does not preserve NI property as we see in [14, Corollary 13] and [7, Proposition 4.2]. If $R[x]$ is NI then R is NI by [7, Proposition 2.4(2)]. Thus we examine some conditions, between local 2-primal and NI properties, over which the polynomial ring can preserve NI property. The class of NI rings is closed under subrings and direct sums by [7, Proposition 2.4(2)]. We will use this fact freely in this note.

A ring R will be called *polynomial-NI* if $R[Y]$ is NI for any finite set Y of commuting indeterminates over R. Locally 2-primal rings are polynomial-NI by [6, Theorem 3.2]. Any polynomial-NI ring is NI by [7, Proposition 2.4(2)]. However these implications are irreversible by the following examples.

We use $\oplus$ to denote the direct sum. Let R be an algebra (with or without identity) over a commutative ring S. Due to Dorroh [3], the *Dorroh extension* of R by S is the Abelian group $R \oplus S$ with multiplication given by $(r_1, s_1)(r_2, s_2) = (r_1 r_2 + s_1 r_2 + s_2 r_1, s_1 s_2)$ for $r_i \in R$ and $s_i \in S$.

Example 1.1. (1) Let k be a field, $\mathbb{Z}$ be the ring of integers and $\{t_i \mid i \in \mathbb{Z}\}$ be commuting indeterminates over k. According to Ram [12, Example 3.2], set

$$A = k[\{t_i\}_{i \in \mathbb{Z}}]/(\{t_{n_1} t_{n_2} t_{n_3} \mid n_3 - n_2 = n_2 - n_1 > 0\}) \text{ and } R = A[x; \sigma],$$

the skew polynomial ring with one indeterminate x over A, where σ is the k-automorphism of A satisfying $\sigma(t_i) = t_{i+1}$ for all $i \in \mathbb{Z}$. Then $\sum_{i \in \mathbb{Z}} t_i x R = \text{L-rad}(R) = N^*(R)$ by Marks [11, Example 2.2]. Set $I = \sum_{i \in \mathbb{Z}} t_i x R$, then $R/I \cong A \oplus kx \oplus kx^2 \oplus \cdots$ is reduced by the computation in [11, Example 2.2]. Note $R[X]/I[X] \cong (R/I)[X]$, and so $R[X]/I[X]$ is reduced (hence NI). Next $I = \text{L-rad}(R)$ implies that I is locally 2-primal by [6, Proposition 2.2] as a ring without identity, and moreover $I[X]$ is locally 2-priaml (hence NI) by [6, Theorem 3.2]. Now $R[X]$ is NI by [7, Proposition 2.4(1)], concluding that R is polynomial-NI. However R is not locally 2-primal by [6, Example 1.5].

(2) The construction of the following ring R is due to Smoktunowicz [14]. Let $\mathbb{Q}$ be the field of all rational numbers and $\bar{A}$ be the algebra of polynomials with zero constant terms in non-commuting indeterminates x, y, z over $\mathbb{Q}$. Then $\bar{A}$ can be denumerated, say $\bar{A} = \{f_1, f_2, \ldots\}$. Let I be the ideal of $\bar{A}$ generated by $\{f_i^{10m_i+1} \mid i = 1, 2, \ldots\}$ and $R = \bar{A}/I$. Then R is a nil ring. Next consider the Dorroh extension of R by $\mathbb{Q}$, say D. Then D is NI by [7, Proposition 4.2] since R is nil (so NI), but the polynomial ring over D is not NI by [14, Theorem 12]. Thus D is not polynomial-NI.

From Example 1.1(2), we have $N(D[x]) \subsetneq N_*(D)[x]$ since $\dfrac{D}{N_*(D)}[x] \cong \dfrac{D[x]}{N_*(D)[x]}$ and $D/N_*(D)$ is a reduced ring. We will see useful conditions equivalent to the equality of $N(D[x]) = N_*(D)[x]$. The degree of a polynomial f is denoted by $\deg f$.

Theorem 1.2. *Let R be a ring and X be any set of commuting indeterminates over R. Then the following conditions are equivalent:*

(1) *R is polynomial-NI;*

(2) *$N^*(R[Y]) = N(R[Y]) = N^*(R)[Y] = N(R)[Y]$ for any finite set Y of commuting indeterminates over R;*

(3) *$N^*(R[X]) = N(R[X]) = N^*(R)[X] = N(R)[X]$;*

(4) *Every minimal strongly prime ideal of $R[X]$ is completely prime;*
(5) *$R[X]/N^*(R[X])$ is a subdirect product of domains;*
(6) *$R[X]/N^*(R[X])$ is reduced;*
(7) *$R[X]$ is NI.*

Proof. (1)$\Rightarrow$(2): Let R be polynomial-NI. Then $N(R[Y]) = N^*(R[Y])$ for any finite set Y of commuting indeterminates over R. Since $N(R) \subseteq N(R[Y])$ and $N(R[Y])$ is an ideal of $R[Y]$, $N(R)$ is also an ideal of R and so $N(R) = N^*(R)$. Then $R/N^*(R)$ is reduced, and hence we have $N(R[Y]) \subseteq N^*(R)[Y]$ from $\dfrac{R[Y]}{N^*(R)[Y]} \cong \dfrac{R}{N^*(R)}[Y]$. But since $N(R[Y])$ is an ideal of $R[Y]$ and $N^*(R) \subseteq N(R[Y])$, we also get $N^*(R)[Y] \subseteq N(R[Y])$, entailing $N^*(R[Y]) = N(R[Y]) = N^*(R)[Y] = N(R)[Y]$.

(2)$\Rightarrow$(3): Suppose that $N^*(R[Y]) = N(R[Y]) = N^*(R)[Y] = N(R)[Y]$ for any finite set Y of commuting indeterminates over R. Then R is NI by the argument in the proof of (1)$\Rightarrow$(2). Let X be a nonempty set (possibly infinite) of commuting indeterminates over R. We first obtain $N(R[X]) \subseteq N^*(R)[X]$ from the fact that $\dfrac{R[X]}{N^*(R)[X]} \cong \dfrac{R}{N^*(R)}[X]$ is reduced. Next we will show $N(R[X]) \supseteq N^*(R)[X]$. Let $f(X) = \sum_{i=1}^{n} a_i P_i \in N^*(R)[X]$, where $a_i \in N^*(R)$ for all i and each P_i is a finite product of indeterminates in X. Then there exists a finite subset X_0 of X such that $f(X) \in N^*(R)[X_0]$. By the condition (2), we get $N^*(R[X_0]) = N(R[X_0])$ and so $f(X)$ is nilpotent. This yields $N(R[X]) \supseteq N^*(R)[X]$, so we now have $N(R[X]) = N^*(R)[X] = N(R)[X]$. But $N^*(R)[X]$ is an ideal of $R[X]$, so this entails $N(R[X]) = N^*(R[X])$. Thus $N^*(R[X]) = N(R[X]) = N^*(R)[X] = N(R)[X]$.

(3)$\Rightarrow$(4) is proved by [5, Corollary 13] since $R[X]$ is NI under the condition (3).

(4)$\Rightarrow$(5), (5)$\Rightarrow$(6), (6)$\Rightarrow$(7), and (7)$\Rightarrow$(1) are obvious. $\square$

Corollary 1.3. *A ring R is polynomial-NI if and only if so is $R[X]$ for any set X of commuting indeterminates over R.*

Proof. Let R be a polynomial-NI ring. Consider the polynomial ring $R[X]$. Next consider the polynomial ring $R[X][X_1]$ for given any set X_1 of commuting indeterminates over $R[X]$. Then $R[X_2]$ is the polynomial ring with the set X_2 of commuting indeterminates over R, where $X_2 = X \cup X_1$. Note $R[X_2] = R[X][X_1]$. Since R is polynomial-NI, $R[X_2]$ is NI by Theorem 1.2. This implies that $R[X]$ is polynomial-NI by Theorem 1.2. The converse is obvious. $\square$

From Theorem 1.2 and Corollary 1.3, we can obtain that a ring R is polynomial-NI if and only if $R[X]$ is polynomial-NI if and only if $R[X]$ is NI.

Considering Theorem 1.2, it is naturally asked whether a ring R is polynomial-NI when $R[x]$ is NI, i.e., the set X of indeterminates is a singleton. We do not know the answer but will find a condition under which the answer is affirmative.

The *index of nilpotency* of a nilpotent element x in a ring R is the least positive integer n such that $x^n = 0$. The *index of nilpotency* of a subset I of R is the supremum of the indices of nilpotency of all nilpotent elements in I. If such a supremum is finite, then I is said to be *of bounded index of nilpotency*.

Proposition 1.4. *Let R be a ring of bounded index of nilpotency. Then the following conditions are equivalent:*

(1) *R is NI;*

(2) *R is polynomial-NI;*

(3) *R is locally 2-primal;*

(4) *R is 2-primal;*

(5) *$R[X]$ is polynomial-NI for any set X of commuting indeterminates over R;*

(6) *$R[X]$ is NI for any set X of commuting indeterminates over R;*

(7) *$R[x]$ is polynomial-NI with x an indeterminate over R;*

(8) *$R[x]$ is NI with x an indeterminate over R.*

Proof. The equivalences of the conditions (1), (2), (3), and (4) are obtained from [7, Proposition 1.4], noting that 2-primal rings are locally 2-primal and locally 2-primal rings are polynomial-NI. Moreover Birkenmeier et al. showed that polynomial rings over 2-primal rings are also 2-primal in [2, Proposition 2.6], so we get (4)⇒(5). (5)⇒(6), (5)⇒(7), (7)⇒(8), (6)⇒(1), and (8)⇒(1) are obvious. □

Corollary 1.5. *Let R be a left or right Goldie ring. Then R is NI if and only if R is polynomial-NI if and only if R is locally 2-primal if and only if R is 2-primal if and only if $R[X]$ is NI if and only if $R[x]$ is NI.*

Proof. If R is left or right Goldie, then $N(R) = N^*(R)$ is nilpotent by [10] and so we get the equivalences from Proposition 1.4. □

Let T be a 2-primal ring of bounded index of nilpotency, $n \geq 1$, and $R_n = U_{2^n}(T)$. Define a map $\sigma : R_n \to R_{n+1}$ by $A \mapsto \begin{pmatrix} A & 0 \\ 0 & A \end{pmatrix}$, then R_n can be considered as a subring of R_{n+1} via σ (i.e., $A = \sigma(A)$ for $A \in R_n$). Let R be the direct limit of the direct system (R_n, σ_{ij}), where $\sigma_{ij} = \sigma^{j-i}$. Then R is a locally 2-primal ring (hence polynomial-NI) but not 2-primal by [6, Example 1.2]. Analyzing this ring R, we can become aware fact that every finite subset of R generates a subring of bounded index of nilpotency. This condition can be found in many kinds of ordinary ring theoretic studies.

Theorem 1.6. *Let R be a ring such that every finite subset of R generates a subring of bounded index of nilpotency. Then R is polynomial-NI if and only if $R[x]$ is NI.*

Proof. It suffices to show the sufficiency by the definition. Suppose that $R[x]$ is NI. Apply the proof of $(1) \Rightarrow (2)$ in Theorem 1.2 to obtain $N^*(R[x]) = N(R[x]) = N^*(R)[x] = N(R)[x]$, using $R[x]$ in place of $R[Y]$.

We first compute the case of $R[x, y]$ for $x, y \in X$. Note that $N(R[x][y]) \subseteq N^*(R)[x][y]$ because $\frac{R[x]}{N^*(R)[x]}[y] \cong \frac{R[x][y]}{N^*(R)[x][y]}$ is reduced. For the proof of the converse inclusion, we apply the proof of [1, Theorem 2]. Let $f(y) = f_0 + f_1 y + \cdots + f_n y^n \in N^*(R)[x][y]$, i.e., $f_i \in N^*(R)[x]$ for all i. Here let C be the set of all coefficients of all f_i's, and S be the subring of R generated by C. Then $S \subseteq N^*(R)$ clearly, and S is of bounded index of nilpotency by hypothesis. This implies that $S[x]$ is nil of bounded index of nilpotency by [9, Theorem 9]. Say that l is the upper bound of indices of $S[x]$.

Set m be the maximum in $\{deg\, f_0, deg\, f_1, \ldots, deg\, f_n\}$, where the degree is considered as a polynomial in x and the degree of the zero polynomial is taken to be 0. Next let

$$k = nml \text{ and define } f(x^k) = f_0 + f_1 x^k + \cdots + f_n x^{nk}.$$

Then $f(x^k)$ is contained in $N^*(R)[x]$ and the structure of coefficients of $f(x^k)$ is equal to one of $f(y)$. From $N(R[x]) = N^*(R)[x]$, $f(x^k)$ is nilpotent and so $f(x^k)^\ell = 0$. Note that every coefficient in the expansion of $f(x^k)^\ell$ is equal to one of $f(y)^\ell$, by the construction of k. This yields $f(y)^l = 0$ and $N(R[x][y]) \supseteq N^*(R)[x][y]$, entailing $N(R[x][y]) = N^*(R)[x][y]$. This also implies $N^*(R[x][y]) = N(R[x][y]) = N^*(R)[x][y] = N(R)[x][y]$.

Proceeding in this manner, we inductively obtain that $N^*(R[Y]) = N(R[Y]) = N^*(R)[Y] = N(R)[Y]$ for any finite set Y of commuting indeterminates over R. Thus R is polynomial-NI. $\square$

Corollary 1.7. *Let R be a ring such that every finite subset of R generates a subring of bounded index of nilpotency. Then the following conditions are equivalent:*

(1) R is polynomial-NI;

(2) $R[X]$ is polynomial-NI for any set X of commuting indeterminates over R;

(3) $R[x]$ is polynomial-NI with x an indeterminate over R;

(4) $R[X]$ *is NI for any set X of commuting indeterminates over R;*
(5) $R[x]$ *is NI.*

Proof. By Corollary 1.3 and Theorem 1.6. □

In fact we do not know whether the preceding result holds without the condition that every finite subset generates a subring of bounded index of nilpotency.

Question. Let R be a ring such that $R[x]$ is NI. Then is R polynomial-NI?

2. Basic Properties of Polynomial-NI Rings

In this section we study basic properties and examples of polynomial-NI rings.

Lemma 2.1. (1) [7, Proposition 2.4(1)] *Let R be a ring and I be a proper ideal of R. If R/I and I are both NI then so is R.*
(2) [7, Proposition 2.4(2)] *The class of NI rings is closed under subrings and direct sums.*

In the following we can obtain similar results for polynomial-NI rings.

Proposition 2.2. (1) *Let R be a ring and I be a proper ideal of R. If R/I and I are both polynomial-NI then so is R, where I is considered as a ring without identity.*
(2) *The class of polynomial-NI rings is closed under subrings and direct sums, where subrings and direct sums are considered as rings possibly without identity.*

Proof. (1) Suppose that R/I and I are both polynomial-NI. Then $\dfrac{R[X]}{I[X]}$ $\cong \dfrac{R}{I}[X]$ and $I[X]$ are both NI by Theorem 1.2; hence $R[X]$ is NI by Lemma 2.1(1), proving that R is polynomial-NI.
(2) Let R be a polynomial-NI ring and S be a subring of R. Then $R[x]$ is NI and so $S[x]$ is also NI by Lemma 2.1(2), concluding that S is polynomial-NI.

Next suppose that R_i is a polynomial-NI ring for each i in a nonempty index set I, and let D be the direct sum of R_i's. Note $D[X] \cong \oplus_{i \in I} R_i[x]$. Since each $R_i[X]$ is NI, $\oplus_{i \in I} R_i[x]$ is also NI by (2) and so $D[X]$ is NI, concluding that D is polynomial-NI. □

As the converse of Proposition 2.2(1), it is also natural to ask whether factor rings of polynomial-NI rings are polynomial-NI. But the answer is negative by the following.

Example 2.3. Let R be the ring of quaternions with integer coefficients. Then R is a domain, so polynomial-NI. However, for any odd prime integer q, the ring R/qR is isomorphic to the 2 by 2 matrix ring over the field $\mathbb{Z}_q$ of integers modulo q, by the argument in [4, Exercise 2A]. Thus R/qR cannot be polynomial-NI.

Next we study the form of a kind of basic finite polynomial-NI ring. In this note we characterize the minimal polynomial-NI rings, where the "minimal" means the smallest cardinality. We denote the Galois field of order p^n by $GF(p^n)$.

Proposition 2.4. *If R is a minimal noncommutative polynomial-NI ring, then R is isomorphic to $U_2(GF(2))$.*

Proof. Let R be a minimal noncommutative finite ring. Then R is polynomial-NI if and only if locally 2-primal by Proposition 1.4. Thus R is isomorphic to $U_2(GF(2))$ by [6, Theorem 1.7]. $\square$

NI rings, polynomial-NI rings, and 2-primal rings are equal by Proposition 1.4 when they are minimal noncommutative rings.

$L_n(R)$ denotes the n by n lower triangular matrix ring over given a ring R.

Proposition 2.5. *Given a ring R the following conditions are equivalent:*

(1) *R is polynomial-NI;*
(2) *$U_n(R)$ is polynomial-NI for $n \geq 2$;*
(3) *$L_n(R)$ is polynomial-NI for $n \geq 2$.*

Proof. First note $U_n(R)[Y] \cong U_n(R[Y])$ for any finite set Y of commuting indeterminates over R. Let

$$I = \{m \in U_n(R) \mid \text{the diagonal entries of } m \text{ are all zero}\}.$$

Then I is a nilpotent ideal of $U_n(R)$, so I is polynomial-NI. Moreover $U_n(R)/I \cong \oplus_{i=1}^n R_i$ is polynomial-NI by Proposition 2.2(2), where $R_i = R$ for all i. Thus $U_n(R)$ is also polynomial-NI by Proposition 2.2(1). (2)$\Rightarrow$(1) is obtained by Proposition 2.2(2). The proof of (2)$\Leftrightarrow$(1) is similar. $\square$

Recall that $Mat_n(A)$ ($n \geq 2$) cannot be (polynomial-) NI for any ring A.

Acknowledgments. The authors thank the referee for very careful reading of the manuscript and valuable suggestions that improved the paper by much. This work was supported by Basic Science Research Program through the National Research Foundation of Korea

(NRF) funded by the Ministry of Education, Science and Technology (No. 20110004745).

References

[1] D.D. Anderson, V. Camillo, *Armendariz rings and Gaussian rings*, Comm. Algebra **26** (1998), 2265-2272.

[2] G.F. Birkenmeier, H.E. Heatherly and E.K. Lee, *Completely prime ideals and associated radicals*, Proc. Biennial Ohio State-Denison Conference 1992, edited by S.K. Jain and S.T. Rizvi, World Scientific, New Jersey (1993), 102-129.

[3] J.L. Dorroh, *Concerning adjunctins to algebras*, Bull. Amer. Math. Soc. **38** (1932), 85-88.

[4] K.R. Goodearl, R.B. Warfield, JR. *An Introduction to Noncommutative Noetherian Rings*, Cambridge University Press, 1989.

[5] C.Y. Hong, T.K. Kwak, *On minimal strongly prime ideals*, Comm. Algebra **28** (2000), 4867-4878.

[6] C.Y. Hong, H.K. Kim, N.K. Kim, T.K. Kwak, Y. Lee, K.S. Park, *Rings whose nilpotent elements form a Levitzki radical ring*, Comm. Algebra **35** (2007), 1379-1390.

[7] S.U. Hwang, Y.C. Jeon and Y. Lee, *Structure and topological conditions of NI rings*, J. Algebra **302** (2006), 186-199.

[8] N.K. Kim, Y. Lee, S.J. Ryu, *An ascending condition on Wedderburn radicals*, Comm. Algebra **34** (2006), 37-50.

[9] A.A. Klein, *Rings of bounded index*, Comm. Algebra **12** (1984), 9-21.

[10] C. Lanski, *Nil subrings of Goldie rings are nilpotent*, Canad. J. Math. **21** (1969), 904-907.

[11] G. Marks, *On 2-primal Ore extensions*, Comm. Algebra **29** (2001), 2113-2123.

[12] J. Ram, *On the semisimplicity of skew polynomial rings*, Proc. Amer. Math. Soc. **90** (1984), 347-351.

[13] L.H. Rowen, *Ring Theory*, Academic Press, Inc., San Diego, 1991.

[14] A. Smoktunowicz, *Polynomial rings over nil rings need not be nil*, J. Algebra **233** (2000), 427-436.

[15] G.Y. Shin, *Prime ideals and sheaf representation of a pseudo symmetric ring*, Trans. Amer. Math. Soc. **184** (1973), 43-60.

DEPARTMENT OF MATHEMATICS EDUCATION
PUSAN NATIONAL UNIVERSITY
PUSAN 609-735, KOREA
E-mail address: jchan@pusan.ac.kr

DEPARTMENT OF MATHEMATICS EDUCATION
PUSAN NATIONAL UNIVERSITY
PUSAN 609-735, KOREA
E-mail address: ylee@pusan.ac.kr

DEPARTMENT OF MATHEMATICS EDUCATION
PUSAN NATIONAL UNIVERSITY
PUSAN 609-735, KOREA
E-mail address: aeroyang@hanmail.net

Proceedings of the Sixth China-Japan-Korea
International Conference on Ring Theory
June 27-July 2, 2011 Suwon, Korea

THE GALOIS MAP AND ITS INDUCED MAPS

GEORGE SZETO AND LIANYONG XUE*

ABSTRACT. Let B be a Galois extension of B^G with Galois group G such that B^G is a separable C^G-algebra where C is the center of B, $J_g = \{b \in B \mid bx = g(x)b \text{ for each } x \in B\}$ for $g \in G$, $\alpha : H \longrightarrow B^H$ the Galois map for a subgroup H of G, $\beta : H \longrightarrow \alpha(H)C$, and $\gamma : H \longrightarrow V_B(\alpha(H))$ where $V_B(\alpha(H))$ is the commutator subring of $\alpha(H)$ in B. Relations between α, β, and γ are obtained, and several conditions are given for a one-to-one Galois map α.

1. Introduction

Let F be a field Galois extension of F^G with Galois group G. The fundamental theorem states that the Galois map $\alpha : H \longrightarrow F^H$ is a one-to-one correspondence between the set of subgroups of G and the set of separable subfields of F over F^G. This was generalized to an indecomposable commutative ring Galois extension ([1]) and a partial Galois extension ([4]). Moreover, Galois algebras satisfying the fundamental theorem were classified ([9]). Let B be a noncommutative ring Galois extension of B^G with Galois group G, C is the center of B, $J_g = \{b \in B \mid bx = g(x)b$ for each $x \in B\}$ for $g \in G$. It was shown that if $J_g \neq \{0\}$ for each $g \in G$, then the Galois map $\alpha : H \longrightarrow B^H$ is one-to-one but not necessarily onto ([10]). Let B be a Galois extension of B^G with Galois group G. The condition that $J_g \neq \{0\}$ for each $g \in G$ is satisfied by all Hirata separable Galois extensions with Galois group G, central Galois algebras with Galois group G, and Galois extensions with an inner Galois group G ([10]). On the other hand, for all commutative ring Galois extensions with Galois group G, $J_g = \{0\}$ for each non-identity $g \in G$. In general, there are Galois extensions with Galois group G such that $J_h \neq \{0\}$ and $J_l = \{0\}$ for some non-identities $h, l \in G$. We denote $S_G = \{g \in G \mid J_g \neq \{0\}\}$ and $T_G = \{g \in G \mid J_g = \{0\}\}$. Then $G = S_G \cup T_G$. Since any subgroup H of G is also a Galois group of B over B^H, $H = S_H \cup T_H$. Let $\overline{H} = \{L \mid L$ is a subgroup of G and $S_L = S_H\}$. We define two maps

2010 Mathematics Subject Classification : 13B05.

Keywords : Galois extensions, Galois maps, separable extensions, Azumaya algebras.

*Corresponding author.

induced by α: (1) $\beta : H \longrightarrow \alpha(H)C$ and (2) $\gamma : H \longrightarrow V_B(\alpha(H))$ where $V_B(\alpha(H))$ is the commutator subring of $\alpha(H)$ in B. In section 3, we shall show that the maps, induced by β and γ, $\overline{\beta} : \overline{H} \longrightarrow \alpha(H)C$ and $\overline{\gamma} : \overline{H} \longrightarrow V_B(\alpha(H))$ are one-to-one for a Galois extension B of B^G with Galois group G such that B^G is a separable C^G-algebra where C is the center of B. In section 4, we shall show some relations between α, β, and γ, and give some conditions for a one-to-one Galois map α. This generalizes the results in [10].

2. Basic Definitions and Notations

Let B be a ring with 1, G a finite automorphism group of B, B^G the set of elements in B fixed under each element in G, and A a subring of B with the same identity 1. We call B a Galois extension of B^G with Galois group G if there exist elements $\{a_i, b_i$ in B, $i = 1, 2, ..., m\}$ for some integer m such that $\sum_{i=1}^m a_i g(b_i) = \delta_{1,g}$ for each $g \in G$ ([2]). A ring B is called a Galois algebra over R if B is a Galois extension of R such that R is contained in the center C of B, and B is called a central Galois algebra if B is a Galois extension of its center C ([8]). We call B a separable extension of A if there exist $\{a_i, b_i$ in B, $i = 1, 2, ..., m\}$ for some integer m such that $\sum a_i b_i = 1$, and $\sum ba_i \otimes b_i = \sum a_i \otimes b_i b$ for all b in B where $\otimes$ is over A ([3]). In particular, B is called an Azumaya algebra if it is a separable extension over its center. A ring B is called a Hirata separable extension of A if $B \otimes_A B$ is isomorphic to a direct summand of a finite direct sum of B as a B-bimodule ([7]). We call B is a Hirata separable Galois extension of B^G with Galois group G if it is a Hirata separable and Galois extension of B^G with Galois group G ([7]).

Throughout this paper, B is a Galois extension of B^G with Galois group G, $\alpha : H \longrightarrow B^H$ the Galois map for a subgroup H of G, C is the center of B, $J_g = \{b \in B \mid bx = g(x)b$ for each $x \in B\}$ for $g \in G$, and $V_B(A)$ the commutator subring of A in B for a subring A of B.

3. Maps Induced by the Galois Map

By keeping the definitions and notations in section 2, in this section, we assume that B is a Galois extension of B^G with Galois group G such that B^G is a separable C^G-algebra. For a subgroup H of G, let $S_H = \{g \in H \mid J_g \neq \{0\}\}$, $T_H = \{g \in H \mid J_g = \{0\}\}$, and $\overline{H} = \{L \mid L$ is a subgroup of G and $S_L = S_H\}$. We define two maps induced by the Galois map α: (1) $\beta : H \longrightarrow \alpha(H)C$ and (2) $\gamma : H \longrightarrow V_B(\alpha(H))$. We shall show that $\overline{\beta} : \overline{H} \longrightarrow \alpha(H)C$ and $\overline{\gamma} : \overline{H} \longrightarrow V_B(\alpha(H))$ are one-to-one from the set $\{\overline{H} \mid H$ is a subgroup of $G\}$ to the separable subalgebras of B over C. This derives some conditions for a one-to-one Galois map α in section 4. We begin with some properties proved by T. Kanzaki.

Lemma 3.1. *([5], Proposition 1) If B is a Galois extension of B^G with Galois group G, then $V_B(B^G) = \oplus \sum_{g \in G} J_g$.*

Lemma 3.2. *([6], Proposition 3.1) If B is a Galois extension of B^G with Galois group G such that B^G is a separable C^G-algebra, then B^H is a separable C^G-algebra for each subgroup H of G.*

Lemma 3.3. *Let B be a Galois extension of B^G with Galois group G such that B^G is a separable C^G-algebra. Then $\overline{\gamma} : \overline{H} \longrightarrow V_B(\alpha(H))$ for a subgroup H of G is well defined.*

Proof. Let $L \in \overline{H}$. It suffices to show that $V_B(\alpha(L)) = V_B(\alpha(H))$. In fact, since $L \in \overline{H}$, $S_L = S_H$ by the definition of $\overline{H}$. Hence $\oplus \sum_{l \in L} J_l = \oplus \sum_{l \in S_L} J_l = \oplus \sum_{h \in S_H} J_h = \oplus \sum_{h \in H} J_h$. This implies $V_B(B^L) = V_B(B^H)$ by Lemma 3.1; that is, $V_B(\alpha(L)) = V_B(\alpha(H))$. $\qquad \square$

Lemma 3.4. *Let B be a Galois extension of B^G with Galois group G such that B^G is a separable C^G-algebra. Then $\overline{\beta} : \overline{H} \longrightarrow \alpha(H)C$ for a subgroup H of G is well defined.*

Proof. Let $L \in \overline{H}$. It suffices to show that $\alpha(L)C = \alpha(H)C$. In fact, since $L \in \overline{H}$, by Lemma 3.3, $V_B(\alpha(L)) = V_B(\alpha(H))$. Hence $V_B(\alpha(L)C) = V_B(\alpha(H)C)$. Moreover, by Lemma 3.2, $\alpha(L)$ $(= B^L)$ and $\alpha(H)$ $(= B^H)$ are separable C^G-algebras, so $\alpha(L)C$ and $\alpha(H)C$ are separable C-algebras. Since B is a Galois extension of B^G with Galois group G such that B^G is a separable C^G-algebra, B is an Azumaya algebra over C. Therefore, by the commutator theorem for Azumaya algebras ([3], Theorem 4.3, page 57), $V_B(\alpha(L)C) = V_B(\alpha(H)C)$ implies that $\alpha(L)C = V_B(V_B(\alpha(L)C)) = V_B(V_B(\alpha(H)C)) = \alpha(H)C$. This completes the proof. $\qquad \square$

Lemma 3.5. *Let B be a Galois extension of B^G with Galois group G, $S_G = \{g \in G \mid J_g \neq \{0\}\}$, and $f : A \longrightarrow \oplus \sum_{g \in A} J_g$ for a subset A of S_G. Then f is a one-to-one map from the set of nonempty subsets of S_G to the set of direct summands of $V_B(B^G)$.*

Proof. We first show that f restricted on the set of singleton subsets of S_G is one-to-one. In fact, let $g, h \in S_G$ such that $f(\{g\}) = f(\{h\})$, that is, $J_g = J_h$. Since $g, h \in S_G$, $J_g = J_h \neq \{0\}$. Hence there exists a nonzero element $b \in J_g = J_h$ such that $bx = g(x)b = h(x)b$; and so $(g(x) - h(x))b = 0$ for all $x \in B$. Thus $(x - (g^{-1}h)(x))g^{-1}(b) = 0$ for all $x \in B$. Since B is a Galois extension of B^G with Galois group G, there exist elements $\{a_i, b_i$ in B, $i = 1, 2, ..., m\}$ for some integer m such that $\sum_{i=1}^{m} a_i g(b_i) = \delta_{1,g}$ for each $g \in G$. Taking x as b_i in the above equation, we get $(b_i - (g^{-1}h)(b_i))g^{-1}(b) = 0$; and so $a_i(b_i - (g^{-1}h)(b_i))g^{-1}(b) = 0$

for each $i = 1, 2, \ldots, m$. Hence $\sum_{i=1}^{m} a_i\big(b_i - (g^{-1}h)(b_i)\big)g^{-1}(b) = 0$. Therefore,

$$g^{-1}(b) = \Big(\sum_{i=1}^{m} a_i b_i\Big)g^{-1}(b) = \Big(\sum_{i=1}^{m} a_i(g^{-1}h)(b_i)\Big)g^{-1}(b) = \delta_{1,g^{-1}h}\, g^{-1}(b).$$

Since $b \neq 0$, $g^{-1}(b) \neq 0$. Thus $g^{-1}h = 1$, $g = h$, that is, $\{g\} = \{h\}$. Next, let A, D be two nonempty subsets of S_G such that $f(A) = f(D)$, that is, $\oplus \sum_{g \in A} J_g = \oplus \sum_{h \in D} J_h$. Then for each $l \in A$, we claim that $l \in D$. In fact, suppose that $l \notin D$. Since $\oplus \sum_{g \in S_G} J_g = (\oplus \sum_{h \in D} J_h) \oplus (\oplus \sum_{h \notin D} J_h)$ by the early part of the proof, we have that $J_l \cap (\oplus \sum_{h \in D} J_h) = \{0\}$ for $l \notin D$. This contradicts to $J_l \subset \oplus \sum_{g \in A} J_g = \oplus \sum_{h \in D} J_h$ for $l \in A$. Thus $l \in D$, that is, $A \subset D$. Similarly, $D \subset A$. Therefore $A = D$. This completes the proof. $\square$

Now we show the main theorem in this section.

Theorem 3.6. *Let B be a Galois extension of B^G with Galois group G such that B^G is a separable C^G-algebra. Then $\overline{\beta} : \overline{H} \longrightarrow \alpha(H)C$ and $\overline{\gamma} : \overline{H} \longrightarrow V_B(\alpha(H))$ are one-to-one.*

Proof. (i) Let $\overline{\beta}(\overline{H}) = \overline{\beta}(\overline{L})$. Then $B^H C = \alpha(H)C = \overline{\beta}(\overline{H}) = \overline{\beta}(\overline{L}) = \alpha(L)C = B^L C$. Taking commutators in B, we have $V_B(B^H C) = V_B(B^L C)$, and so $V_B(B^H) = V_B(B^H C) = V_B(B^L C) = V_B(B^L)$. Hence $\oplus \sum_{h \in S_H} J_h = \oplus \sum_{l \in S_L} J_l$ by Lemma 3.1. Thus, $S_H = S_L$ by Lemma 3.5. Therefore $\overline{H} = \overline{L}$. This implies that $\overline{\beta}$ is one-to-one.

(ii) Let $\overline{\gamma}(\overline{H}) = \overline{\gamma}(\overline{L})$. Then $\oplus \sum_{h \in S_H} J_h = V_B(B^H) = V_B(\alpha(H)) = \overline{\gamma}(\overline{H}) = \overline{\gamma}(\overline{L}) = V_B(\alpha(L)) = V_B(B^L) = \oplus \sum_{l \in S_L} J_l$. Hence $S_H = S_L$ by Lemma 3.5 again; and so $\overline{H} = \overline{L}$. This implies that $\overline{\gamma}$ is one-to-one. $\square$

4. The Galois Map

Throughout this section, we assume that B is a Galois extension of B^G with Galois group G such that B^G is a separable C^G-algebra. Keeping the notations in section 3, let α be the Galois map, $S_H = \{g \in H \mid J_g \neq \{0\}\}$, $T_H = \{g \in H \mid J_g = \{0\}\}$ for a subgroup H of G, $\beta : H \longrightarrow \alpha(H)C$, and $\gamma : H \longrightarrow V_B(\alpha(H))$. In this section, we shall show several conditions under which α is one-to-one. We begin with a relation between β and γ.

Lemma 4.1. *β is one-to-one if and only if γ is one-to-one.*

Proof. ($\Longrightarrow$) Let $\gamma(H) = \gamma(L)$ for some subgroups H and L of G. Then $V_B(\alpha(H)) = V_B(\alpha(L))$, that is, $V_B(B^H) = V_B(B^L)$. Hence $V_B(B^H C) = V_B(B^L C)$. Since B^H and B^L are separable C^G-algebras by Lemma 3.2, $B^H C$ and $B^L C$ are separable C-algebras of the Azumaya C-algebra B.

Thus, $B^H C = V_B(V_B(B^H C)) = V_B(V_B(B^L C)) = B^L C$ by the double centralizer property for Azumaya algebras ([3], Theorem 4.3, page 57), that is, $\beta(H) = \beta(L)$. But β is one-to-one, so $H = L$. This implies that γ is one-to-one.

($\Longleftarrow$) Let $\beta(H) = \beta(L)$ for some subgroups H and L of G. Then $\alpha(H)C = \alpha(L)C$. Hence $V_B(\alpha(H)) = V_B(\alpha(H)C) = V_B(\alpha(L)C) = V_B(\alpha(L))$. Thus $\gamma(H) = \gamma(L)$. Since γ is one-to-one by hypothesis, $H = L$. This implies that β is one-to-one. $\qquad\square$

Theorem 4.2. *If either β or γ is one-to-one, then α is one-to-one.*

Proof. Assume β is one-to-one. Let $\alpha(H) = \alpha(L)$ for some subgroups H and L of G. Then $B^H = B^L$, and so $B^H C = B^L C$. But then $\beta(H) = \beta(L)$. Since β is one-to-one, $H = L$. Thus α is one-to-one. Also, assume γ is one-to-one. Then β is one-to-one by Lemma 4.1; and so α is one-to-one. $\qquad\square$

We recall that $\overline{H} = \{L \mid L \text{ is a subgroup of } G \text{ and } S_L = S_H\}$ for a subgroup H of G as defined in section 3. Next we give a sufficient condition for α being one-to-one.

Theorem 4.3. *If $\overline{H} = \{H\}$, a singleton, for each subgroup H of G, then α is one-to-one.*

Proof. Let H and L be subgroups of G such that $\alpha(H) = \alpha(L)$. Then $B^H = B^L$, and so $B^H C = B^L C$. Hence $\overline{\beta}(\overline{H}) = \overline{\beta}(\overline{L})$ by Lemma 3.4. Since $\overline{\beta}$ is one-to-one by Theorem 3.6, $\overline{H} = \overline{L}$. By hypothesis, $\overline{H} = \{H\}$ and $\overline{L} = \{L\}$, so $H = L$. Thus α is one-to-one. $\qquad\square$

Next is another sufficient condition for α being one-to-one.

Theorem 4.4. *Let $\langle S_H \rangle$ be the subgroup of G generated by the elements in S_H for a subgroup H of G. If $\langle S_H \rangle = H$ for each subgroup H of G, then α is one-to-one.*

Proof. Let H and L be subgroups of G such that $\alpha(H) = \alpha(L)$. Then $B^H = B^L$. Taking commutator subrings in B, we have $\oplus \sum_{h \in H} J_h = V_B(B^H) = V_B(B^L) = \oplus \sum_{l \in L} J_l$ by Lemma 3.1. Hence $\oplus \sum_{h \in S_H} J_h = \oplus \sum_{l \in S_L} J_l$. Thus, $S_H = S_L$ by Lemma 3.5; and so $\langle S_H \rangle = \langle S_L \rangle$. Therefore, $H = \langle S_H \rangle = \langle S_L \rangle = L$ by hypothesis. This proves that α is one-to-one. $\qquad\square$

Remark 4.5. Theorem 4.4. holds for any Galois extensions B over B^G which is not necessarily a separable C^G-algebra.

Theorem 4.4 generalizes Theorem 3.4 in [10].

Corollary 4.6. *([10], Theorem 3.4) If $J_g \neq \{0\}$ for each $g \in G$, then α is one-to-one.*

Proof. Since $J_g \neq \{0\}$ for each $g \in G$, $\langle S_H \rangle = S_H = H$ for each subgroup H of G. Hence α is one-to-one by Theorem 4.4. $\square$

Acknowledgments. This paper was written under the support of a Caterpillar Fellowship at Bradley University. The authors would like to thank Caterpillar Inc. for the support.

References

[1] S.U. Chase, D.K. Harrison and A. Rosenberg, Galois theory and Galois cohomology of commutative rings, Memoirs Amer. Math. Soc. No. 52, 1965.

[2] F.R. DeMeyer, *Galois theory in separable algebras over commutative rings*, Illinois J. Math. **10** (1966), 287-295.

[3] F.R. DeMeyer and E. Ingraham, Separable algebras over commutative rings, Volume 181, Springer Verlag, Berlin, Heidelberg, New York, 1971.

[4] M. Dokuchaev, M. Ferrero and A. Paques, *Partial action and Galois theory*, Journal of Pure and Applied Algebra **208** (2007), 77-87.

[5] T. Kanzaki, *On Galois algebra over a commutative ring*, Osaka J. Math. **2** (1965), 309-317.

[6] T. Kanzaki, *On Galois extension of rings*, Nagoya J. Math. **27** (1966), 43-49.

[7] K. Sugano, *On a special type of Galois extensions*, Hokkaido J. Math. **9** (1980), 123-128.

[8] G. Szeto and L. Xue, *The structure of Galois algebras*, Journal of Algebra **237**(1) (2001), 238-246.

[9] G. Szeto and L. Xue, *On Galois algebras satisfying the fundamental theorem*, Communications in Algebra **35**(12) (2007), 3979-3985.

[10] G. Szeto and L. Xue, *On Galois extensions with a one-to-one Galois map*, International Journal of Algebra **5**(17) (2011), 801-807.

DEPARTMENT OF MATHEMATICS
BRADLEY UNIVERSITY
PEORIA, IL 61625, U.S.A.
E-mail address: szeto@bradley.edu

DEPARTMENT OF MATHEMATICS
BRADLEY UNIVERSITY
PEORIA, IL 61625, U.S.A.
E-mail address: lxue@bradley.edu

Proceedings of the Sixth China-Japan-Korea
International Conference on Ring Theory
June 27-July 2, 2011 Suwon, Korea

NOTES ON WEAKLY d-KOSZUL MODULES

JIAFENG LÜ* AND XIAOLAN YU

ABSTRACT. It was proved in [5] that each weakly d-Koszul module
M possesses a natural filtration of graded submodules $0 = U_0 \subset
U_1 \subset \cdots \subset U_{p-1} \subset U_p = M$ such that all quotients U_{i+1}/U_i
are d-Koszul modules. This paper continues the study of weakly
d-Koszul modules. In particular, we have $P_n = \bigoplus_{i=1}^{p} P_n^i$ for
all $n \geq 0$, where $\mathcal{P}_*^i \to U_i/U_{i-1} \to 0$ and $\mathcal{P}_* \to M \to 0$ are
the corresponding minimal graded projective resolutions, which
implies easily that $pd(M) = \max_i\{pd(U_i/U_{i-1})\}$ and that the
finitistic dimension conjecture is true in the category of weakly
d-Koszul modules under certain conditions.

1. Introduction

As an extension of d-Koszul modules (c.f. [1]-[4]) and weakly
Koszul modules (c.f. [7]), the authors developed the notion of *weakly
d-Koszul module* and proved that each weakly d-Koszul module can be
approximated by d-Koszul modules in [5]. More precisely, let A be a
d-Koszul algebra, M an arbitrary finitely generated graded A-module,
and $\{S_{d_1}, S_{d_2}, \cdots, S_{d_p}\}$ a set of minimal homogeneous generating spaces
of M. We may assume that $S_{d_i} \subseteq M_{d_i}$, $d_i \in \mathbb{N}$ for $1 \leq i \leq p$, and
$d_1 < d_2 < \cdots < d_p$. Then it is easy to see that M admits the following
natural filtration:

$$\mathcal{F}M: \quad 0 = U_0 \subset U_1 \subset \cdots \subset U_{p-1} \subset U_p = M,$$

where $U_1 = \langle S_{d_1} \rangle$, $U_2 = \langle S_{d_1}, S_{d_2} \rangle$, $\cdots$, $U_p = \langle S_{d_1}, S_{d_1}, \cdots, S_{d_p} \rangle$. The
main result of [5] is to prove that M is a weakly d-Koszul module if and
only if U_i/U_{i-1} are d-Koszul modules for all $1 \leq i \leq p$.

This paper is a continuous work of weakly d-Koszul modules. It is
well known that both d-Koszul modules and weakly d-Koszul modules
are originally defined in terms of their minimal graded projective resolutions. Therefore, a natural ideal is to find some relationships between
the minimal graded projective resolutions of M and these U_i/U_{i-1}'s.

2010 Mathematics Subject Classification : 16S37, 16W50.

Keywords : d-Koszul algebras, d-Koszul modules, weakly d-Koszul modules.

*Corresponding author.

In Section 2, we mainly discuss the relationships between the minimal graded projective resolutions of M and those U_i/U_{i-1}. Besides, we also give an equivalent characterization for weakly d-Koszul modules in terms of the filtration of their related projective complexes. The followings are the main results:

Theorem 1.1. *Let M be a weakly d-Koszul module and $\mathcal{F}M$ its natural graded submodule filtration. Put $K_i = U_i/U_{i-1}$ for $i = 1, 2, \cdots, p$. Let $\mathcal{P}_* \to M \to 0$ and $\mathcal{P}_*^i \to K_i \to 0$ be the corresponding minimal graded projective resolutions. Then for all $n \geq 0$ we have*

$$P_n \cong \bigoplus_{i=1}^{p} P_n^i,$$

where P_n and P_n^i are the nth terms of the resolution $\mathcal{P}_*$ and $\mathcal{P}_*^i$, respectively.

Theorem 1.2. *Let $M \in gr(A)$ and $\cdots \to P_i \to P_{i-1} \to \cdots \to P_1 \to P_0 \to M \to 0$ be the minimal projective resolution of M. Let $\mathcal{P}_*$ denote the complex: $\cdots \to P_i \to P_{i-1} \to \cdots \to P_1 \to P_0 \to 0$. Then M is a weakly d-Koszul module if and only if the complex $\mathcal{P}_*$ has a filtration $\mathcal{F}\mathcal{P}_* : \quad 0 = \mathcal{P}_*^0 \subset \mathcal{P}_*^1 \subset \cdots \subset \mathcal{P}_*^{p-1} \subset \mathcal{P}_*^p = \mathcal{P}_*$, such that the complex $\mathcal{P}_*^j/\mathcal{P}_*^{j-1}: \quad \cdots \to P_i^j/P_i^{j-1} \to P_{i-1}^j/P_{i-1}^{j-1} \to \cdots \to P_1^j/P_1^{j-1} \to P_0^j/P_0^{j-1} \to 0$ has only one non-zero d-Koszul homology K_j at P_0^j/P_0^{j-1} for each $1 \leq j \leq p$. Moreover, $P_i = \bigoplus_{j=1}^{p} P_i^j/P_i^{j-1}$ and in this case M has a filtration $\mathcal{F}M : 0 = U_0 \subset U_1 \subset \cdots \subset U_{p-1} \subset U_p = M$ such that $U_j/U_{j-1} \cong K_j$ and all K_j are d-Koszul modules.*

The last section is devoted to give some easy applications of Theorem 1.1. More precisely, using the notations of Theorem 1.1, we investigate the relationships of the projective dimensions of the weakly d-Koszul module M and the quotients K_i of its fixed filtration. Moreover, we show that the finitistic dimension conjecture is true in the category of weakly d-Koszul modules under certain conditions. The following is the main result of Section 3:

Theorem 1.3. *Using the notations of Theorem 1.1, we have the following statements:*

(1) Let M be a weakly d-Koszul module. Then $pd(M) = \max\{pd(K_i) : i = 1, 2, \cdots, p\}$, where "pd" means projective dimension.

(2) The finitistic dimension conjecture is true in $\mathcal{F}\mathcal{W}\mathcal{K}^d(A)$, where $\mathcal{F}\mathcal{W}\mathcal{K}^d(A)$ denotes the category of weakly d-Koszul modules over a finite dimensional d-Koszul algebra A.

Now we give some notations.

Throughout, $\Bbbk$ denotes an arbitrary field, $\mathbb{N}$ and $\mathbb{Z}$ denote the sets of natural numbers and integers, respectively. The positively graded $\Bbbk$-algebra $A = \bigoplus_{i \geq 0} A_i$ will be called *standard* provided

(1) $A_0 = \Bbbk \times \cdots \times \Bbbk$, a finite product of $\Bbbk$;

(2) $A_i \cdot A_j = A_{i+j}$ for all $0 \leq i, j < \infty$; and

(3) each A_i is of finite dimension as a $\Bbbk$-space.

The graded Jacobson radical of the standard graded algebra A is obvious $\bigoplus_{i \geq 1} A_i$. Let $Gr(A)$ and $gr(A)$ denote the categories of graded A-modules and finitely generated graded A-modules, respectively.

2. The Proofs of Theorems 1.1 and 1.2

We begin with

Definition 2.1. *(c.f. [2], [5]) Let A be a standard graded algebra and $M \in gr(A)$. We call M a d-Koszul module if there exists a minimal graded projective resolution of M*

$$\cdots \longrightarrow Q_n \longrightarrow \cdots \longrightarrow Q_1 \longrightarrow Q_0 \longrightarrow M \longrightarrow 0$$

and a fixed integer s such that for each $n \geq 0$, Q_n is generated in degree $\delta(n, s)$, where the set function $\delta : \mathbb{N} \times \mathbb{Z} \to \mathbb{Z}$ is defined by

$$\delta(n, s) = \begin{cases} \frac{nd}{2} + s, & \text{if } n \text{ is even,} \\ \frac{(n-1)d}{2} + 1 + s, & \text{if } n \text{ is odd,} \end{cases}$$

and $d \geq 2$ an integer, $n \in \mathbb{N}$. In particular, if the trivial A-module A_0 is a d-Koszul module, then we call A a d-Koszul algebra.

Definition 2.2. *(c.f. [5]) Let A be a d-Koszul algebra. We say that $M \in gr(A)$ is a weakly d-Koszul module if there exists a minimal graded projective resolution of M:*

$$\cdots \longrightarrow Q_i \xrightarrow{f_i} \cdots \longrightarrow Q_1 \xrightarrow{f_1} Q_0 \xrightarrow{f_0} M \longrightarrow 0$$

such that for all $i, k \geq 0$, $J^k \ker f_i = J^{k+1} Q_i \cap \ker f_i$ if i is even and $J^k \ker f_i = J^{k+d-1} Q_i \cap \ker f_i$ if i is odd.

Lemma 2.3. *Let $0 \longrightarrow K \longrightarrow M \longrightarrow N \longrightarrow 0$ be an exact sequence of finitely generated graded A-modules. Then $JK = K \cap JM$ if and only if we have the following commutative diagram with exact rows and columns*

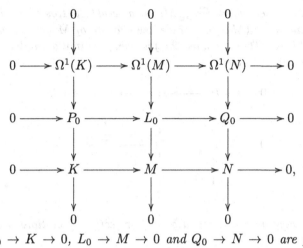

where $P_0 \to K \to 0$, $L_0 \to M \to 0$ and $Q_0 \to N \to 0$ are projective covers.

Proof. ($\Rightarrow$) Clearly, we obtain the exact sequence

$$0 \longrightarrow K/JK \longrightarrow M/JM \longrightarrow N/JN \longrightarrow 0.$$

Note that for any finitely generated graded A-module X, $A \otimes_{A_0} X/JX \longrightarrow X \longrightarrow 0$ is a projective cover. Now set

$$P_0 := A \otimes_{A_0} K/JK, \quad L_0 := A \otimes_{A_0} M/JM \quad \text{and} \quad Q_0 := A \otimes_{A_0} N/JN.$$

We have the following exact sequence

$$0 \longrightarrow P_0 \longrightarrow L_0 \longrightarrow Q_0 \longrightarrow 0$$

since A_0 is semisimple. Therefore, we have the desired diagram in terms of "Snake Lemma".

($\Leftarrow$) Suppose that we have the above commutative diagram. Note that the projective cover of a module is unique up to isomorphisms. We may assume that

$$P_0 := A \otimes_{A_0} K/JK, \quad L_0 := A \otimes_{A_0} M/JM \quad \text{and} \quad Q_0 := A \otimes_{A_0} N/JN.$$

From the middle row of the diagram, we have the following exact sequence

$$0 \longrightarrow A \otimes_{A_0} K/JK \longrightarrow A \otimes_{A_0} M/JM \longrightarrow A \otimes_{A_0} N/JN \longrightarrow 0.$$

Thus, we have the following short exact sequence as A_0-modules

$$0 \longrightarrow K/JK \longrightarrow M/JM \longrightarrow N/JN \longrightarrow 0$$

since A_0 is semisimple, which implies $JK = K \cap JM$. $\qquad\square$

Lemma 2.4. *Let $M = \bigoplus_{i \geq 0} M_i$ be a weakly d-Koszul module with $M_0 \neq 0$ and $K := \langle M_0 \rangle$, the graded submodule of M generated by M_0. Set $N := M/K$. Then we have the following commutative diagram with exact rows and columns*

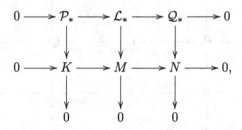

where the columns are minimal graded projective resolutions and for all $n \geq 0$, we have $L_n = P_n \oplus Q_n$.

Proof. By Theorem 3.1 of [5], we have $JK = K \cap JM$. By Lemma 2.3, we have the same commutative diagram as in Lemma 2.3, which implies easily that $L_0 = P_0 \oplus Q_0$ since the split exact sequence

$$0 \longrightarrow P_0 \longrightarrow L_0 \longrightarrow Q_0 \longrightarrow 0.$$

Clearly, we have the following commutative diagram with exact rows and columns

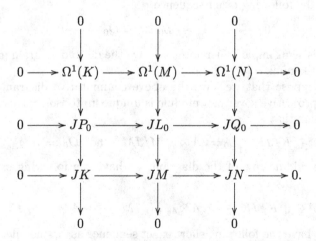

Apply the functor $A/J \otimes_A -$ to the above diagram, note that M and N are weakly d-Koszul modules, we obtain the following commutative diagram

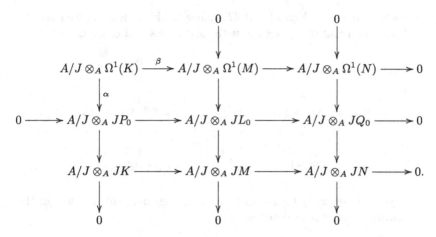

By Theorem 3.1 of [5], K is a d-Koszul module. Thus we have that $J\Omega^1(K) = \Omega^1(K) \cap J^2 P_0$, which implies that α is a monomorphism. Hence β is also a monomorphism and we have $J\Omega^1(K) = \Omega^1(K) \cap J\Omega^1(M)$. Similarly, we get the following commutative diagram with exact rows and columns

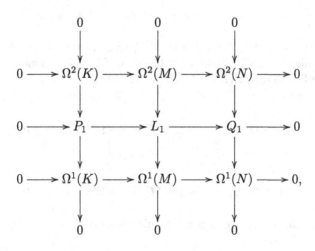

where P_1, Q_1 and L_1 are graded projective covers of $\Omega^1(K)$, $\Omega^1(M)$ and $\Omega^1(N)$ respectively, and $\Omega^2(K)$, $\Omega^2(M)$ and $\Omega^2(N)$ are the second syzygies of K, M and N respectively. Of course, $L_1 = P_1 \oplus Q_1$ since the exact sequence

$$0 \longrightarrow P_1 \longrightarrow L_1 \longrightarrow Q_1 \longrightarrow 0$$

and Q_1 is a graded projective module.

Note that K is d-Koszul and M, N are weakly d-Koszul, we have the following commutative diagram with exact rows and columns

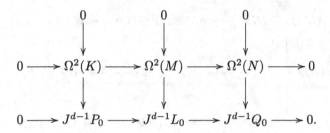

$$0 \longrightarrow \Omega^2(K) \longrightarrow \Omega^2(M) \longrightarrow \Omega^2(N) \longrightarrow 0$$

$$0 \longrightarrow J^{d-1}P_0 \longrightarrow J^{d-1}L_0 \longrightarrow J^{d-1}Q_0 \longrightarrow 0.$$

Apply the functor $A/J \otimes_A -$ to the above diagram, similarly, we get the following commutative diagram

$$A/J \otimes_A \Omega^2(K) \xrightarrow{\delta} A/J \otimes_A \Omega^2(M) \longrightarrow A/J \otimes_A \Omega^2(N) \longrightarrow 0$$

$$0 \longrightarrow A/J \otimes_A J^{d-1}P_0 \longrightarrow A/J \otimes_A J^{d-1}L_0 \longrightarrow A/J \otimes_A J^{d-1}Q_0 \longrightarrow 0.$$

Further we have $J\Omega^2(K) = \Omega^2(K) \cap J^d P_1$ since K is a d-Koszul module, which implies that γ is a monomorphism. So δ is also a monomorphism, which implies that $J\Omega^2(K) = \Omega^2(K) \cap J\Omega^2(M)$.

Now replace

$$0 \longrightarrow \Omega^1(K) \longrightarrow \Omega^1(M) \longrightarrow \Omega^1(N) \longrightarrow 0$$

by

$$0 \longrightarrow \Omega^2(K) \longrightarrow \Omega^2(M) \longrightarrow \Omega^2(N) \longrightarrow 0$$

and repeat the above argument, we finish the proof. $\square$

Corollary 2.5. *Let* $M = \bigoplus_{i \geq 0} M_i$ *be a weakly d-Koszul module with its natural filtration of graded submodules*

$$\mathcal{F}M : 0 = U_0 \subset U_1 \subset \cdots \subset U_{p-1} \subset U_p = M.$$

Then for each j, we have the following commutative diagram with exact rows and columns

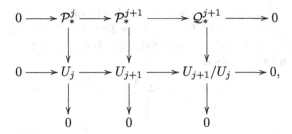

where the columns are minimal graded projective resolutions.

Proof. It is immediate from Lemma 2.4. □

Now we can prove Theorem 1.1.

Proof. Consider the following exact sequence

$$0 \longrightarrow U_1 \longrightarrow M \longrightarrow M/U_1 \longrightarrow 0.$$

By Lemma 2.4, we have the following commutative diagram with exact rows and columns

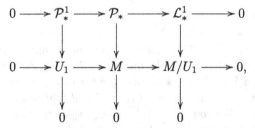

where $\mathcal{P}_*^1$, $\mathcal{P}_*$ and $\mathcal{L}_*^1$ are the minimal graded projective resolutions of U_1, M and M/U_1, respectively. Clearly, $\mathcal{P}_* = \mathcal{P}_*^1 \oplus \mathcal{L}_*^1$. Setting $W = M/U_1$. Then $\langle W_{d_2} \rangle = U_2/U_1 = K_2$. Consider the following exact sequence

$$0 \longrightarrow K_2 \longrightarrow W \longrightarrow W/K_2 \longrightarrow 0.$$

By Lemma 2.4 again, we have the following commutative diagram with exact rows and columns

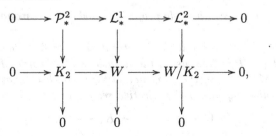

where $\mathcal{P}_*^2$, $\mathcal{L}_*^1$ and $\mathcal{L}_*^2$ are the minimal graded projective resolution of K_2, W and W/K_2, respectively. Clearly, $\mathcal{L}_*^1 = \mathcal{P}_*^2 \oplus \mathcal{L}_*^2$. Repeating the above argument and by induction, we finally get the following commutative diagram with exact rows and columns

$$
\begin{array}{ccccccccc}
& & 0 & \longrightarrow & \mathcal{P}_*^{p-1} & \longrightarrow & \mathcal{L}_*^{p-1} & \longrightarrow & \mathcal{P}_*^p & \longrightarrow & 0 \\
& & & & \downarrow & & \downarrow & & \downarrow & & \\
& & 0 & \longrightarrow & K_{p-1} & \longrightarrow & X & \longrightarrow & K_p & \longrightarrow & 0. \\
& & & & \downarrow & & \downarrow & & \downarrow & & \\
& & & & 0 & & 0 & & 0 & &
\end{array}
$$

Therefore, we have $\mathcal{P}_* \cong \bigoplus_{i=1}^p \mathcal{P}_*^i$ and we are done. □

We now give a concrete example to explain Theorem 1.1.

Example 2.6. *Let Γ be the following quiver:*

$$
\bullet^1 \xrightarrow{u_1} \bullet^2 \xrightarrow{u_2} \bullet^3 \xrightarrow{u_3} \bullet^4.
$$
$$
\Big\downarrow u_4
$$
$$
\bullet^5
$$

Set $A = \mathbb{k}\Gamma/(u_1u_2u_3)$. Then under a routine computation, A is a 3-Koszul algebra. Let $e_1, \cdots, e_5$ be the idempotents of A corresponding to the vertices and $V = \mathbb{k}v_0 \oplus \mathbb{k}v_1$ be a graded vector space with basis v_0 and v_1. Assume that the degree of v_0 is 0 and that of v_1 is 1. Define a left A_0-module action on V as follows: $e_4 \cdot v_0 = v_0$ and $e_i \cdot v_0 = 0$ for $i \neq 4$; $e_5 \cdot v_1 = v_1$ and $e_i \cdot v_1 = 0$ for $i \neq 5$. Let

$$
M = \frac{A \otimes_{A_0} V}{\langle u_2 \otimes_{A_0} u_3 \otimes_{A_0} v_0 - u_4 \otimes_{A_0} v_1 \rangle}.
$$

Now it is not hard to check that M is a weakly 3-Koszul module. Now under a routine computation, M possesses the following minimal projective resolution:

$$
0 \longrightarrow P_2[2] \longrightarrow P_4 \oplus (P_5[1]) \longrightarrow M \longrightarrow 0,
$$

where P_i is the indecomposable projective module corresponding to vertex i.

Obviously, M has the following natural filtration $0 = U_0 \subseteq U_1 \subseteq U_2 = M$, where $U_1 = P_4$. Now using the notations of Theorem 1.1, $K_1 = P_4$ and $K_2 = M/P_4 \cong S_5[1]$. It is not difficult to check that K_1 and K_2 admit the following minimal projective resolutions:

$$
0 \longrightarrow P_4 \longrightarrow K_1 \longrightarrow 0
$$

and

$$0 \longrightarrow P_2[2] \longrightarrow P_5[1] \longrightarrow K_2 \longrightarrow 0.$$

Therefore, the projective resolution of M is the direct sum of the resolutions of K_1 and K_2.

Now we are ready to prove Theorem 1.2.

Proof. ($\Rightarrow$) If $M = \bigoplus_{i \geq 0} M_i$ is a weakly d-Koszul module with its natural filtration

$$\mathcal{F}M : 0 = U_0 \subset U_1 \subset \cdots \subset U_{p-1} \subset U_p = M.$$

By Lemma 2.5, we get the following commutative diagram with exact columns,

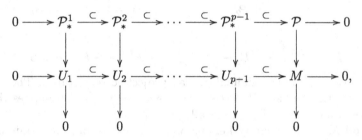

which induces minimal graded projective resolutions of $U_j/U_{j-1} = K_j$ for each $1 \leq j \leq p$,

$$\cdots \to P_i^j/P_i^{j-1} \to P_{i-1}^j/P_{i-1}^{j-1} \to \cdots \to P_1^j/P_1^{j-1} \to P_0^j/P_0^{j-1} \to K_j \to 0.$$

Thus, each K_j is a d-Koszul module.

($\Leftarrow$) If $M \in gr(A)$ has the minimal projective resolution $\cdots \to P_i \to P_{i-1} \to \cdots \to P_1 \to P_0 \to M \to 0$ and the complex $\mathcal{P}_* : \cdots \to P_i \to P_{i-1} \to \cdots \to P_1 \to P_0 \to 0$ has a filtration as in the theorem. Then it can be checked that for each $1 \leq j \leq p$, $\mathcal{P}_j : \cdots \to P_i^j \to P_{i-1}^j \to \cdots \to P_1^j \to P_0^j \to 0$ has only one non-zero homology, say U_j, at P_0^j. Therefore, the filtration of the complex $\mathcal{P}_*$ induces a filtration of the module M. Moreover, for each $1 \leq j \leq p$, we have $K_j = U_j/U_{j-1}$ is a d-Koszul module. Then M is a weakly d-Koszul module. $\square$

3. The Proof of Theorem 1.3

As an application of Theorem 1.1, we will prove Theorem 1.3 in this section.

Lemma 3.1. *Let A be a standard graded algebra and M a finitely generated graded A-module. Then the length l of a minimal graded projective resolution of M equals to $pd(M)$.*

Proof. Note that a minimal graded projective resolution is in particular a graded projective resolution, which implies that $pd(M) \leq l$. But if there would be a minimal resolution of length $< l$, then we have $Ext^l_A(M, A_0) \cong Tor^A_l(A_0, M) = 0$, which is a contradiction. $\square$

Lemma 3.2. *Let M be a weakly d-Koszul module. Using the notations of Theorem 1.1. Then $pd(M) = \max\{pd(K_i) : i = 1, 2, \cdots, p\}$, where "pd" means projective dimension.*

Proof. Let M be a weakly d-Koszul module and K_i $(i = 1, 2, \cdots, p)$ be the related d-Koszul modules. Let

$$\mathcal{P}_* : \quad \cdots \longrightarrow P_2 \longrightarrow P_1 \longrightarrow P_0 \longrightarrow M \longrightarrow 0$$

and

$$\mathcal{P}^i_* : \quad \cdots \longrightarrow P^i_2 \longrightarrow P^i_1 \longrightarrow P^i_0 \longrightarrow K_i \longrightarrow 0$$

be the corresponding minimal graded projective resolutions. Then by Theorem 1.1, we have $P_n \cong \bigoplus_{i=1}^p P^i_n$ for all $n \geq 0$, which implies that $P_n = 0$ if and only if $P^i_n = 0$ for all $i = 1, 2, \cdots, p$.

If $pd(M) = \infty$, then by Lemma 3.1, there exists an infinite minimal graded projective resolution of M

$$\cdots \longrightarrow P_n \longrightarrow \cdots \longrightarrow P_2 \longrightarrow P_1 \longrightarrow P_0 \longrightarrow M \longrightarrow 0.$$

Note that the minimal graded projective resolution of a module is unique up to isomorphisms. Thus at least one of

$$\cdots \longrightarrow P^i_n \longrightarrow \cdots \longrightarrow P^i_2 \longrightarrow P^i_1 \longrightarrow P^i_0 \longrightarrow K_i \longrightarrow 0$$

is infinite, which implies $pd(M) = \max\{pd(K_i) : i = 1, 2, \cdots, p\}$.

If $pd(M) = n < \infty$, then by Lemma 3.1, there exists a minimal graded projective resolution of M

$$0 \longrightarrow P_n \longrightarrow \cdots \longrightarrow P_2 \longrightarrow P_1 \longrightarrow P_0 \longrightarrow M \longrightarrow 0,$$

which implies that K_i possess the following minimal projective resolutions

$$0 \longrightarrow P^i_n \longrightarrow \cdots \longrightarrow P^i_2 \longrightarrow P^i_1 \longrightarrow P^i_0 \longrightarrow K_i \longrightarrow 0$$

such that at least one of P^i_n isn't zero, which implies $pd(M) = \max\{pd(K_i) : i = 1, 2, \cdots, p\}$ as well. $\square$

We also give an example to illustrate Lemma 3.2.

Example 3.3. *Let A and M be the same as Example 2.6. Then by Example 2.6, M, K_1 and K_2 admit the following minimal projective resolutions*

$$0 \longrightarrow P_2[2] \longrightarrow P_4 \oplus (P_5[1]) \longrightarrow M \longrightarrow 0,$$

$$0 \longrightarrow P_4 \longrightarrow K_1 \longrightarrow 0$$

and

$$0 \longrightarrow P_2[2] \longrightarrow P_5[1] \longrightarrow K_2 \longrightarrow 0,$$

where P_i is the indecomposable projective module corresponding to vertex i. Thus, $pd(M) = 1$, $pd(K_1) = 0$ and $pd(K_2) = 1$, which implies that $pd(M) = \max\{pd(K_1),\ pd(K_2)\}$.

It is well known that the finitistic dimension conjecture is one of the most important and interesting conjectures in the representation theory of Artin algebras. First let us recall the contents of the conjecture.

Let Λ be an arbitrary Artin R-algebra, where R is a commutative Artin ring with identity. Let $\mathrm{mod}(\Lambda)$ be the category of finitely generated Λ-modules and

$$\mathbf{B}(\Lambda) := \{M \in \mathrm{mod}(\Lambda)|pd_\Lambda(M) < \infty\}.$$

The famous finitistic dimension conjecture says that $\sup\{pd_\Lambda(M)|M \in \mathbf{B}(\Lambda)\} < \infty$. It is too far to solve the conjecture completely and it still remains open now. Therefore it is also interesting to find certain subcategories in which the finitistic dimension conjecture holds.

Lemma 3.4. *Let A be a finite dimensional d-Koszul algebra. Then the finitistic dimension conjecture holds in the category of d-Koszul modules.*

Proof. It is immediate from Theorem 4.5 of [3]. $\qquad\qquad\square$

Lemma 3.5. *The finitistic dimension conjecture is true in $\mathcal{FWK}^d(A)$, where $\mathcal{FWK}^d(A)$ denotes the category of weakly d-Koszul modules over a finite dimensional d-Koszul algebra A.*

Proof. Let $M \in \mathcal{FWK}^d(A)$ and $0 = \mathcal{M}_1 \subset \mathcal{M}_2 \subset \cdots \subset \mathcal{M}_{m-1} \subset \mathcal{M}_m = M$ its special fixed graded submodule filtration, the same as in Theorem 1.1. Set $K_i := \mathcal{M}_i/\mathcal{M}_{i-1}$ for $i = 1,\ 2,\ \cdots,\ m$. First note that for each $M \in \mathcal{FWK}^d(A)$ with finite projective dimension. Then by Theorem 1.3, we have $pd(M) = \max\{pd(K_i) : i = 1, 2, \cdots, m\}$, which implies that $pd(K_i) < \infty$ for all $i = 1, 2, \cdots, m$. Therefore, we have

$$\sup\{pd(M) < \infty|M \in \mathcal{FWK}^d(A)\}$$

$$= \sup\{\max\{pd(K_1),\ pd(K_2),\ \cdots,\ pd(K_m)\}\} < \infty.$$

That is, the finitistic dimension conjecture is true in $\mathcal{FWK}^d(A)$. $\qquad\square$

Now Theorem 1.3 is immediate from Lemmas 3.2 and 3.5.

Acknowledgments. The first author was supported by National Natural Science Foundation of China (11001245 and 11101288), Natural Science Foundation of Zhejiang Province (Y6110323), Zhejiang Province Department of Education Fund (Y201016432) and Zhejiang Innovation Project (T200905).

References

[1] R. Berger, *Koszulity for nonquadratic algebras*, J. Algebra **239** (2001), 705-734.

[2] E. L. Green, E. N. Marcos, R. Martínez-Villa and Pu Zhang, *D-Koszul algebras*, J. Pure Appl. Algebra **193** (2004), 141-162.

[3] E. L. Green, R. Martínez-Villa, I. Reiten, ϕ. Solberg and D. Zacharia, *On modules with linear presentations*, J. Algebra **205**(2) (1998), 578-604.

[4] J.-W. He and D.-M. Lu, *Higher Koszul Algebras and A-infinity Algebras*, J. Algebra **293** (2005), 335-362.

[5] J.-F. Lü, J.-W. He and D.-M. Lu, *On modules with d-Koszul towers*, Chinese Ann. Mathematics **28**(2) (2007), 231-238.

[6] J.-F. Lü and G.-J. Wang, *Weakly d-Koszul modules*, Vietnam J. Mathematics **34**(3) (2006), 341-351.

[7] R. Martínez-Villa and D. Zacharia, *Approximations with modules having linear resolutions*, J. Algebra **266** (2003), 671-697.

[8] C. A. Weibel, *An Introduction to Homological Algebra*, Cambridge Studies in Avanced Mathematics, **38**, Cambridge Univ. Press, 1995.

DEPARTMENT OF MATHEMATICS
ZHEJIANG NORMAL UNIVERSITY
JINHUA, ZHEJIANG, 321004, P.R. CHINA
E-mail address: jiafenglv@zjnu.edu.cn

DEPARTMENT OF MATHEMATICS
HANGZHOU NORMAL UNIVERSITY
HANGZHOU, ZHEJIANG, 310036, P.R. CHINA
E-mail address: xlyu@hznu.edu.cn

Proceedings of the Sixth China-Japan-Korea
International Conference on Ring Theory
June 27-July 2, 2011 Suwon, Korea

AN EXTENSION OF RINGS AND HOCHSCHILD
2-COCYCLES

M. TAMER KOŞAN, TSIU-KWEN LEE, AND YIQIANG ZHOU*

ABSTRACT. The focus of this paper is on a ring construction
$H_n(R; \sigma)$ based on a given ring R and a Hochschild 2-cocycle σ.
This construction is a unified generalization of the ring $R[x]/(x^{n+1})$
and the Hochschild extension $H_\sigma(R, R)$. Here we discuss when the
ring $H_n(R; \sigma)$ is reversible, symmetric, Armendariz, abelian and
uniquely clean, respectively. Several known results of $R[x]/(x^{n+1})$
and $H_\sigma(R, R)$ are extended to $H_n(R; \sigma)$, and new examples of reversible, symmetric and Armendariz rings are given.

1. The ring $H_n(R; \sigma)$

Let R be an associative ring not necessarily having unity and let
M be a bimodule over R. A $\mathbb{Z}$-bilinear map $\sigma : R \times R \to M$ is called a
Hochschild 2-cocycle if

$$\sigma(ab, c) - \sigma(a, bc) = a\sigma(b, c) - \sigma(a, b)c$$

for all $a, b, c \in R$. The Hochschild extension of R by M, denoted
$H_\sigma(R, M)$, is the ring whose abelian group is $R \oplus M$, with multiplication

$$(a, x)(b, y) = \big(ab, ay + xb + \sigma(a, b)\big) \quad \text{(see [3], [10])}.$$

If $\sigma = 0$, the ring $H_0(R, M)$ is called the trivial extension of R by M in
the literature. In particular, $H_0(R, R) \cong R[x]/(x^2)$. The following ring
construction contains a subring which is a unified generalization of both
$H_\sigma(R, R)$ and $R[x]/(x^{n+1})$.

Let $\mathbb{T}_n(R)$ be the $n \times n$ upper triangular matrix ring over R. If
$(a_{ij}) \in \mathbb{T}_n(R)$, we write $[a; (a_{ij})]$ for
$\begin{pmatrix} a & a_{11} & a_{12} & \cdots & a_{1n} \\ 0 & a & a_{22} & \cdots & a_{2n} \\ 0 & 0 & a & \cdots & a_{3n} \\ \vdots & \vdots & \vdots & \ddots & \vdots \\ 0 & 0 & 0 & \cdots & a \end{pmatrix}$ and

2010 Mathematics Subject Classification : 16S99.

Keywords : Extension of rings, Hochschild 2-cocycle, reversible ring, symmetric
ring, Armendariz ring.

*Corresponding author.

write $rE_{1,n+1}$ for the $(n+1) \times (n+1)$ matrix whose $(1, n+1)$-entry is r and all other entries are 0. We write $J(R)$ for the Jacobson radical of R, and $U(R)$ for the set of units of R if R has unity.

Proposition 1.1. *Let* $\sigma : R \times R \to R$ *be a Hochschild 2-cocycle and* $n \geq 1$. *Then*

$$B_n(R; \sigma) := \{ [a; (a_{ij})] : a \in R, (a_{ij}) \in \mathbb{T}_n(R) \}$$

is a ring, with usual addition of matrices and with multiplication $*$ given by, for $A = [a; (a_{ij})]$ *and* $B = [b; (b_{ij})]$,

$$A * B = AB + \sigma(a, b)E_{1,n+1},$$

where AB is the usual multiplication of matrices. If R has unity 1, then $B_n(R; \sigma)$ *has unity* $[1; -\sigma(1, 1)E_{1,n+1}]$ *and the following hold:*

(1) $J\big(B_n(R; \sigma)\big) = \{ [a; (a_{ij})] : a \in J(R) \}$.
(2) $U\big(B_n(R; \sigma)\big) = \{ [a; (a_{ij})] : a \in U(R) \}$.

Proof. We verify the associative law and leave the rest for the readers. Let $A = [a; (a_{ij})]$, $B = [b; (b_{ij})]$ and $C = [c; (c_{ij})]$. Then

$$
\begin{aligned}
(A * B) * C &= \big(AB + \sigma(a, b)E_{1,n+1}\big) * C \\
&= \big(AB + \sigma(a, b)E_{1,n+1}\big)C + \sigma(ab, c)E_{1,n+1} \\
&= ABC + \big(\sigma(a, b)c + \sigma(ab, c)\big)E_{1,n+1}
\end{aligned}
$$

and

$$
\begin{aligned}
A * (B * C) &= A * \big(BC + \sigma(b, c)E_{1,n+1}\big) \\
&= A\big(BC + \sigma(b, c)E_{1,n+1}\big) + \sigma(a, bc)E_{1,n+1} \\
&= ABC + \big(a\sigma(b, c) + \sigma(a, bc)\big)E_{1,n+1}.
\end{aligned}
$$

So $(A * B) * C = A * (B * C)$ because σ is a Hochschild 2-cocycle. $\square$

If $\sigma = 0$, $B_n(R; 0)$ is a subring of $\mathbb{T}_{n+1}(R)$. The focus of this paper is on the following subring of $B_n(R; \sigma)$.

Proposition 1.2. *Let* $\sigma : R \times R \to R$ *be a Hochschild 2-cocycle and* $n \geq 1$. *Then*

$$
H_n(R; \sigma) := \left\{ \begin{pmatrix} a_0 & a_1 & a_2 & \cdots & a_n \\ 0 & a_0 & a_1 & \cdots & a_{n-1} \\ 0 & 0 & a_0 & \cdots & a_{n-2} \\ \vdots & \vdots & \vdots & \ddots & \vdots \\ 0 & 0 & 0 & \cdots & a_0 \end{pmatrix} : a_i \in R \text{ for } 0 \leq i \leq n \right\}
$$

is a subring of $B_n(R; \sigma)$. *We identify* $H_n(R; \sigma)$ *with the set*

$$\{ (a_0, \ldots, a_n) : a_i \in R \text{ for } i = 0, \ldots, n \}.$$

Thus, $H_n(R;\sigma)$ is a ring with addition defined componentwise and multiplication given by

$$(a_0,\ldots,a_n)(b_0,\ldots,b_n) = (c_0,\ldots,c_n)$$

where, for $i = 0,\ldots,n-1$,

$$c_i = a_0 b_i + a_1 b_{i-1} + \cdots + a_i b_0$$
$$c_n = a_0 b_n + a_1 b_{n-1} + \cdots + a_n b_0 + \sigma(a_0, b_0).$$

If R has unity 1 then $H_n(R;\sigma)$ has unity $(1,\ldots,-\sigma(1,1))$ and the following hold:

(1) $J\big(H_n(R;\sigma)\big) = \big\{(a_0,\ldots,a_n) \in H_n(R;\sigma) : a_0 \in J(R)\big\}$.
(2) $U\big(H_n(R;\sigma)\big) = \big\{(a_0,\ldots,a_n) \in H_n(R;\sigma) : a_0 \in U(R)\big\}$.

We note that $H_1(R;\sigma) = H_\sigma(R,R)$ and that for $\sigma = 0$, $H_n(R;0) \cong R[x]/(x^{n+1})$. From now on, $\sigma : R \times R \to R$ is always a Hochschild 2-cocycle.

2. Reversible and symmetric rings

A ring R is called reduced if it contains no nonzero nilpotent elements. For a reduced ring R, $a_1 a_2 \cdots a_n = 0$ in R implies that $a_{\rho(1)} a_{\rho(2)} \cdots a_{\rho(n)} = 0$ for any permutation ρ on $\{1, 2, \ldots, n\}$, and $ab = 0$ in R implies that $aRb = bRa = 0$. These properties of a reduced ring will be used later without explicit mention.

A ring R is called reversible if $ab = 0$ in R implies $ba = 0$ (see [4]), while R is said to be symmetric if $abc = 0$ in R implies $acb = 0$ (see [8]). Reduced rings are reversible and symmetric. A reversible ring with unity need not be symmetric (see [2] or [11, Example 5]). A symmetric ring with unity is always reversible, but a symmetric ring without unity need not be reversible (see [11, Example 1] or [10]).

For a reduced ring R with unity, $R[x]/(x^{n+1})$ is reversible by [7, Theorem 2.5] and is symmetric by [5, Theorem 2.3]. On the other hand, for a reduced ring R with or without unity, $H_\sigma(R,I)$ is symmetric and reversible for any ideal I of R by [10, Theorem 1.1]. Note that subrings of a symmetric (or reversible) ring are still symmetric (or reversible). Since $H_\sigma(R,I)$ embeds in $H_\sigma(R,R)$ as a subring, $H_\sigma(R,R)$ being symmetric (or reversible) always implies that $H_\sigma(R,I)$ is symmetric (or reversible) for every ideal I of R. Theorems 2.4 and 2.5 below extend [10, Theorem 1.1], [5, Theorem 2.3] and [7, Theorem 2.5].

Lemma 2.1. *If $ab = bc = 0$ in R, then $\sigma(a,b)c = a\sigma(b,c)$.*

Proof. This is obvious. □

Lemma 2.2. *Let R be a reduced ring. If $(a_0 + a_1 x + \cdots + a_n x^n)(b_0 + b_1 x + \cdots + b_n x^n) = 0$ in $R[x]/(x^{n+1})$, then $a_i b_j = 0$ for all i, j with $0 \le i + j \le n$.*

Proof. This is Claim 2 in the proof of [9, Theorem 20].   □

Lemma 2.3. *Let R be a ring. Then $(a_0, \ldots, a_n)(b_0, \ldots, b_n) = 0$ in $H_n(R; \sigma)$ if and only if $(a_0 + a_1 x + \cdots + a_{n-1} x^{n-1})(b_0 + b_1 x + \cdots + b_{n-1} x^{n-1}) = 0$ in $R[x]/(x^n)$ and $\sum_{i=0}^n a_i b_{n-i} + \sigma(a_0, b_0) = 0$.*

Proof. This is by the definition of multiplication in $H_n(R; \sigma)$.   □

Theorem 2.4. *Let R be a reduced ring. Then $H_n(R; \sigma)$ is reversible for each $n \ge 1$.*

Proof. Suppose $\alpha\beta = 0$ in $H_n(R; \sigma)$, where $\alpha = (a_0, \ldots, a_n)$ and $\beta = (b_0, \ldots, b_n)$. Then, by Lemmas 2.2 and 2.3,

$$(2.1) \qquad a_i b_j = 0 \text{ for all } i, j \text{ with } 0 \le i + j \le n - 1$$

and

$$(2.2) \qquad c := a_0 b_n + a_1 b_{n-1} + \cdots + a_n b_0 + \sigma(a_0, b_0) = 0.$$

Note that (2.1) clearly shows that $(\sum_{i=0}^{n-1} b_i x^i)(\sum_{i=0}^{n-1} a_i x^i) = 0$ in $R[x]/(x^n)$. So by Lemma 2.3, to show $\beta\alpha = 0$, it suffices to prove that

$$d := b_0 a_n + b_1 a_{n-1} + \cdots + b_n a_0 + \sigma(b_0, a_0) = 0.$$

For any $0 < i < n$, by (2.1), we have

$$
\begin{aligned}
0 &= b_{n-i}(c a_i) \\
&= b_{n-i}\big[\big(a_0 b_n + a_1 b_{n-1} + \cdots + a_{i-1} b_{n-i+1} + a_i b_{n-i} + \sigma(a_0, b_0) \big) a_i \big] \\
&= b_{n-i}[a_i b_{n-i} + \sigma(a_0, b_0)] a_i \\
&= b_{n-i}[a_i b_{n-i} a_i + a_0 \sigma(b_0, a_i)] \\
&= b_{n-i} a_i b_{n-i} a_i.
\end{aligned}
$$

So, $b_{n-i} a_i = 0$ for $i = 1, \ldots, n - 1$. It follows that

$$c = a_0 b_n + a_n b_0 + \sigma(a_0, b_0) = 0 \text{ and } d = b_0 a_n + b_n a_0 + \sigma(b_0, a_0).$$

Thus,

$$
\begin{aligned}
a_0 d &= (a_0 b_n + \sigma(a_0, b_0)) a_0 = c a_0 = 0, \\
d b_0 &= b_0(a_n b_0 + \sigma(a_0, b_0)) = b_0 c = 0.
\end{aligned}
$$

Now $d^3 = d[(b_0 a_n + b_n a_0 + \sigma(b_0, a_0))d] = d(\sigma(b_0, a_0)d) = d(b_0 \sigma(a_0, d)) = 0$, so $d = 0$.   □

Theorem 2.5. *Let R be a reduced ring. Then $H_n(R; \sigma)$ is symmetric for each $n \ge 1$.*

Proof. Suppose $\alpha\beta\gamma = 0$ in $H_n(R; \sigma)$, where $\alpha = (a_0, \ldots, a_n)$, $\beta = (b_0, \ldots, b_n)$ and $\gamma = (c_0, \ldots, c_n)$. For convenience, let $a = a_0, b = b_0, c = c_0$. Then, by Lemma 2.3,

$$(2.3) \qquad u := \sum_{i+j+k=n} a_i b_j c_k + \sigma(a, b)c + \sigma(ab, c) = 0$$

and

$$(2.4) \qquad \left(\sum_{i=0}^{n-1} a_i x^i\right)\left(\sum_{i=0}^{n-1} b_i x^i\right)\left(\sum_{i=0}^{n-1} c_i x^i\right) = 0 \text{ in } R[x]/(x^n).$$

By Lemma 2.2, it follows from (2.4) that

$$(a_0 b_i + a_1 b_{i-1} + \cdots + a_i b_0)c_j = 0 \text{ for all } i, j \text{ with } 0 \le i + j \le n - 1,$$

which implies (because R is reduced)

$$(2.5) \qquad a_i b_j c_k = 0 \text{ for all } i, j, k \text{ with } 0 \le i + j + k \le n - 1.$$

Note that (2.5) clearly shows that

$$\left(\sum_{i=0}^{n-1} a_i x^i\right)\left(\sum_{i=0}^{n-1} c_i x^i\right)\left(\sum_{i=0}^{n-1} b_i x^i\right) = 0 \text{ in } R[x]/(x^n).$$

So by Lemma 2.3, to show $\alpha\gamma\beta = 0$, it suffices to prove that

$$v := \sum_{i+j+k=n} a_i c_j b_k + \sigma(a, c)b + \sigma(ac, b) = 0.$$

Claim 1. $vab = vac = vbc = 0$.

Proof of Claim 1. In view of (2.5), one obtains $0 = uab = [abc_n + \sigma(ab, c)]ab$, so

$$0 = [abc_n + \sigma(ab, c)]ba = ab[c_n ba + \sigma(c, ba)];$$

hence $ba[c_n ba + \sigma(c, ba)] = 0$. Thus, in view of (2.5),

$$\begin{aligned}
bva &= b[ac_n b + \sigma(a, c)b + \sigma(ac, b)]a \\
&= bac_n ba + [b\sigma(a, c)]ba + [b\sigma(ac, b)]a \\
&= bac_n ba + [\sigma(ba, c) - \sigma(b, ac) + \sigma(b, a)c]ba \\
&\qquad + [\sigma(bac, b) - \sigma(b, acb) + \sigma(b, ac)b]a \\
&= bac_n ba + \sigma(ba, c)ba \\
&= ba[c_n ba + \sigma(c, ba)] = 0.
\end{aligned}$$

One also obtains

$$
\begin{aligned}
0 = cua &= c[ab_nc + \sigma(a,b)c + \sigma(ab,c)]a \\
&= cab_nca + [c\sigma(a,b)]ca + [c\sigma(ab,c)]a \\
&= cab_nca + [\sigma(ca,b) - \sigma(c,ab) + \sigma(c,a)b]ca \\
&\quad + [\sigma(cab,c) - \sigma(c,abc) + \sigma(c,ab)c]a \\
&= cab_nca + \sigma(ca,b)ca \\
&= [cab_n + \sigma(ca,b)]ca,
\end{aligned}
$$

so $0 = [cab_n + \sigma(ca,b)]ac = ca[b_nac + \sigma(b,ac)]$, implying that $ac[b_nac + \sigma(b,ac)] = 0$. Hence

$$
vac = [acb_n + \sigma(ac,b)]ac = ac[b_nac + \sigma(b,ac)] = 0.
$$

We also have

$$
\begin{aligned}
0 = bcu &= bc[a_nbc + \sigma(a,b)c + \sigma(ab,c)] \\
&= bca_nbc + [bc\sigma(a,b)]c + [bc\sigma(ab,c)] \\
&= bca_nbc + [\sigma(bca,b) - \sigma(bc,ab) + \sigma(bc,a)b]c \\
&\quad + [\sigma(bcab,c) - \sigma(bc,abc) + \sigma(bc,ab)c] \\
&= bca_nbc + \sigma(bc,a)bc \\
&= [bca_n + \sigma(bc,a)]bc,
\end{aligned}
$$

implying $0 = [bca_n + \sigma(bc,a)]cb = bc[a_ncb + \sigma(a,cb)$; so $cb[a_ncb + \sigma(a,cb)] = 0$. Thus,

$$
\begin{aligned}
cbv &= cb[a_ncb + \sigma(a,c)b + \sigma(ac,b)] \\
&= cb[a_ncb + \sigma(a,cb) + a\sigma(c,b)] \\
&= cb[a_ncb + \sigma(a,cb)] = 0.
\end{aligned}
$$

Therefore, $vab = vac = vbc = 0$.

Claim 2. $v\delta = v\eta = 0$, where $\delta = \sigma(a,c)b + \sigma(ac,b)$ and $\eta = \sigma(a,b)c + \sigma(ab,c)$.

Proof of Claim 2. In view of Claim 1, one obtains

$$
\begin{aligned}
v\delta av &= v[\sigma(a,c)b + \sigma(ac,b)]av = v[\sigma(ac,b)av] \\
&= v[ac\sigma(b,av)] = 0,
\end{aligned}
$$

so $v\delta a = 0$; and hence

$$
\begin{aligned}
v\delta^2 v &= v\delta[\sigma(a,c)b + \sigma(ac,b)]v = v\delta[a\sigma(c,b) + \sigma(a,cb)]v \\
&= [v\delta\sigma(a,cb)]v = \sigma(v\delta,a)cbv = 0,
\end{aligned}
$$

implying $v\delta = 0$. We also have

$$
\begin{aligned}
v\eta av &= v[\sigma(a,b)c + \sigma(ab,c)]av = v[\sigma(ab,c)av] \\
&= v[ab\sigma(c,av)] = 0,
\end{aligned}
$$

so $v\eta a = 0$; and hence

$$v\eta^2 v = v\eta[\sigma(a,b)c + \sigma(ab,c)]v = v\eta[a\sigma(b,c) + \sigma(a,bc)]v$$
$$= [v\eta\sigma(a,bc)]v = \sigma(v\eta,a)bcv = 0,$$

implying $v\eta = 0$.

For any integer l with $0 \le l \le n$, let $P(l)$ denote the condition that

$$P(l): \qquad va_ib_j = va_ic_j = vb_ic_j = 0 \text{ for all } i,j \text{ with } i+j = l.$$

Thus $P(0)$ holds by Claim 1. We now assume that $0 < l \le n$ and $P(s)$ holds for all $0 \le s < l$, i.e.,

$$(2.6) \qquad va_ib_j = va_ic_j = vb_ic_j = 0 \text{ for all } i,j \text{ with } i+j < l.$$

We next show that $P(l)$ holds. That is

$$(2.7) \qquad va_ib_j = va_ic_j = vb_ic_j = 0 \text{ for all } i,j \text{ with } i+j = l.$$

Let us fix non-negative integers s,t with $s+t = l$.

Claim 3. $(b_sc_t)(a_ib_jc_k) = 0$ for all i,j,k with $i+j+k = n$.

Proof of Claim 3. If $i < n - l$ then $i + s + t < n$; so $a_ib_sc_t = 0$ by (2.5) and hence $(b_sc_t)(a_ib_jc_k)v = 0$.

Now suppose $i = n-l$, so $s+t = j+k = l$. If $s \ne j$, then either $s > j$ or $s < j$; this implies that either $j + t < l$ or $s + k < l$, so $b_jc_tv = 0$ or $b_sc_kv = 0$ by (2.6). Hence $(b_sc_t)(a_{n-l}b_jc_k)v = 0$ if $s \ne j$. This clearly implies that $(b_sc_t)(a_{n-l}b_jc_k)v = 0$ if $t \ne k$.

To see $(b_sc_t)(a_{n-l}b_sc_t)v = 0$, note that,

$$0 = (b_sc_t)uv = (b_sc_t)\left(\sum_{i+j+k=n} a_ib_jc_k + \eta \right)v = (b_sc_t)(a_{n-l}b_sc_t)v.$$

It follows by Claims 2 and 3 that

$$(b_sc_t)v^2 = (b_sc_t)\left(\sum_{i+j+k=n} a_ic_jb_k + \delta \right)v = 0. \text{ So } (b_sc_t)v = 0.$$

Similar arguments show that $(a_sb_t)v = 0 = (a_sc_t)v$. So (2.7) holds. By induction principle, we have proved that

$$(2.8) \qquad va_ib_j = va_ic_j = vb_ic_j = 0 \text{ for all } i,j \text{ with } i+j \le n.$$

It now follows from (2.8) and Claim 2 that

$$v^2 = \left(\sum_{i+j+k=n} a_ic_jb_k + \delta \right)v = 0;$$

so $v = 0$. $\qquad\square$

A ring R is semicommutative if $ab = 0$ in R implies $aRb = 0$. Reversible rings are semicommutative (see [8]). By [7], for $\sigma = 0$, $B_2(R; 0)$ is semicommutative but not reversible (so not symmetric) for any reduced ring R with unity, and $B_n(R; 0)$ (for $n \geq 3$) is never semicommutative (for any ring R with unity).

Proposition 2.6. *If R is a reduced ring, then $B_2(R; \sigma)$ is semicommutative.*

Proof. Let $\alpha\beta = 0$ in $B_2(R; \sigma)$ where $\alpha = \begin{pmatrix} a & a_1 & a_3 \\ 0 & a & a_2 \\ 0 & 0 & a \end{pmatrix}$ and $\beta = \begin{pmatrix} b & b_1 & b_3 \\ 0 & b & b_2 \\ 0 & 0 & b \end{pmatrix}$, and let $\gamma := \begin{pmatrix} r & r_1 & r_3 \\ 0 & r & r_2 \\ 0 & 0 & r \end{pmatrix} \in B_2(R; \sigma)$. Because R is reduced, $\alpha\beta = 0$ gives

(2.9) $$ab = ab_1 = ab_2 = a_1b = a_2b = 0$$

and

(2.10) $$u := ab_3 + a_1b_2 + a_3b + \sigma(a, b) = 0.$$

Thus, by (2.9),

$$\alpha\gamma\beta = \begin{pmatrix} arb & arb_1 + (ar_1 + a_1r)b & v \\ 0 & arb & arb_2 + (ar_2 + a_2r)b \\ 0 & 0 & arb \end{pmatrix} = \begin{pmatrix} 0 & 0 & v \\ 0 & 0 & 0 \\ 0 & 0 & 0 \end{pmatrix}$$

where

$$v := arb_3 + ar_1b_2 + a_1rb_2 + ar_3b + a_1r_2b + a_3rb + \sigma(a, r)b + \sigma(ar, b)$$
$$= arb_3 + a_1rb_2 + a_3rb + \sigma(a, r)b + \sigma(ar, b).$$

So, to show $\alpha\gamma\beta = 0$, it suffices to show that $v = 0$. We notice that

$$va = [arb_3 + a_1rb_2 + a_3rb + \sigma(a, r)b + \sigma(ar, b)]a$$
$$= arb_3a + \sigma(ar, b)a = ar[b_3a + \sigma(b, a)] = 0,$$

because $a[b_3a + \sigma(b, a)] = [ab_3 + \sigma(a, b)]a = ua = 0 \cdot a = 0$, and that

$$bv = b[arb_3 + a_1rb_2 + a_3rb + \sigma(a, r)b + \sigma(ar, b)]$$
$$= b[a_3rb + \sigma(a, r)b + \sigma(ar, b)]$$
$$= b[a_3rb + a\sigma(r, b) + \sigma(a, rb)]$$
$$= ba_3rb + b\sigma(a, rb)$$
$$= [ba_3 + \sigma(b, a)]rb = 0,$$

because $[ba_3 + \sigma(b,a)]b = b[a_3b + \sigma(a,b)] = bu = b \cdot 0 = 0$. Moreover, by (2.10),

$$-va_1b_2v = v[ab_3 + a_3b + \sigma(a,b)]v$$
$$= v\sigma(a,b)v = va\sigma(b,v) = 0,$$

so $a_1b_2v = 0$. Therefore,

$$v^3 = v[arb_3 + a_1rb_2 + a_3rb + \sigma(a,r)b + \sigma(ar,b)]v$$
$$= v\sigma(ar,b)v = v(ar)\sigma(b,v) = 0.$$

So $v = 0$, as required.   □

3. Armendariz rings

A ring R is called Armendariz (resp., Armendariz of power series type) if, whenever $(\Sigma_{i\geq 0}a_ix^i)(\Sigma_{j\geq 0}b_jx^j) = 0$ in $R[x]$ (resp., in $R[[x]]$), $a_ib_j = 0$ for all i and j. It was proved by Anderson and Camillo in [1] that, for $n \geq 1$, R is a reduced ring iff $R[x]/(x^{n+1})$ is Armendariz.

Theorem 3.1. *Let R be a reduced ring and $n \geq 1$. Then $H_n(R;\sigma)$ is an Armendariz ring of power series type. In particular, $H_n(R;\sigma)$ is an Armendariz ring.*

Proof. Let $S = H_n(R;\sigma)$ and let

$$(3.1) \qquad (\alpha_0 + \alpha_1y + \cdots)(\beta_0 + \beta_1y + \cdots) = 0 \quad \text{in } S[[y]].$$

We need to show that $\alpha_i\beta_j = 0$ in S for all i and j. Write, for $i = 0, 1, \ldots$,

$$\alpha_i = \left(a_0^{(i)}, a_1^{(i)}, \ldots, a_n^{(i)}\right), \quad \beta_i = \left(b_0^{(i)}, b_1^{(i)}, \ldots, b_n^{(i)}\right)$$

and let

$$f_i = a_0^{(i)} + a_1^{(i)}x + \cdots + a_{n-1}^{(i)}x^{n-1}$$
$$g_i = b_0^{(i)} + b_1^{(i)}x + \cdots + b_{n-1}^{(i)}x^{n-1}$$

be in $R[x]/(x^n)$. Note that

$$H_n(R;\sigma) \to R[x]/(x^n), \quad (a_0, \ldots, a_n) \mapsto a_0 + \cdots + a_{n-1}x^{n-1},$$

is an onto homomorphism of rings; thus, it follows from (3.1) that

$$(f_0 + f_1y + \cdots)(g_0 + g_1y + \cdots) = 0 \quad \text{in } \left(R[x]/(x^n)\right)[[y]].$$

Since $R[x]/(x^n)$ is an Armendariz ring of power series type by [9, Theorem 20], it follows that $f_ig_j = 0$ in $R[x]/(x^n)$ for all i and j. Hence Lemma 2.2 gives, for any i and j,

$$(3.2) \qquad a_s^{(i)}b_t^{(j)} = 0 \quad \text{for all } s,t \text{ with } 0 \leq s+t \leq n-1.$$

For convenience, we write, for any i and j,

$$X_{i,j} = a_0^{(i)}b_n^{(j)} + a_1^{(i)}b_{n-1}^{(j)} + \cdots + a_n^{(i)}b_0^{(j)} + \sigma\big(a_0^{(i)}, b_0^{(j)}\big).$$

Because of (3.2), it is always true that

$$\big(a_0^{(i)}+a_1^{(i)}x+\cdots+a_{n-1}^{(i)}x^{n-1}\big)\big(b_0^{(j)}+b_1^{(j)}x+\cdots+b_{n-1}^{(j)}x^{n-1}\big) = 0 \ \text{in } R[x]/(x^n);$$

so, by Lemma 2.3,

(3.3) $\alpha_i\beta_j = 0$ if and only if $X_{i,j} = 0$.

We now assign a total order "$\leq$" to the set of the ordered pairs (i,j) of non-negative integers:

- If $i + j < s + t$, then $(i,j) < (s,t)$.
- For $i + j = s + t$, $(i,j) < (s,t)$ iff $i < s$.

We have $\alpha_0\beta_0 = 0$ by (3.1). We assume that $(0,0) < (s,t)$ and $\alpha_i\beta_j = 0$ for all $(i,j) < (s,t)$, and we prove next that $\alpha_s\beta_t = 0$. By (3.3), we assume that

(3.4) $X_{i,j} = 0$ for all $(i,j) < (s,t)$

and we want to show that $X_{s,t} = 0$.

By (3.1), $\sum_{i+j=s+t} \alpha_i\beta_j = 0$, i.e., $\sum_{i+j=s+t} X_{i,j} = 0$; so by (3.4) we have

(3.5) $X_{s,t} + X_{s+1,t-1} + \cdots + X_{s+t,0} = 0.$

Claim 1. $X_{s,t}a_0^{(s)} = 0.$

Proof of Claim 1. For any i, j with $i + j = s + t$ and $i > s$, we show that $X_{i,j}a_0^{(s)} = 0$; thus Claim 1 follows by (3.5). Because $X_{s,j} = 0$ by (3.4),

$$a_0^{(s)}\big(b_n^{(j)}a_0^{(s)} + \sigma(b_0^{(j)}, a_0^{(s)})\big)$$
$$= \big(a_0^{(s)}b_n^{(j)} + \sigma(a_0^{(s)}, b_0^{(j)})\big)a_0^{(s)}$$
$$= X_{s,j}a_0^{(s)} = 0 \cdot a_0^{(s)} = 0;$$

so

$$X_{i,j}a_0^{(s)}a_0^{(s)} = [a_0^{(i)}b_n^{(j)} + \cdots + a_n^{(i)}b_0^{(j)} + \sigma(a_0^{(i)}, b_0^{(j)})]a_0^{(s)}a_0^{(s)}$$
$$= [(a_0^{(i)}b_n^{(j)} + \sigma(a_0^{(i)}, b_0^{(j)}))a_0^{(s)}]a_0^{(s)}$$
$$= [a_0^{(i)}(b_n^{(j)}a_0^{(s)} + \sigma(b_0^{(j)}, a_0^{(s)}))]a_0^{(s)}$$
$$= a_0^{(i)}[(b_n^{(j)}a_0^{(s)} + \sigma(b_0^{(j)}, a_0^{(s)}))a_0^{(s)}]$$
$$= a_0^{(i)} \cdot 0 = 0.$$

This shows that $X_{i,j}a_0^{(s)} = 0$.

Claim 2. $X_{s,t}(a_k^{(s)} b_{n-k}^{(t)}) = 0$ or $b_{n-k}^{(t)} X_{s,t} a_k^{(s)} = 0$ for all k with $0 < k < n$.

Proof of Claim 2. For any i,j with $i+j = s+t$ and $i > s$, we show that $b_{n-k}^{(t)} X_{i,j} a_k^{(s)} = 0$; thus Claim 2 follows by (2.15). Because $X_{s,j} = 0$ by (2.14), $0 = b_{n-k}^{(t)} X_{s,j} a_k^{(i)} = b_{n-k}^{(t)} a_k^{(s)} b_{n-k}^{(j)} a_k^{(i)}$; so

$$
\begin{aligned}
b_{n-k}^{(t)} X_{i,j} a_k^{(s)} &= b_{n-k}^{(t)} [a_0^{(i)} b_n^{(j)} + \cdots + a_n^{(i)} b_0^{(j)} + \sigma(a_0^{(i)}, b_0^{(j)})] a_k^{(s)} \\
&= b_{n-k}^{(t)} a_k^{(i)} b_{n-k}^{(j)} a_k^{(s)} + b_{n-k}^{(t)} \sigma(a_0^{(i)}, b_0^{(j)}) a_k^{(s)} \\
&= b_{n-k}^{(t)} a_k^{(i)} b_{n-k}^{(j)} a_k^{(s)} + b_{n-k}^{(t)} a_0^{(i)} \sigma(b_0^{(j)}, a_k^{(s)}) \\
&= b_{n-k}^{(t)} a_k^{(i)} b_{n-k}^{(j)} a_k^{(s)} + 0 = 0.
\end{aligned}
$$

Claim 3. $b_0^{(t)} X_{s,t} = 0$.

Proof of Claim 3. We first show that

$$(3.6) \qquad b_0^{(k)} X_{i,j} = 0 \text{ for all } i,j,k \text{ with } i+j = s+t \text{ and } k < j.$$

In fact, because

$$
\begin{aligned}
[b_0^{(k)} a_n^{(i)} + \sigma(b_0^{(k)}, a_0^{(i)})] b_0^{(k)} &= b_0^{(k)} [a_n^{(i)} b_0^{(k)} + \sigma(a_0^{(i)}, b_0^{(k)})] \\
&= b_0^{(k)} X_{i,k} \\
&= b_0^{(k)} \cdot 0 \\
&= 0,
\end{aligned}
$$

one obtains

$$
\begin{aligned}
b_0^{(k)} b_0^{(k)} X_{i,j} &= b_0^{(k)} b_0^{(k)} [a_0^{(i)} b_n^{(j)} + \cdots + a_n^{(i)} b_0^{(j)} + \sigma(a_0^{(i)}, b_0^{(j)})] \\
&= b_0^{(k)} [b_0^{(k)} a_n^{(i)} b_0^{(j)} + b_0^{(k)} \sigma(a_0^{(i)}, b_0^{(j)})] \\
&= b_0^{(k)} [b_0^{(k)} a_n^{(i)} + \sigma(b_0^{(k)}, a_0^{(i)})] b_0^{(j)} \\
&= 0 \cdot b_0^{(j)} = 0;
\end{aligned}
$$

showing that $b_0^{(k)} X_{i,j} = 0$.

For any i,j with $i+j = s+t$ and $i > s$, we have

$$
\begin{aligned}
b_0^{(t)} X_{i,j} &= b_0^{(t)} [a_0^{(i)} b_n^{(j)} + \cdots + a_n^{(i)} b_0^{(j)} + \sigma(a_0^{(i)}, b_0^{(j)})] \\
&= b_0^{(t)} [a_n^{(i)} b_0^{(j)} + \sigma(a_0^{(i)}, b_0^{(j)})] \\
&= [b_0^{(t)} a_n^{(i)} + \sigma(b_0^{(t)}, a_0^{(i)})] b_0^{(j)} \in Rb_0^{(j)}.
\end{aligned}
$$

Thus, there exist $r_1, \ldots, r_t \in R$ such that

$$
-b_0^{(t)} X_{s,t} = b_0^{(t)}(X_{s+1,t-1} + \cdots + X_{s+t,0}) = r_1 b_0^{(t-1)} + r_2 b_0^{(t-2)} + \cdots + r_t b_0^{(0)}.
$$

It follows from (3.6) that $-b_0^{(t)} X_{s,t} X_{s,t} = \left(r_1 b_0^{(t-1)} + r_2 b_0^{(t-2)} + \cdots + r_t b_0^{(0)}\right) X_{s,t} = 0$; this shows that $b_0^{(t)} X_{s,t} = 0$.

Now by Claims 1-3, we have

$$
\begin{aligned}
(X_{s,t})^3 &= X_{s,t}[a_0^{(s)} b_n^{(t)} + \cdots + a_n^{(s)} b_0^{(t)} + \sigma\left(a_0^{(s)}, b_0^{(t)}\right)] X_{s,t} \\
&= X_{s,t} \sigma\left(a_0^{(s)}, b_0^{(t)}\right) X_{s,t} \\
&= X_{s,t} a_0^{(s)} \sigma\left(b_0^{(t)}, X_{s,t}\right) = 0,
\end{aligned}
$$

showing that $X_{s,t} = 0$. By induction principle, $X_{i,j} = 0$ for all i and j. The proof is now complete. $\qquad\square$

We do not know whether $H_n(R; \sigma)$ being Armendariz implies that R is reduced. By [6], for $\sigma = 0$, $B_2(R; 0)$ is Armendariz for any reduced ring R and $B_n(R)$ (for $n \geq 3$) is never Armendariz (for any ring R).

Proposition 3.2. Let R be a reduced ring. Then $B_2(R; \sigma)$ is an Armendariz ring of power series type. In particular, $B_2(R; \sigma)$ is an Armendariz ring.

Proof. Suppose

$$(3.7) \qquad (\alpha_0 + \alpha_1 y + \cdots)(\beta_0 + \beta_1 y + \cdots) = 0 \quad \text{in } B_2(R; \sigma)[[y]]$$

with $\alpha_i = \begin{pmatrix} a^{(i)} & a_1^{(i)} & a_3^{(i)} \\ 0 & a^{(i)} & a_2^{(i)} \\ 0 & 0 & a^{(i)} \end{pmatrix}$ and $\beta_i = \begin{pmatrix} b^{(i)} & b_1^{(i)} & b_3^{(i)} \\ 0 & b^{(i)} & b_2^{(i)} \\ 0 & 0 & b^{(i)} \end{pmatrix}$ for $i \geq 0$. We

need to show that $\alpha_i \beta_j = 0$ in $B_2(R; \sigma)$ for all i and j. Note that

$$\begin{pmatrix} a & a_1 & a_3 \\ 0 & a & a_2 \\ 0 & 0 & a \end{pmatrix} \mapsto a + a_1 x, \qquad \begin{pmatrix} a & a_1 & a_3 \\ 0 & a & a_2 \\ 0 & 0 & a \end{pmatrix} \mapsto a + a_2 x$$

are two homomorphisms from $B_2(R; \sigma)$ to $R[x]/(x^2)$; thus, it follows from (3.7) that

$$\left[(a^{(0)} + a_1^{(0)} x) + (a^{(1)} + a_1^{(1)} x) y + \cdots\right]\left[(b^{(0)} + b_1^{(0)} x) + (b^{(1)} + b_1^{(1)} x) y + \cdots\right]$$
$$= 0$$

$$\left[(a^{(0)} + a_2^{(0)} x) + (a^{(1)} + a_2^{(1)} x) y + \cdots\right]\left[(b^{(0)} + b_2^{(0)} x) + (b^{(1)} + b_2^{(1)} x) y + \cdots\right]$$
$$= 0$$

in $\left(R[x]/(x^2)\right)[[y]]$. Since $R[x]/(x^2)$ is an Armendariz ring of power series type by [9, Theorem 20], it follows that

$$(3.8) \quad \left[a^{(i)} + a_1^{(i)} x\right]\left[b^{(j)} + b_1^{(j)} x\right] = 0 \quad \text{and} \quad \left[a^{(i)} + a_2^{(i)} x\right]\left[b^{(j)} + b_2^{(j)} x\right] = 0$$

in $R[x]/(x^2)$ for all i and j. Hence, by Lemma 2.2, (3.8) gives, for any i and j,

$$(3.9) \qquad a^{(i)} b^{(j)} = a^{(i)} b_1^{(j)} = a^{(i)} b_2^{(j)} = a_1^{(i)} b^{(j)} = a_2^{(i)} b^{(j)} = 0.$$

Because of (3.9), for any i and j,

$$\alpha_i \beta_j = 0 \text{ iff } X_{i,j} := a^{(i)} b_3^{(j)} + a_1^{(i)} b_2^{(j)} + a_3^{(i)} b^{(j)} + \sigma\big(a^{(i)}, b^{(j)}\big) = 0.$$

As done in the proof of Theorem 3.1, we assign the same total order "$\leq$" to the set of ordered pairs of non-negative integers. Since $\alpha_0 \beta_0 = 0$ by (3.7), we have $X_{0,0} = 0$. We assume that $(0,0) < (s,t)$ and $X_{i,j} = 0$ for all $(i,j) < (s,t)$, and we prove next that $X_{s,t} = 0$. The same arguments proving Claims 1-3 in the proof of Theorem 3.1 show that

$$X_{s,t} a^{(s)} = 0, \ X_{s,t} a_1^{(s)} b_2^{(t)} = 0, \ \text{and} \ b^{(t)} X_{s,t} = 0.$$

Therefore,

$$\begin{aligned}
(X_{s,t})^3 &= X_{s,t} [a^{(s)} b_3^{(t)} + a_1^{(s)} b_2^{(t)} + a_3^{(s)} b^{(t)} + \sigma\big(a^{(s)}, b^{(t)}\big)] X_{s,t} \\
&= X_{s,t} \sigma\big(a^{(s)}, b^{(t)}\big) X_{s,t} \\
&= X_{s,t} a^{(s)} \sigma\big(b^{(t)}, X_{s,t}\big) = 0,
\end{aligned}$$

showing that $X_{s,t} = 0$. By induction principle, $X_{i,j} = 0$ for all i and j. $\qquad\qquad\square$

4. Abelian rings and uniquely clean rings

Throughout this section, all rings are assumed to have unity. A ring R is called abelian if every idempotent of R is central.

Lemma 4.1. *Let* $\sigma : R \times R \to R$ *be a Hochschild 2-cocycle. Then* $e\sigma(e, e) = \sigma(e, e)e$ *for all* $e^2 = e \in R$.

Proof. This is because that $0 = \sigma(ee, e) - \sigma(e, ee) = e\sigma(e, e) - \sigma(e, e)e$. $\qquad\square$

Lemma 4.2. *Let* $S = B_n(R; \sigma)$. *The following hold:*

(1) *If* $e^2 = e \in R$, *then* $[e; (1 - 2e)\sigma(e, e)E_{1,n+1}]$ *is an idempotent of* S.

(2) *If* R *is abelian then every idempotent of* S *is of the form as given in* (1).

Proof. (1) This is an easy calculation because of Lemma 4.1.

(2) Suppose that $\alpha = [a; (a_{ij})]$ is an idempotent of S. Then $\alpha^2 = \alpha$ gives

$$a = a^2,$$

$$a_{ii} = aa_{ii} + a_{ii}a \text{ for } i = 1, \ldots, n,$$

$$a_{ij} = aa_{ij} + a_{ii}a_{i+1,j} + \cdots + a_{j,j-1}a_{jj} + a_{ij}a \text{ for } i < j \text{ with } (i,j) \neq (1,n),$$

$$a_{1n} = aa_{1n} + a_{11}a_{2n} + \cdots + a_{1,n-1}a_{nn} + a_{1n}a + \sigma(a,a).$$

Since a is a central idempotent and $1 - 2a \in U(R)$, it easily follows that $a_{1n} = (1 - 2a)\sigma(a,a)$, and $a_{ij} = 0$ for all $1 \leq i \leq j \leq n$ with $(i,j) \neq (1,n)$. □

Proposition 4.3. *Let $n \geq 1$. The following are equivalent:*

(1) R is abelian.

(2) $\mathrm{H}_n(R;\sigma)$ is abelian.

(3) $\mathrm{B}_n(R;\sigma)$ is abelian.

Proof. (1) $\Rightarrow$ (3). Let $\alpha^2 = \alpha \in S := \mathrm{B}_n(R;\sigma)$ and let $\beta \in S$ be arbitrary with a the entry on the diagonal. By Lemma 4.2, $\alpha = [e; (1 - 2e)\sigma(e,e)E_{1,n+1}]$ where $e^2 = e \in R$. Since e is central, direct calculation shows that

$$\alpha\beta = \beta\alpha \Longleftrightarrow (1 - 2e)\sigma(e,e)a + \sigma(e,a) = a(1 - 2e)\sigma(e,e) + \sigma(a,e)$$

$$\Longleftrightarrow \sigma(e,e)a - a\sigma(e,e) + (1 - 2e)\sigma(e,a) - (1 - 2e)\sigma(a,e) = 0$$

(by multiplying the equality by unit $1 - 2e$).

But we have

$$\sigma(e,e)a - a\sigma(e,e) + (1 - 2e)\sigma(e,a) - (1 - 2e)\sigma(a,e)$$

$$= [\sigma(e,ea) - \sigma(e,a) + e\sigma(e,a)] - [\sigma(ae,e) - \sigma(a,e) + \sigma(a,e)e]$$

$$+ (1 - 2e)\sigma(e,a) - (1 - 2e)\sigma(a,e)$$

$$= \sigma(e,ea) - \sigma(ae,e) + e\sigma(a,e) - e\sigma(e,a)$$

$$= \sigma(e,ae) - \sigma(ea,e) + e\sigma(a,e) - \sigma(e,a)e = 0.$$

So α is central in S.

(3) $\Rightarrow$ (2). This is clear because $\mathrm{H}_n(R;\sigma)$ is a subring of $\mathrm{B}_n(R;\sigma)$.

(2) $\Rightarrow$ (1). Let $e^2 = e \in R$ and let $a \in R$. Write $\alpha = [e; (1 - 2e)\sigma(e,e)E_{1,n+1}]$ and $\beta = aI_{n+1}$ where I_{n+1} is the $(n+1) \times (n+1)$ identity matrix. Then α is an idempotent of S by Lemma 4.2, so it is central by hypothesis. Hence $\alpha\beta = \beta\alpha$. It follows that $ea = ae$. □

A ring is called clean (uniquely clean) if every element can be (uniquely) written as the sum of an idempotent and a unit. It is known that, for an ideal $I \subseteq J(R)$, a ring R is clean iff R/I is clean and idempotents lift modulo I. It follows quickly that, for any $n \geq 1$, R is clean iff

$H_n(R; \sigma)$ is clean. It is known that a ring R is uniquely clean iff so is $R[[x]]$ ([12, Corollary 10]). Because the image of a uniquely clean ring is again uniquely clean ([12, Theorem 22]), it follows that R is uniquely clean iff so is $R[x]/(x^{n+1})$. Moreover, if $\sigma = 0$ and R is uniquely clean then $B_n(R; 0)$ is uniquely clean by [12, Example 8]. Here we have the following observation.

Proposition 4.4. *Let $n \geq 1$. Then R is uniquely clean iff $H_n(R; \sigma)$ is uniquely clean iff $B_n(R; \sigma)$ is uniquely clean.*

Proof. Let $S = H_n(R; \sigma)$ and let $T = B_n(R; \sigma)$. The following facts hold:

  (1) If one of the rings R, S, T is clean then the same is true of the others.
  (2) If one of the rings R, S, T is abelian then the same is true of the others (by Proposition 4.3).
  (3) $R/J(R) \cong S/J(S) \cong T/J(T)$.

Now the claim follows by [12, Theorem 20]. □

We want to point out that $H_n(R; \sigma)$ and $R[x]/(x^{n+1})$ are not isomorphic in general.

Example 4.5. We use an example of Ohnuki, Takeda and Yamagata. Let $K = \mathbb{Z}_2(u, v)$ be the field of rational functions in two variables u and v over $\mathbb{Z}_2$. Let $K[X, Y]$ be the polynomial algebra in variables X and Y over K, $L = K[X, Y]/(X^2 - u, Y^2 - v)$ and x, y be the residue classes of X, Y in L, respectively. Define $\sigma : L \times L \to L$ by

$$\sigma(x^l y^m, x^p y^q) = lq x^{l+p} y^{m+q},$$

where $l, m, p, q \in \{0, 1\}$. By [13, Example 5.6, p.669], σ is a Hochschild 2-cocycle. Note that $\sigma(x^0 y^1, x^1 y^1) = 0$ and $\sigma(x^1 y^1, x^0 y^1) = xv \neq 0$. But it is easy to see that $H_\sigma(L, L)$ being commutative would imply that σ is symmetric (i.e., $\sigma(a, b) = \sigma(b, a)$ for all $a, b \in L$). Thus $H_\sigma(L, L)$ is not commutative, and hence $H_\sigma(L, L) \not\cong L[t]/(t^2)$ (where $L[t]$ is the polynomial ring in variable t over L) because the latter is commutative.

We conclude by giving some rings R for which $H_n(R; \sigma) \cong R[x]/(x^{n+1})$ or $H_\sigma(R, M) \cong H_0(R, M)$.

Lemma 4.6. *Let M be a bimodule over a ring R and let $\sigma : R \times R \to M$ be a Hochschild 2-cocycle. Then $H_\sigma(R, M) \cong H_0(R, M)$ provided $\sigma(r, s) = \sigma(r, 1)s$ for all $r, s \in R$.*

Proof. If $\sigma(r, s) = \sigma(r, 1)s$ for all $r, s \in R$, then it can be easily verified that $\theta : H_0(R, M) \to H_\sigma(R, M)$, $(r, x) \mapsto (r, x - \sigma(r, 1))$, is the required isomorphism. □

In the situations where $R = \mathbb{Z}$ or $R = \mathbb{Z}_n$, it is always true that $\sigma(r, s) = \sigma(r, 1)s$ for all $r, s \in R$. So the next two corollaries follow immediately.

Corollary 4.7. *Let $\sigma : \mathbb{Z} \times \mathbb{Z} \to M$ be a Hochschild 2-cocycle where M is an abelian group. Then $\mathrm{H}_\sigma(\mathbb{Z}, M) \cong \mathrm{H}_0(\mathbb{Z}, M)$.*

Corollary 4.8. *Let $\sigma : \mathbb{Z}_n \times \mathbb{Z}_n \to M$ be a Hochschild 2-cocycle where M is a bimodule over $\mathbb{Z}_n$. Then $\mathrm{H}_\sigma(\mathbb{Z}_n, M) \cong \mathrm{H}_0(\mathbb{Z}_n, M)$.*

Proposition 4.9. *Let M be a bimodule over a Boolean ring R and let $\sigma : R \times R \to M$ be a Hochschild 2-cocycle. Then $\mathrm{H}_\sigma(R, M) \cong \mathrm{H}_0(R, M)$ iff*

$$\sigma(a, b) = a\sigma(b, b) + \sigma(a, a)b + \sigma(ab, ab)$$

for all $a, b \in R$.

Proof. Let $S = \mathrm{H}_0(R, M)$ and let $T = \mathrm{H}_\sigma(R, M)$.

($\Leftarrow$). It is by hypothesis that the map

$$\theta : S \to T, \quad (a, x) \mapsto \big(a, x + \sigma(a, a)\big),$$

is the required isomorphism.

($\Rightarrow$). Suppose that $\theta : S \to T$ is an isomorphism. For $a \in R$, let $\theta(a, 0) = (a', y)$. Since $(a, 0)$ is an idempotent in S, $(a', y) = (a', y)^2 = (a', \sigma(a', a'))$. So

$$\theta(a, 0) = (a', \sigma(a', a')).$$

Since $\theta(J(S)) = J(T) = (0) \oplus M$, the map $a \mapsto a'$ is an automorphism of R. Moreover, for all $a, b \in R$, it follows from $\theta(ab, 0) = \theta(a, 0)\theta(b, 0)$ that

$$\sigma(a'b', a'b') = a'\sigma(b', b') + \sigma(a', a')b' + \sigma(a', b').$$

Replacing a' by a and b' by b in the equality gives that, for all $a, b \in R$,

$$\sigma(ab, ab) = a\sigma(b, b) + \sigma(a, a)b + \sigma(a, b).$$

$\square$

Proposition 4.10. *If R is a Boolean ring and $\sigma : R \times R \to R$ is a 2-cocycle, then $\mathrm{H}_n(R; \sigma) \cong R[x]/(x^{n+1})$.*

Proof. Let $a, b \in R$. We have

$$a\sigma(a, b) + \sigma(a, a)b = \sigma(a, b) + \sigma(a, ab)$$
$$a\sigma(b, a) + \sigma(a, b)a = \sigma(ab, a) + \sigma(a, ba)$$
$$b\sigma(a, a) + \sigma(b, a)a = \sigma(ba, a) + \sigma(b, a).$$

Adding up the left-handed sides, respectively the right-handed sides, of the three equalities gives

$$(4.1) \qquad\qquad \sigma(a, b) = \sigma(b, a).$$

We also have

$$a\sigma(b, ab) + \sigma(a, b)ab = \sigma(ab, ab) + \sigma(a, ab)$$
$$\sigma(b, b)a + b\sigma(b, a) = \sigma(b, ba) + \sigma(b, a).$$

Multiplying the last equality by a gives

$$\sigma(b, b)a + ab\sigma(b, a) = a\sigma(b, ba) + a\sigma(b, a).$$

Now, because of (4.1), adding up the left-handed sides, respectively the right-handed sides, of the first, the fifth, and the seventh equalities yields

$$a\sigma(a, b) + \sigma(a, a)b + a\sigma(b, ab) + \sigma(b, b)a$$
$$= \sigma(a, b) + \sigma(ab, ab) + a\sigma(b, ab) + a\sigma(b, a).$$

That is,

(4.2) $$\sigma(a, b) = a\sigma(b, b) + \sigma(a, a)b + \sigma(ab, ab).$$

Now it is by (4.1) and (4.2) that the mapping $H_n(R; \sigma) \to R[x]/(x^{n+1})$ given by $(a_0, \ldots, a_{n-1}, a_n) \mapsto (a_0, \ldots, a_{n-1}, a_n + \sigma(a_0, a_0))$, is the required isomorphism.   □

Acknowledgments. Part of the work was carried out when the third author was visiting Gebze Institute of Technology sponsored by TUBITAK. He gratefully acknowledges the support from TUBITAK and hospitality from the host university. The research of the second author was supported by NSC of Taiwan and by NCTS of Taipei, and that of the third author by a Discovery Grant from NSERC of Canada.

References

[1] D.D. Anderson and V. Camillo, *Armendariz rings and Gaussian rings*, Comm. Alg. **26**(7)(1998), 2265-2272.

[2] D.D. Anderson and V. Camillo, *Semigroups and rings whose zero products commute*, Comm. Alg. **27**(6)(1999), 2847-2852.

[3] H. Cartan and S. Eilenberg, *Homological Algebra*, Princeton Landmarks in Mathematics, 1973. Originally published in 1956.

[4] P.M. Cohn, *Reversible rings*, Bull. London Math. Soc. **31**(6)(1999), 641-648.

[5] C. Huh, H.K. Kim, N.K. Kim and Y. Lee, *Basic examples and extensions of symmetric rings*, J. Pure Appl. Alg. **202**(2005), 154-167.

[6] N.K. Kim and Y. Lee, *Armendariz rings and reduced rings*, J. Alg. **223**(2000), 477-488.

[7] N.K. Kim and Y. Lee, *Extensions of reversible rings*, J. Pure Appl. Alg. **185**(2003), 207-223.

[8] J. Lambek, *On the representation of modules by sheaves of factor modules*, Canad. Math. Bull. **14**(3)(1971), 359-368.

[9] T-K. Lee and Y. Zhou, *A unified approach to the Armendariz property for polynomial rings and power series rings*, Colloq. Math. **113**(1)(2008), 151-168.

[10] H. Lin and C-C. Xi, *On Hochschild extensions of reduced and clean rings*, Comm. Alg. **36**(2008), 388-396.
[11] G. Marks, *Reversible and symmetric rings*, J. Pure Appl. Alg. **174**(2002), 311-318.
[12] W.K. Nicholson and Y. Zhou, *Rings in which elements are uniquely the sum of an idempotent and a unit*, Glasgow Math. J. **46**(2004), 227-236.
[13] A. Skowronsk and K. Yamagata, *Selfinjective algebras of quasitilted type*, Trends in representation theory of algebras and related topics, 639-708, EMS Ser. Congr. Rep., Eur. Math. Soc., Zurich, 2008.

DEPARTMENT OF MATHEMATICS
GEBZE INSTITUTE OF TECHNOLOGY
GEBZE/KOCAELI, TURKEY
E-mail address: mtkosan@gyte.edu.tr

DEPARTMENT OF MATHEMATICS
NATIONAL TAIWAN UNIVERSITY, TAIPEI 106, TAIWAN
MEMBER OF MATHEMATICS DIVISION (TAIPEI OFFICE)
NATIONAL CENTER FOR THEORETICAL SCIENCES
E-mail address: tklee@math.ntu.edu.tw

DEPARTMENT OF MATHEMATICS AND STATISTICS
MEMORIAL UNIVERSITY OF NEWFOUNDLAND
ST. JOHN'S, NFLD A1C 5S7, CANADA
E-mail address: zhou@mun.ca

Proceedings of the Sixth China-Japan-Korea
International Conference on Ring Theory
June 27-July 2, 2011 Suwon, Korea

WHEN DO THE DIRECT SUMS OF MODULES
INHERIT CERTAIN PROPERTIES?

GANGYONG LEE, S. TARIQ RIZVI*, AND COSMIN ROMAN

ABSTRACT. It is of obvious interest to know whether an algebraic
property of modules is preserved by direct sums of such modules.
In this paper we provide a survey of this question for various
classes of modules of interest. The question of inheritance of a
property by direct sums of modules has been explored for the
classes of (quasi-)injective modules and some of their generalizations as a motivation for further work. In the main part of this
paper we provide latest results and developments on this question
for the related classes of Baer, quasi-Baer, and Rickart modules.
Examples are provided that delimit our results and explain the
notions. Some open problems are listed at the end of the paper.

1. Introduction

For a long time, algebraists have been interested in finding out when
do (certain) properties of modules (or of other algebraic structures) go
over to finite or infinite direct sums of modules (or such structures)?
Among other things, this quest has led one to the conditions needed for
the property of (quasi-)injectivity of modules and some of its generalizations to go to direct sums of such modules. In this survey paper,
starting from the notion of (quasi-)injectivity, we will consider this question about direct sums for some well-known classes of modules which
generalize injective modules or are related to the notion of injectivity by
other means. In particular, we will also consider this direct sum question
for the related classes of Baer, quasi-Baer, and Rickart modules. While
the question of when do the direct sums of modules with a property
P inherit the property P has been satisfactorily settled for the classes
of (quasi-)injective, (quasi-)continuous, and FI-extending modules, the
problem of a satisfactory characterization of when is a direct sum of
extending modules, extending, remains an open problem. There have

2010 Mathematics Subject Classification : 16D40, 16D50, 16D70, 16D80, 16S50.

Keywords : Direct sums, Rickart modules, Baer modules, (FI-)extending modules,
(quasi-) continuous modules, annihilators, idempotents, endomorphism rings.

*Corresponding author.

been a number of attempts to solve this open problem but with limited success. By an interesting result of Chatters and Khuri, a ring is right extending right nonsingular if and only if it is Baer and right cononsingular. This result on rings was extended to a module theoretic setting by Rizvi and Roman after the introduction of relevant notions for modules. That development has allowed us to connect the study of the class of extending modules to that of Baer modules and vice versa, under suitable conditions. In this paper, we will consider relevant properties of the classes of Baer, quasi-Baer, and Rickart modules needed in our study. We will discuss conditions needed for a direct sum of Baer, quasi-Baer, and Rickart modules to be Baer, quasi-Baer, and Rickart, respectively. A major part of the paper will be devoted to results related to these latter classes of modules in an attempt to bring the reader up to date on the latest developments in this newer area of research. It is hoped that some of these results and other related investigations may lead to a satisfactory answer for the question of the direct sum problem for the class of extending modules mentioned earlier. It will also be of interest if a solution to any of the general open problems listed at the end of the paper can be obtained. As a general observation, we will note from the results presented that for a direct sum to inherit a property under our consideration, one often needs some sort of 'relative conditions' between each of its direct summands. In addition, it will be seen that often for a property $\mathcal{P}$ to go to 'infinite' direct sums of modules with $\mathcal{P}$, one may require some kind of finiteness or additional conditions.

After this introduction, in Section 2 we consider injective modules and some of their generalizations. Connections between the classes of injective, quasi-injective, (quasi-)continuous, extending, and FI-extending modules are explored. Explicit examples are given which show that direct sums of modules belonging to each of these classes (except for the class of FI-extending modules) do not belong to these respective classes, in general. Necessary and sufficient conditions and other results for the direct sum problem are provided.

Our focus in Section 3 is on the related notions of Baer, Quasi-Baer, and Rickart rings and modules. Connections between extending and Baer properties are mentioned for rings via Chatters-Khuri's result (Theorem 3.1) and for modules via Theorem 3.4. The basic properties of these notions are discussed and highlighted. It is shown that in general direct sums of Baer, Quasi-Baer, and Rickart modules do not inherit each of these properties. This is the main topic of our discussions in the remainder of this paper.

Section 4 is devoted to results on direct sums of Baer and quasi-Baer modules. The direct sum problem of Baer modules is quite difficult. It is shown that the direct sum of Baer modules inherits the Baer property only under special conditions. For the quasi-Baer module case, it is shown that every free module over a quasi-Baer ring is quasi-Baer. More generally, any direct sum of copies of a quasi-Baer module is quasi-Baer.

Section 5 consists of latest developments on the direct sum problem in the theory of Rickart modules. Relative Rickart and relative C_2 condition are introduced to study the direct sums of Rickart modules. We use these conditions to obtain specific results for direct sums of Rickart modules to be Rickart.

In Section 6, we consider the question of when are free R-modules Rickart or Baer. A number of well-known classes of rings R are characterized via Rickart or Baer properties of certain classes of free R-modules. In particular, a ring R is right hereditary (resp., right semihereditary) iff every (resp., every finitely generated) free right R-module is Rickart. A ring R is right hereditary and semiprimary (resp., right semihereditary and left Π-coherent) iff every (resp., every finitely generated) free right R-module is Baer. As a consequence, a new characterization of a Prüfer domain R is obtained via Rickart (or Baer) property of its finitely generated free modules. An example of a module M is included showing that $M^{(n)}$ is a Baer module while $M^{(n+1)}$ is not Baer. It is shown that a ring R is von Neumann regular iff every finitely generated free R-module is Rickart with C_3 condition.

Throughout this paper, R is a ring with unity and M is a unital right R-module. For a right R-module M, $S = End_R(M)$ denotes the endomorphism ring of M; thus M can be viewed as a left S- right R-bimodule. For $\varphi \in S$, $Ker\varphi$ and $Im\varphi$ stand for the kernel and the image of φ, respectively. The notations $N \subseteq M$, $N \leq M$, $N \unlhd M$, $N \leq^{ess} M$, or $N \leq^{\oplus} M$ mean that N is a subset, a submodule, a fully invariant submodule, an essential submodule, or a direct summand of M, respectively. $M^{(n)}$ denotes the direct sum of n copies of M and $\mathrm{Mat}_n(R)$ denotes an $n \times n$ matrix ring over R. By $\mathbb{C}$, $\mathbb{R}$, $\mathbb{Q}$, $\mathbb{Z}$, and $\mathbb{N}$ we denote the set of complex, real, rational, integer, and natural numbers, respectively. $E(M)$ denotes the injective hull of M and $\mathbb{Z}_n$ denotes the $\mathbb{Z}$-module $\mathbb{Z}/n\mathbb{Z}$.

We also denote $r_M(I) = \{m \in M \mid Im = 0\}$, $r_S(I) = \{\varphi \in S \mid I\varphi = 0\}$ for $\emptyset \neq I \subseteq S$; $r_R(N) = \{r \in R \mid Nr = 0\}$, $l_S(N) = \{\varphi \in S \mid \varphi N = 0\}$ for $N \leq M$.

2. Injectivity and some of its generalizations

We begin with some basic definitions and examples.

Definition 2.1. *Let M and N be right R-modules. M is called N-injective if, $\forall\ N' \le N$ and $\forall\ \varphi : N' \to M$, $\exists\ \overline{\varphi} : N \to M$ such that $\overline{\varphi}|_{N'} = \varphi$. M is said to be quasi-injective if M is M-injective. M is called injective if M is N-injective for all right R-modules N.*

It is easy to see that any vector space is an injective module over its base field, and every semisimple module is always quasi-injective. The following examples motivate this study about direct sums.

Example 2.2. *Let $R = \prod_{i\in\mathbb{N}} \mathbb{F}$ be a product of fields $\mathbb{F}$ and let $M_i = \mathbb{F}_R$. Then $M = \bigoplus_{i\in\mathbb{N}} M_i$ is semisimple and $E(M) = \prod_{i\in\mathbb{N}} \mathbb{F}$, thus M is a quasi-injective R-module which is not injective.*

Example 2.3. *Let $R = \left(\begin{smallmatrix} \mathbb{F} & \mathbb{F} \\ 0 & \mathbb{F} \end{smallmatrix}\right)$ with $\mathbb{F}$ a field. Then $M = \left(\begin{smallmatrix} \mathbb{F} & \mathbb{F} \\ 0 & 0 \end{smallmatrix}\right)$ is an injective R-module and $N = \left(\begin{smallmatrix} 0 & 0 \\ 0 & \mathbb{F} \end{smallmatrix}\right)$ is a quasi-injective R-module. However, $M \oplus N = R$ is not a quasi-injective R-module.*

Example 2.4. *Consider $\mathbb{Z}_p$ and $\mathbb{Z}_{p^2}$, where p is a prime number. Each of these is a quasi-injective $\mathbb{Z}$-module. However, $\mathbb{Z}_p \oplus \mathbb{Z}_{p^2}$ is not a quasi-injective $\mathbb{Z}$-module.*

For the stronger notion of injective modules, the finite direct sums inherit the property.

Theorem 2.5. *$E = \bigoplus_{i=1}^{n} E_i$ is injective iff each E_i is injective for $1 \le i \le n$.*

Next, we introduce some finiteness conditions that will be required for results on infinite direct sums.

Definition 2.6. *Let $\mathcal{I}$ be an index set, and $\{M_\alpha\}_{\alpha\in\mathcal{I}}$ be a family of R-modules.*

(A_1) *For every choice of distinct $\alpha_i \in \mathcal{I}$ ($i \in \mathbb{N}$) and $m_i \in M_{\alpha_i}$, the ascending sequence $(\bigcap_{i\ge n} r_R(m_i))_{n\in\mathbb{N}}$ becomes stationary.*

(A_2) *For every choice of $x \in M_\alpha$ ($\alpha \in \mathcal{I}$) and $m_i \in M_{\alpha_i}$ for distinct $\alpha_i \in \mathcal{I}$ ($i \in \mathbb{N}$) such that $r_R(m_i) \ge r_R(x)$, the ascending sequence $(\bigcap_{i\ge n} r_R(m_i))_{n\in\mathbb{N}}$ becomes stationary.*

(A_3) *For every choice of distinct $\alpha_i \in \mathcal{I}$ ($i \in \mathbb{N}$) and $m_i \in M_{\alpha_i}$, if the sequence $(r_R(m_i))_{i\in\mathbb{N}}$ is ascending, then it becomes stationary.*

Note that $A_1 \Rightarrow A_2 \Rightarrow A_3$: Reverse implications do not hold true in general [36].

For the case of a specific infinite direct sum of injective module we need condition (A_1) to get the following:

Proposition 2.7. *(Proposition 1.10, [36]) $M = \bigoplus_{\alpha\in\mathcal{I}} M_\alpha$ is injective if and only if each M_α is injective and (A_1) holds where $\mathcal{I}$ is an index set.*

For an infinite direct sum of quasi-injective modules to be quasi-injective, we also need the slightly weaker chain condition (A_2):

Proposition 2.8. *(Proposition 1.18, [36]) The following conditions are equivalent for a direct sum decomposition of a module $M = \bigoplus_{\alpha \in \mathcal{I}} M_\alpha$:*

(a) M *is quasi-injective;*

(b) M_α *is quasi-injective and $\bigoplus_{\beta \in \mathcal{I} - \{\alpha\}} M_\beta$ is M_α-injective for every $\alpha \in \mathcal{I}$;*

(c) M_α *is M_β-injective for all $\alpha, \beta \in \mathcal{I}$ and (A_2) holds.*

Corollary 2.9. *(Corollary 1.19, [36]) $\bigoplus_{i=1}^{n} M_i$ is quasi-injective if and only if M_i is M_j-injective for all $1 \leq i, j \leq n$. In particular, $M^{(n)}$ is quasi-injective if and only if M is quasi-injective.*

In recent years the notions of continuous, quasi-continuous, and extending modules have garnered a great deal of interest. The remainder of this section is devoted to the direct sum problem for these classes of modules.

Definition 2.10. *Let M be a right R-module. Consider the following conditions:*

(C_1) *Every submodule of M is essential in a direct summand of M.*

(C_2) *If a submodule L of M is isomorphic to a direct summand of M, then L is a direct summand of M.*

(C_3) *If M_1 and M_2 are direct summands of M such that $M_1 \cap M_2 = 0$, then $M_1 \oplus M_2$ is a direct summand of M.*

A module M is called extending *(or CS) if it satisfies (C_1), M is called* continuous *if it satisfies (C_1) and (C_2), and M is called* quasi-continuous *if it satisfies (C_1) and (C_3). A ring R is called* right extending *or* right *(quasi-)continuous if R_R is extending or (quasi-)continuous, respectively.*

The notion of extending modules was generalized to that of FI-extending modules in [7].

Definition 2.11. *A right R-module M is called* FI-extending *if every fully invariant submodule of M is essential in a direct summand of M. (When $M = R_R$, the fully invariant submodules are precisely the 2-sided ideals of R.) A ring R is right FI-extending if R_R is an FI-extending module, i.e., every ideal of R is essential in a right ideal direct summand of R.*

Large classes of modules and rings are FI-extending, but not necessarily extending (for example, direct sums of uniform modules, the ring of upper triangular matrices over $\mathbb{Z}$, etc; see Example 2.13(v) and [7]).

Remark 2.12. *The hierarchy of the notions we have considered until now, goes as follows: Injective $\Rightarrow$ Quasi-injective $\Rightarrow$ Continuous $\Rightarrow$ Quasi-continuous $\Rightarrow$ Extending $\Rightarrow$ FI-extending.*

Example 2.13. *The following examples show that, in general, the reverse implications in Remark 2.12 are not true.*

  (i) $M = \bigoplus_{i \in \mathbb{N}} \mathbb{F}$ *is a quasi-injective R-module but not an injective R-module where $R = \prod_{i \in \mathbb{N}} \mathbb{F}$ with $\mathbb{F}$ a field. More precisely, the injective hull of M is $E(M) = \prod_{i \in \mathbb{N}} \mathbb{F}$.*

 (ii) *Let R be a ring which has only 3 right ideals but which is not left artinian. Then $M = R_R$ is continuous but not quasi-injective. Since if so, R will be right self-injective and hence quasi-Frobenius, a contradiction. (See page 337 in [17] for an explicit example.)*

(iii) $\mathbb{Z}_{\mathbb{Z}}$ *is quasi-continuous but not continuous ($\mathbb{Z}_{\mathbb{Z}} \cong n\mathbb{Z}$, but $n\mathbb{Z}$ is not a direct summand of $\mathbb{Z}_{\mathbb{Z}}$).*

 (iv) *Let $\mathbb{F}$ be a field and let $R = \left(\begin{smallmatrix} \mathbb{F} & \mathbb{F} \\ 0 & \mathbb{F} \end{smallmatrix}\right)$. Then R_R is an extending module but is not quasi-continuous.*

  (v) *The $\mathbb{Z}$-module $\bigoplus_{i=1}^{\infty} \mathbb{Z}$ is FI-extending but not extending. Further, let $R = \left(\begin{smallmatrix} \mathbb{Z} & \mathbb{Z} \\ 0 & \mathbb{Z} \end{smallmatrix}\right)$. Then R_R is FI-extending but not extending.*

The module $M = R_R$ in Example 2.3 and the module $M = \mathbb{Z}_p \oplus \mathbb{Z}_{p^2}$ in Example 2.4 also exhibit direct sums of (quasi-)continuous modules which are not (quasi-)continuous. Yet, in each of these cases, the direct sum is extending. In the next two examples, we present a situation where similar direct sums are not even extending.

Example 2.14. $\mathbb{Z}_p, \mathbb{Z}_{p^3}$ *are quasi-injective $\mathbb{Z}$-modules, where p is a prime number; consequently, each of these modules is quasi-injective hence (quasi-)continuous, and so extending. However, $\mathbb{Z}_p \oplus \mathbb{Z}_{p^3}$ is not an extending $\mathbb{Z}$-module.*

Example 2.15. *Direct sums of extending modules are not extending in general:*
(i) Let $M = \bigoplus_{i=1}^{\infty} \mathbb{Z}$. Then M is not an extending $\mathbb{Z}$-module, while the domain $\mathbb{Z}$ is uniform and hence extending;
(ii) Let $R = \mathbb{Z}[X]$. Thus R is a commutative Noetherian domain (hence quasi-continuous), but $R \oplus R$ is not an extending R-module.

These examples illustrate that some extra conditions will be required for a direct sum of (quasi-)continuous or extending modules to be (quasi-)continuous or extending, respectively. The next two results address this question for the classes of continuous and quasi-continuous modules, respectively.

Theorem 2.16. *(Theorem 3.16, [36]) The following conditions are equivalent for a module $M = \bigoplus_{\alpha \in \mathcal{I}} M_\alpha$:*

(a) M is continuous;
(b) M is quasi-continuous and the M_α are continuous;
(c) M_α is continuous and M_β-injective for all $\alpha \neq \beta$, and (A_2) holds.

Proof. See also [40] and Theorem 8 in [39]. $\qquad\qquad\qquad\qquad\square$

Theorem 2.17. *(Theorem 2.13, [36]) Let $\{M_\alpha\}_{\alpha\in\mathcal{I}}$ be a family of quasi-continuous modules. Then the following conditions are equivalent:*

(a) $M = \bigoplus_{\alpha\in\mathcal{I}} M_\alpha$ *is quasi-continuous;*
(b) $\bigoplus_{\beta\in\mathcal{I}-\{\alpha\}} M_\beta$ *is M_α-injective for every $\alpha \in \mathcal{I}$;*
(c) M_α *is M_β-injective for all $\alpha \neq \beta \in \mathcal{I}$ and (A_2) holds.*

Proof. See also [40] and Theorem 7 in [39]. $\qquad\qquad\qquad\qquad\square$

Corollary 2.18. *(Theorems 12 and 13, [38]) Let $M = \bigoplus_{i\in\mathcal{I}} M_i$ where $\mathcal{I}$ is an index set. If $\mathcal{I}$ is finite or R is right Noetherian, then M is continuous if and only if each M_i is continuous and M_j-injective for all $j \neq i \in \mathcal{I}$.*

Example 2.19. *(i) Any direct sum of simple modules is quasi-injective, but the direct sum of their injective hulls need not be injective.*
(ii) Let R be a domain, and $A_i = E(R)$ $(i = 0, 1, \dots)$. Then $\bigoplus_{i=0}^\infty A_i$ need not be quasi-continuous. Otherwise, $\bigoplus_{i=1}^\infty A_i$ is A_0-injective, hence $\bigoplus_{i=1}^\infty A_i$ is injective as $R \subset A_0$, thus $E(R)$ is Σ-injective, and R is a right Ore domain.

Proposition 2.20. *(Lemma 7.9, [13]) Let M_1 and M_2 be extending modules. Then $M = M_1 \oplus M_2$ is extending if and only if every closed submodule K of M with $K \cap M_1 = 0$ or $K \cap M_2 = 0$ is a direct summand of M.*

Proposition 2.21. *(Proposition 7.10, [13]) Let $M = M_1 \oplus \cdots \oplus M_n$ such that M_i is M_j-injective for all $1 \leq i \neq j \leq n$. Then M is extending iff each M_i is extending.*

In Proposition 2.21, the assumption that M_i is M_j-injective for all $1 \leq i \neq j \leq n$ is a sufficient but not necessary condition for the direct sum of extending modules to be extending (e.g, $\mathbb{Z} \oplus \mathbb{Z}$ is extending, but $\mathbb{Z}$ is not $\mathbb{Z}$-injective.)

To obtain conditions for a direct sum of extending modules to be extending, Mohamed and Müller renamed a notion studied by Oshiro earlier and called it Ojectivity to honor Oshiro for his contributions to this study.

Definition 2.22. *For M, N right R-modules, N is M-ojective if for any submodule $X \leq M$ and any homomorphism $\varphi : X \to N$, there* *exist decompositions* $M = M_1 \oplus M_2$ *and* $N = N_1 \oplus N_2$ *together with homomorphisms* $\varphi_1 : M_1 \to N_1$ *and* $\varphi_2 : N_2 \to M_2$, *such that* φ_2 *is one-to-one, and for* $x = m_1 + m_2$ *and* $\varphi(x) = n_1 + n_2$ *one has* $n_1 = \varphi_1(m_1)$ *and* $m_2 = \varphi_2(n_2)$. *If* N *is* M-*ojective and* M *is* N-*ojective, we say that* M *and* N *are mutually ojective.*

Example 2.23. *The* $\mathbb{Z}$-*module* $\mathbb{Z}_{p^n} \oplus \mathbb{Z}_{p^{n+1}}$ *is self-ojective but not self-injective.*

Theorem 2.24. *(Theorem 10, [37]) Let* $M = M_1 \oplus M_2$. *Then* M_i *is extending and is* M_j-*ojective for* $1 \leq i \neq j \leq 2$ *if and only if for any closed submodule* N, *we have* $M = N \oplus M_1' \oplus M_2'$ *with* $M_i' \leq M_i$ *for* $1 \leq i \leq 2$.

Recall that a decomposition $M = \bigoplus_{i \in \mathcal{I}} M_i$ is called *exchangeable* if for any summand N of M, we have $M = \bigoplus_{i \in \mathcal{I}} M_i' \oplus N$ with $M_i' \leq M_i$.

Theorem 2.25. *(Theorem 11, [37]) Let* $n \geq 2$ *be an integer and let* $M = \bigoplus_{i=1}^n M_i$. *Then the following conditions are equivalent:*
  (a) M *is extending and the decomposition is exchangeable;*
  (b) M_i *is extending, and* $M_1 \oplus \cdots \oplus M_{i-1}$ *and* M_i *are mutually ojective for* $2 \leq i \leq n$.

Theorem 2.26. *(Theorem 13, [37]) Let* $M = M_1 \oplus \cdots \oplus M_n$, *where the* M_i *are uniform. Then* M *is extending and the decomposition is exchangeable if and only if* M_i *is* M_j-*ojective for all* $i \neq j$.

More recently, another generalization of relative injectivity was introduced, which is a necessary rather than a sufficient condition for a direct sum of extending modules to be extending. This is in contrast to some of the other generalizations we have considered so far.

Definition 2.27. *([48]) We say that* M_2 *is relatively* *-*injective with respect to* M_1 *if,* $\forall K \leq^{ess} M_1$ *and* $\varphi : K \to M_2$, *there exist homomorphisms* $\alpha, \alpha', \beta, \beta'$ *such that the diagram below commutes:*

$$
\begin{array}{ccc}
M_2 & \xrightarrow{\alpha'} & M_1 \\
\uparrow \varphi & & \uparrow \beta' \\
K & \hookrightarrow & M_1, \\
\downarrow \varphi & & \downarrow \beta \\
M_2 & \xrightarrow{\alpha} & M_2
\end{array}
\qquad i.e. \qquad \alpha\varphi = \beta, \alpha'\varphi = \beta'
$$

with $0 \to K$ on the left.

and either (α, β) *or* (α', β') *is nontrivial.*

Example 2.28. $\mathbb{Z}_{p^k}$ *and* $\mathbb{Z}_{p^l}$ *are relative* *-*injective.*

A result due to Osofsky (Corollary 7.4, [13]) states that, for a uniserial module M with unique composition series $0 < U < V < M$, $M \oplus (V/U)$

is not extending, although both summands are uniform (in fact, one is simple). It can be checked that M and V/U are not relatively $*$-injective.

When $M_1 = M_2$ the definition of relative $*$-injectivity yields the following.

Definition 2.29. *We say that M is quasi-$*$-injective if, $\forall\, K \leq^{ess} M$, $\forall\, \varphi : K \to M$, $\exists\, \alpha \neq 0$, $\beta \neq 0$ such that $\alpha\varphi = \beta$ on K:*

$$
\begin{array}{ccc}
0 & \to & K & \to & M \\
& & \downarrow \varphi & & \downarrow \beta \\
& & M & \overset{\alpha}{\to} & M
\end{array}
$$

Example 2.30. $\mathbb{Z}_{\mathbb{Z}}$ *is quasi-$*$-injective, but it is not quasi-injective.*

We have the following result which is of interest to our present investigations.

Proposition 2.31. *If $M = M_1 \oplus M_2$ is extending then M_i is extending and relatively M_j-$*$-injective for any $1 \leq i \neq j \leq 2$.*

We mention that this condition is, however, not sufficient for direct sums of extending modules to inherit the property.

Finally, we consider the class of FI-extending modules. Recall that every fully invariant submodule of such a module is essential in a direct summand. The class of fully invariant submodules of a module includes some of its most well-known submodules such as its socle, its singular submodule, the Jacobson radical, or the second singular submodule etc. One interesting property of the class of FI-extending modules, is that it is *closed under direct sums without any additional requirements*. This also provides a motivation for the study of this notion.

Theorem 2.32. *(Theorem 1.3, [7]) Every direct sum of FI-extending modules is always FI-extending.*

Corollary 2.33. *(Corollary 1.4, [7]) Let M be a direct sum of extending (e.g., uniform) modules. Then M is FI-extending.*

In view of Corollary 2.33, we obtain that in any direct sum of extending modules, every fully invariant submodule will always be essential in a direct summand without any additional requirements.

So far we have discussed some of the cases when specific direct sums of modules with a property generalizing injectivity, inherit that property. We conclude this section by a result of Matlis and Papp showing that if the direct sum of any family of injective R-modules is injective then the ring has to be right noetherian.

Theorem 2.34. *(Theorem 1.11, [36]) R is right noetherian if and only if $\bigoplus_{i \in \mathcal{I}} E_i$ is injective for any index set $\mathcal{I}$ and any family of injective modules $\{E_i\}_{i \in \mathcal{I}}$.*

3. Baer, quasi-Baer, and Rickart modules

Kaplansky introduced the notion of Baer rings in 1955 [26] which was extended to that of quasi-Baer rings by Clark in 1967 [12]. These two notions have their roots in functional analysis. A ring R is called (*quasi-*)*Baer* if the right annihilator of any nonempty subset (two-sided ideal) of R is generated by an idempotent as a right ideal. Examples of Baer rings include (any product of) domains, the Boolean ring of all subsets of a given set, and the ring of all bounded operators on a Hilbert space. Examples of quasi-Baer rings include any prime ring and any n-by-n upper triangular matrix ring over a domain. A prime ring with a nonzero singular ideal is quasi-Baer but not Baer. Similarly, an n-by-n upper triangular matrix ring over a domain which is not a division ring is quasi-Baer but not Baer, e.g., $\left(\begin{smallmatrix} \mathbb{Z} & \mathbb{Z} \\ 0 & \mathbb{Z} \end{smallmatrix} \right)$.

In 1980, Chatters and Khuri proved a very useful result. Recall that a ring R is called *right nonsingular* if $\{t \in R \mid r_R(t) \leq^{ess} R_R\} = 0$. R is called *right cononsingular* if any right ideal, with zero left annihilator, is essential in R_R.

Theorem 3.1. *(Theorem 2.1, [9]) A ring R is right nonsingular, right extending if and only if R is a right cononsingular Baer ring.*

The notions of Baer and quasi-Baer rings were extended to a general module theoretic setting using the endomorphism ring of a module in 2004 [44].

Definition 3.2. *([44]) A right R-module M is called Baer if the right annihilator in M of any nonempty subset of $End_R(M)$ is a direct summand of M.*

Example 3.3. *Every semisimple module is a Baer module. R_R is a Baer module if R is a Baer ring. Every nonsingular injective (or extending) module is Baer. $\mathbb{Z}^{(\mathbb{N})}$ ($\cong \mathbb{Z}[x]$) is a Baer $\mathbb{Z}$-module, while $\mathbb{Z}^{(\mathbb{R})}$ is not a Baer $\mathbb{Z}$-module.*

Recall that a module M is said to have the *strong summand intersection property* (SSIP) if the intersection of any family of direct summands is a direct summand of M. Examples include Baer modules and indecomposable modules. The $\mathbb{Z}$-module $\mathbb{Z}^{(\mathbb{N})}$ has the SSIP, while the $\mathbb{Z}$-module $\mathbb{Z}^{(\mathbb{R})}$ has not the SSIP.

It was proved in [44] that similar to Theorem 3.1, there are close connections between extending modules and Baer modules. A module M

is called $\mathcal{K}$-*nonsingular* if, for all $\varphi \in End_R(M), Ker\varphi \leq^{ess} M$ implies $\varphi = 0$. M is called $\mathcal{K}$-*cononsingular* if, for all $N \leq M, l_S(N) = 0$ implies $N \leq^{ess} M$.

The next useful result explicitly characterizes an extending module in terms of a Baer module analogous to the ring case in Theorem 3.1. This result may help translate results on (direct sums of) Baer modules to those on (direct sums of) extending modules and vice-versa.

Theorem 3.4. *(Theorem 2.12, [44]) A module M is $\mathcal{K}$-nonsingular and extending iff M is a $\mathcal{K}$-cononsingular Baer module.*

Proposition 3.5. *The following statements hold true:*

  (i) *Every direct summand of a Baer module is a Baer module.*
  (ii) *Every Baer module has the SSIP.*
  (iii) *The endomorphism ring of a Baer module is a Baer ring.*
  (iv) *A finitely generated $\mathbb{Z}$-module M is Baer iff M is semisimple or torsion-free.*

Proof. See Theorem 2.17, Proposition 2.22, Theorem 4.1, and Proposition 2.19 in [44]. □

A module M is said to be *retractable* if, for every $0 \neq N \leq M$, $Hom_R(M, N) \neq 0$. Note that every free R-module is retractable. The next result shows that the property of retractability passes to arbitrary direct sums of copies of a retractable module.

Lemma 3.6. *(Lemma 2.8, [47]) Let $\{M_i\}_{i \in \mathcal{I}}$ be a class of retractable modules. Then $\bigoplus_{i \in \mathcal{I}} M_i$ is retractable.*

We recall a result of Khuri (Theorem 3.2, [29]).

Theorem 3.7. *Let M_R be nonsingular and retractable. Then $End_R(M)$ is a right extending ring if and only if M is a extending module.*

The evident close connections between Baer and extending modules (Theorem 3.4) suggest that a similar result of Theorem 3.7 could possibly hold true for the case of Baer modules. In the following we show that this is the case:

Proposition 3.8. *(Proposition 4.6, [44]) Let M be a retractable module. If the endomorphism ring of M is a Baer ring then M is a Baer module.*

Theorem 3.9. *(Proposition 2.22, [44] and Theorem 2.5, [47]) The following conditions are equivalent for a module M:*

  (a) *M is a Baer module;*
  (b) *M has the SSIP and $Ker\varphi \leq^{\oplus} M$ for all $\varphi \in S$;*
  (c) *$End_R(M)$ is a Baer ring and M is quasi-retractable.*

Recall that a module M is called *quasi-retractable* if, for any left ideal I of $End_R(M)$ and $0 \neq r_M(I)$, $Hom_R(M, r_M(I)) \neq 0$. Note that every retractable module is quasi-retractable. Next, the following example (due to Chatters) exhibits an R-module which is quasi-retractable but not retractable.

Example 3.10. *(Example 3.4, [28]) Let K be a subfield of complex numbers $\mathbb{C}$. Let R be the ring $\left[\begin{smallmatrix} K & C \\ 0 & C \end{smallmatrix}\right]$. Then R is a right nonsingular right extending ring. Consider the module $M = eR$ where $e = \left(\begin{smallmatrix} 1 & 0 \\ 0 & 0 \end{smallmatrix}\right)$. Then M is projective, extending, and nonsingular (as it is a direct summand of R), hence is Baer by Theorem 3.4. Thus M is quasi-retractable, by Theorem 3.9. But M is not retractable, since the endomorphism ring of M, which is isomorphic to K, consists of isomorphisms and the zero endomorphism; on the other hand, M is not simple, and retractability implies that there exist nonzero endomorphisms which are not onto.*

Proposition 3.11. *(Proposition 4.6, [44]) The endomorphism ring of a free module F_R is a Baer ring if and only if F_R is a Baer module.*

Definition 3.12. *([44]) A right R-module M is called* quasi-Baer *if for all $N \trianglelefteq M$, $l_S(N) = Se$ for some $e^2 = e \in S = End_R(M)$.*

Example 3.13. *All semisimple modules are quasi-Baer; all Baer and quasi-Baer rings are quasi-Baer modules, viewed as modules over themselves. Baer modules are obviously quasi-Baer modules. Every finitely generated abelian group is quasi-Baer. Every direct sum of copies of a quasi-Baer module is a quasi-Baer module.*

The class of quasi-Baer modules is strictly larger than that of Baer modules, as the next example will show.

Example 3.14. $\mathbb{Z}^{(\mathbb{R})}$ *is a quasi-Baer module, but is not a Baer module.*

Similar to Theorems 3.1 and 3.4, we show that there are also close connections between the class of FI-extending modules and that of quasi-Baer modules ([44]). A module M is called *FI-$\mathcal{K}$-nonsingular* if, for all $I \trianglelefteq S$ such that $r_M(I) \leq^{ess} eM$ for $e^2 = e \in S$, we have $r_M(I) = eM$. M is called *FI-$\mathcal{K}$-cononsingular* if, for every $N \trianglelefteq^{\oplus} M$ and $N' \trianglelefteq N$ such that $\varphi(N') \neq 0$ for all $\varphi \in End_R(M)$, we have $N' \leq^{ess} N$.

Theorem 3.15. *(Theorem 3.10, [44]) A module M is FI-$\mathcal{K}$-nonsingular and FI-extending iff M is quasi-Baer and FI-$\mathcal{K}$-cononsingular.*

Corollary 3.16. *(Corollary 3.16, [44]) A ring R is right FI-extending and right FI-$\mathcal{K}$-nonsingular if and only if R is quasi-Baer and right FI-$\mathcal{K}$-cononsingular.*

Proposition 3.17. *(Proposition 3.8, [44]) The following equivalences hold true for a module M and $S = End_R(M)$.*

(i) M is FI-$\mathcal{K}$-nonsingular if and only if, for all $I \trianglelefteq S$, $r_M(I) \leq^{ess}$ eM for $e^2 = e \in S$, implies $I \cap Se = 0$;

(ii) M is FI-$\mathcal{K}$-cononsingular if and only if, for all $N \trianglelefteq M$, $r_M(l_S(N))$ $\leq^{\oplus} M$ implies $N \leq^{ess} r_M(l_S(N))$.

Proposition 3.18. *(Theorems 3.17 and 4.1, [44]) The following statements hold true:*

(i) *Every direct summand of a quasi-Baer module is a quasi-Baer module.*

(ii) *The endomorphism ring of a quasi-Baer module is a quasi-Baer ring.*

Example 3.19. *Let $M = \mathbb{Z}_{p^\infty}$, considered as a $\mathbb{Z}$-module. Then it is well-known that $End_{\mathbb{Z}}(M)$ is the ring of p-adic integers. Since the ring of p-adic integers is a commutative domain, it is a (quasi-)Baer ring. However, M is not a (quasi-)Baer module.*

Proposition 3.20. *(Proposition 4.7, [44]) Let M be retractable. Then M is quasi-Baer if and only if $End_R(M)$ is a quasi-Baer ring.*

We remark that Example 3.27 also exhibits that a direct sum of two quasi-Baer modules need not be quasi-Baer, in general.

Next, we consider the notion of a right Rickart ring, introduced independently by Maeda and Hattori in 1960. Maeda defined a ring R to be *right Rickart* if the right annihilator of an element is generated by an idempotent, as a right ideal. Hattori called a ring a *right p.p. ring* if all of its principal right ideals are projective. Examples include von Neumann regular rings, Baer rings, right (semi-)hereditary rings, and $End_R(R^{(\mathcal{I})})$ with R a right hereditary ring and $\mathcal{I}$ an index set. It was discovered that the notion of a right Rickart ring coincides with that of a right p.p. ring. Motivated by the work on Baer modules [44] and the definition of a right Rickart ring we introduce the notion of a Rickart module. We do this by utilizing the endomorphism ring of a module, similar to the case of Baer modules.

Definition 3.21. *([45]) A right R-module M is called Rickart if the right annihilator in M of any single element of $End_R(M)$ is a direct summand of M.*

Example 3.22. *Every semisimple module is a Rickart module. R_R is a Rickart module if R is a right Rickart ring. Every Baer module is a Rickart module. Every projective right R-module over a right hereditary ring R is a Rickart module. The free $\mathbb{Z}$-module $\mathbb{Z}^{(\mathcal{I})}$, for any index set $\emptyset \neq \mathcal{I}$, is Rickart, while $\mathbb{Z}^{(\mathcal{I})}$ is not a Baer $\mathbb{Z}$-module if $\mathcal{I}$ is uncountable. In particular, $\mathbb{Z}^{(\mathbb{N})}$ ($\cong \mathbb{Z}[x]$) is a Rickart (and Baer) $\mathbb{Z}$-module, while*

$\mathbb{Z}^{(\mathbb{R})}$ is a Rickart but not a Baer $\mathbb{Z}$-module. In general, if R is a right hereditary ring which is not Baer then every free R-module is Rickart but not Baer. On the contrary, $\mathbb{Z}_{p^\infty}$ and $\mathbb{Z}_4$ are injective and quasi-injective $\mathbb{Z}$-modules, respectively, neither of which is a Rickart $\mathbb{Z}$-module.

Note that an indecomposable Rickart module is a Baer module by Theorem 3.9. The next example provides another explicit instance when a Rickart module is not a Baer module.

Example 3.23. Let $T = \{(a_n)_{n=1}^\infty \in \prod_{n=1}^\infty \mathbb{Z}_2 \mid a_n$ is eventually constant$\}$ and $I = \{(a_n)_{n=1}^\infty \in \prod_{n=1}^\infty \mathbb{Z}_2 \mid a_n = 0$ eventually$\} = \bigoplus_{n=1}^\infty \mathbb{Z}_2$. Now, consider the ring $R = \left(\begin{smallmatrix} T & T/I \\ 0 & T/I \end{smallmatrix}\right)$ and the idempotent $e = \left(\begin{smallmatrix} (1,1,\dots) & 0+I \\ 0 & 0+I \end{smallmatrix}\right) \in R$. Then $M = eR = \left(\begin{smallmatrix} T & T/I \\ 0 & 0 \end{smallmatrix}\right)$ is a Rickart module, but is not Baer because $r_M \left(\left(\begin{smallmatrix} I & 0 \\ 0 & 0 \end{smallmatrix}\right)\right) = \left(\begin{smallmatrix} 0 & T/I \\ 0 & 0 \end{smallmatrix}\right)$ is not a direct summand of M where $\left(\begin{smallmatrix} I & 0 \\ 0 & 0 \end{smallmatrix}\right) \subseteq End_R(M)$.

Recall that a module M is said to have the *summand intersection property* (SIP) if the intersection of any two direct summands is a direct summand of M. M is said to satisfy D_2 *condition* if, $\forall N \leq M$ with $M/N \cong M' \leq^\oplus M$, we have $N \leq^\oplus M$.

Proposition 3.24. *The following statements hold true:*

  (i) *Every direct summand of a Rickart module is a Rickart module.*
 (ii) *Every Rickart module satisfies D_2 condition.*
(iii) *Every Rickart module is $\mathcal{K}$-nonsingular.*
 (iv) *Every Rickart module has the SIP.*
  (v) *The endomorphism ring of a Rickart module is a right Rickart ring.*

Proof. See Theorem 2.7 and Propositions 2.11, 2.12, 2.16, and 3.2 in [32]. □

Theorem 3.25. *(Proposition 2.11 and Theorem 3.9, [32]) The following conditions are equivalent for a module M and $S = End_R(M)$:*

 (a) *M is a Rickart module;*
 (b) *M satisfies D_2 condition, and $Im\varphi$ is isomorphic to a direct summand of M for any $\varphi \in S$;*
 (c) *S is a right Rickart ring and M is k-local-retractable.*

Recall that a module M is called *k-local-retractable* if $r_M(\varphi) = r_S(\varphi)(M)$ for any $\varphi \in S = End_R(M)$. Note that every free R-module is k-local-retractable.

Proposition 3.26. *(Corollary 5.3, [32]) The endomorphism ring of a free module F_R is a right Rickart ring if and only if F_R is a Rickart module.*

Until now, we have been focused on developing the notions and presenting basic properties of Baer, quasi-Baer, and Rickart modules. As we see from Propositions 3.5(i), 3.18(i), and 3.24(i), every direct summand each of Baer, quasi-Baer, and Rickart modules inherits the respective property. It is, therefore, natural to ask if these properties go to their respective direct sums? The next three examples and the following proposition show that this is not the case, in general. In fact, the Baer and Rickart properties are not inherited by even a direct sum of copies of such modules. In contrast, we will see in Corollary 4.15 that a direct sum of copies of a quasi-Baer module does, in fact, inherit the quasi-Baer property. The focus of our investigations in Sections 4, 5, and 6, will be these direct sum questions.

Example 3.27. *It is easy to see that $\mathbb{Z}$ and $\mathbb{Z}_p$ are both Baer $\mathbb{Z}$-modules where p is a prime number in $\mathbb{N}$. However, the $\mathbb{Z}$-module $M = \mathbb{Z} \oplus \mathbb{Z}_p$ is not Rickart: Consider the endomorphism $\varphi \in End_R(M)$ defined by $\varphi : (m, \overline{n}) \mapsto (0, \overline{m})$, then $Ker\varphi = p\mathbb{Z} \oplus \mathbb{Z}_p \leq^{ess} M$ is not a direct summand of M.*

Example 3.28. *Let $R = \left(\begin{smallmatrix} \mathbb{Z} & \mathbb{Z} \\ 0 & \mathbb{Z} \end{smallmatrix}\right)$, $e_1 = \left(\begin{smallmatrix} 1 & 0 \\ 0 & 0 \end{smallmatrix}\right)$, $e_2 = \left(\begin{smallmatrix} 0 & 0 \\ 0 & 1 \end{smallmatrix}\right)$. Then the modules $e_1 R = \left(\begin{smallmatrix} \mathbb{Z} & \mathbb{Z} \\ 0 & 0 \end{smallmatrix}\right)$ and $e_2 R = \left(\begin{smallmatrix} 0 & 0 \\ 0 & \mathbb{Z} \end{smallmatrix}\right)$ are Baer R-modules (since $End(e_1 R) \cong \mathbb{Z} \cong End(e_2 R)$). But $M = R_R$ is not a Rickart module. Since $End_R(M) \cong R$, the only direct summands of M are: $\left(\begin{smallmatrix} 0 & 0 \\ 0 & 0 \end{smallmatrix}\right)$, $\left(\begin{smallmatrix} \mathbb{Z} & \mathbb{Z} \\ 0 & \mathbb{Z} \end{smallmatrix}\right)$, $\left(\begin{smallmatrix} \mathbb{Z} & \mathbb{Z} \\ 0 & 0 \end{smallmatrix}\right)$ and $\left(\begin{smallmatrix} 0 & n \\ 0 & 1 \end{smallmatrix}\right)\mathbb{Z}$ where $n \in \mathbb{Z}$. Consider $\left(\begin{smallmatrix} 2 & 1 \\ 0 & 0 \end{smallmatrix}\right) \in End_R(M)$. Then $r_M\left(\left(\begin{smallmatrix} 2 & 1 \\ 0 & 0 \end{smallmatrix}\right)\right) = \left(\begin{smallmatrix} 0 & -1 \\ 0 & 2 \end{smallmatrix}\right)\mathbb{Z}$ is not a direct summand of M.*

Example 3.29. *$M = \mathbb{Z}[x]$ is a Baer $\mathbb{Z}[x]$-module, but $M \oplus M$ is not a Rickart $\mathbb{Z}[x]$-module (more details in Example 6.8).*

Our next result extends Example 3.27 to arbitrary modules and motivates our study.

Proposition 3.30. *(Proposition 2.1, [34]) If M is an indecomposable Rickart module which has a nonzero maximal submodule N, then $M \oplus (M/N)$ is not a Baer module, while M and M/N are Baer modules.*

4. Direct sums of Baer and quasi-Baer modules

In this section we will show results on direct sums of Baer and quasi-Baer modules and list conditions which allow for such direct sums to inherit these properties. Recall that a sufficient condition for a finite direct sum of extending modules to be extending is that each direct summand be relatively injective to all others (see [22] or Proposition 7.10 in [13]). We prove that an analogue holds true also for the case of Baer modules.

Definition 4.1. *(Definition 1.3, [47], see also [32]) A module M is called N-Rickart (or relatively Rickart to N) if, for every homomorphism φ : $M \to N$, $Ker\varphi \leq^{\oplus} M$.*

Theorem 4.2. *(Theorem 3.19, [47]) Let $\{M_i\}_{1 \leq i \leq n}$ be a class of Baer modules where $n \in \mathbb{N}$. Assume that, for any $i \neq j$, M_i and M_j are relative Rickart and relative injective. Then $\bigoplus_{i=1}^{n} M_i$ is a Baer module.*

The preceding result was improved further by reducing the requirement of relative injectivity to a smaller subset of a finite index set $\mathcal{I}$ in [34] as follows.

Proposition 4.3. *(Proposition 2.14, [34]) Assume that there exists an ordering $\mathcal{I} = \{1, 2, \ldots, n\}$ for a class of Baer R-modules $\{M_i\}_{i \in \mathcal{I}}$ such that M_i is M_j-injective for all $i < j \in \mathcal{I}$. Then $\bigoplus_{i=1}^{n} M_i$ is a Baer module if and only if M_i is M_j-Rickart for all $i, j \in \mathcal{I}$.*

Using Proposition 4.3, we obtain the following useful consequence. First recall that in view of Theorem 3.4, every nonsingular extending module is Baer.

Theorem 4.4. *(Theorem 2.16, [34]) Let M be a nonsingular extending module. Then M and $E(M)$ are relatively Rickart to each other and $E(M) \oplus M$ is a Baer module.*

Remark 4.5. *In the hypothesis of Theorem 4.4, it suffices to have that $E(M)$ be $\mathcal{K}$-nonsingular instead of M to be nonsingular. Since the $\mathcal{K}$-nonsingularity of $E(M)$ is inherited by M (see Proposition 2.18, [46]), the hypothesis of Theorem 4.4 can be improved to "if M is extending and $E(M)$ is $\mathcal{K}$-nonsingular then $E(M) \oplus M$ is a Baer module".*

Next example shows that the extending condition for the module M in Theorem 4.4 is not superfluous.

Example 4.6. *Let $A = \prod_{n=1}^{\infty} \mathbb{Z}_2$. Then the ring A is commutative, von Neumann regular, and Baer. Consider $R = \{(a_n)_{n=1}^{\infty} \in A \mid a_n$ is eventually constant$\}$, a subring of A. Then R is a von Neumann regular ring which is not a Baer ring (see Example 7.54, [30]). Note that $M = R_R$ is not extending, but is a nonsingular Rickart module. On the other hand, the injective hull, $E(M) = A$, is an injective Rickart R-module. In this case, $E(M)$ is M-injective and M-Rickart, but M is not $E(M)$-Rickart: For $\varphi = (1, 0, 1, 0, \ldots, 1, 0, \ldots) \in Hom_R(M, E(M))$, $Ker\varphi$ is not a direct summand of M. Thus, $E(M) \oplus M$ is not a Rickart module.*

The nonsingular condition for the module M in Theorem 4.4 is not superfluous (and cannot be weakened to $\mathcal{K}$-nonsingularity) as the next example shows.

Example 4.7. *Consider the $\mathbb{Z}$-module $M = \mathbb{Z}_p$ where p is a prime number. Then M is not nonsingular but is $\mathcal{K}$-nonsingular extending. However, $E(M) = \mathbb{Z}_{p^\infty}$ is not a Rickart $\mathbb{Z}$-module. Thus, $E(M) \oplus M = \mathbb{Z}_{p^\infty} \oplus \mathbb{Z}_p$ is not a Rickart $\mathbb{Z}$-module.*

Corollary 4.8. *(Corollary 2.20, [34]) If M is a nonsingular extending module then $E(M)^{(n)} \oplus M$ is a Baer module for any $n \in \mathbb{N}$.*

Definition 4.9. *A module M is called (finitely) Σ-Rickart if every (finite) direct sum of copies of M is a Rickart module. A (finitely) Σ-Baer module and a (finitely) Σ-extending module are defined similarly.*

We remark that every right (semi)hereditary ring R is precisely (finitely) Σ-Rickart as a right R-module (see Theorems 6.2 and 6.16). Also, if M is a finitely generated retractable module and if $End_R(M)$ is a right (semi)hereditary ring then M is a (finitely) Σ-Rickart module (see Proposition 6.22 and Corollary 6.29).

The next corollary provides a rich source of examples of Baer modules (hence, of Rickart modules).

Corollary 4.10. *(Corollary 2.22, [34]) If M is a nonsingular finitely Σ-extending module, then M and $E(M)$ are finitely Σ-Baer modules, and $E(M)^{(m)} \oplus M^{(n)}$ is a Baer (hence, Rickart) module for any $m, n \in \mathbb{N}$.*

An explicit application of Theorem 4.4 and Corollary 4.10 is exhibited in the next examples.

Example 4.11. *Let $R = \prod_{n=1}^{\infty} \mathbb{Z}_2$ and $M = \bigoplus_{n=1}^{\infty} \mathbb{Z}_2$ as a right R-module. Since M is a nonsingular finitely Σ-extending module, $R^{(m)} \oplus M^{(n)} = E(M)^{(m)} \oplus M^{(n)}$ is a Baer R-module for any $m, n \in \mathbb{N}$ by Corollary 4.10. (Compare to Example 5.11.)*

Example 4.12. *Consider $M = \mathbb{Z}^{(n)}$ as a right $\mathbb{Z}$-module for any $n \in \mathbb{N}$. Then M is a nonsingular extending $\mathbb{Z}$-module and $E(M) = \mathbb{Q}^{(n)}$. Thus, from Theorem 4.4, $E(M) \oplus M = \mathbb{Q}^{(n)} \oplus \mathbb{Z}^{(n)}$ is a Baer $\mathbb{Z}$-module. In particular, $\mathbb{Q}^{(m)} \oplus \mathbb{Z}^{(n)}$ is a Baer $\mathbb{Z}$-module for $m, n \in \mathbb{N}$. We remark that $\mathbb{Z}$ is a nonsingular finitely Σ-extending $\mathbb{Z}$-module.*

Note that for any $n \in \mathbb{N}$, $\mathbb{Z}^{(n)}$ is an extending and Baer $\mathbb{Z}$-module, $\mathbb{Z}^{(\mathbb{N})}$ is a Baer but not an extending $\mathbb{Z}$-module (Page 56, [13]), and $\mathbb{Z}^{(\mathbb{R})}$ is a Rickart but neither a Baer nor an extending $\mathbb{Z}$-module (Remark 2.28, [32]).

Theorem 4.13. *(Theorem 3.18, [44]) Let M_1 and M_2 be quasi-Baer modules. If we have the property $\psi(x) = 0 \ \forall \ 0 \neq \psi \in Hom_R(M_i, M_j)$ implies $x = 0$ ($i \neq j$, $i, j = 1, 2$) then $M_1 \oplus M_2$ is quasi-Baer.*

Proposition 4.14. *(Proposition 3.19, [44]) $\bigoplus_{i \in \mathcal{I}} M_i$ is quasi-Baer if M_i is quasi-Baer and subisomorphic to (i.e. isomorphic to a submodule of) M_j for all $i \neq j \in \mathcal{I}$ where $\mathcal{I}$ is an index set.*

Corollary 4.15. *A module M is quasi-Baer if and only if $M^{(\mathcal{I})}$ is a quasi-Baer module for any nonempty index set $\mathcal{I}$.*

Corollary 4.16. *(Corollary 3.20, [44]) A free module over a quasi-Baer ring is a quasi-Baer module.*

We can point out now that we have a general method of producing quasi-Baer modules that are not Baer modules.

Example 4.17. *An infinitely generated free module M over a non-Dedekind commutative domain R is not a Baer R-module. On the other hand, since M is a free R-module over a quasi-Baer ring R, M is a quasi-Baer module.*

Next, we provide a complete characterization for an arbitrary direct sum of (quasi-)Baer modules to be (quasi-)Baer, provided that each module is fully invariant in the direct sum (see Proposition 2.4.15 in [49]).

Proposition 4.18. *(Proposition 3.20, [47]) Let $M_i \trianglelefteq \bigoplus_{j \in \mathcal{I}} M_j$, $\forall\, i \in \mathcal{I}$, $\mathcal{I}$ is an arbitrary index set. Then $\bigoplus_{j \in \mathcal{I}} M_j$ is a (quasi-)Baer module if and only if M_i is a (quasi-)Baer module for all $i \in \mathcal{I}$.*

5. Direct sums of Rickart modules

In this section we provide and exemplify a number of relative conditions for Rickart modules. These conditions are then used to obtain results for direct sums of Rickart modules to inherit the Rickart property. Recall that a module M is called N-*Rickart* (or *relatively Rickart to N*) if, for every homomorphism $\varphi : M \to N$, $Ker\varphi \leq^{\oplus} M$. In particular, a right R-module M is Rickart iff M is M-Rickart.

Our next characterization extends Proposition 3.24(i).

Theorem 5.1. *(Theorem 2.6, [34]) Let M and N be right R-modules. Then M is N-Rickart if and only if for any direct summand $M' \leq^{\oplus} M$ and any submodule $N' \leq N$, M' is N'-Rickart.*

Proposition 5.2. *(Proposition 2.9, [34]) Let $\{M_i\}_{i \in \mathcal{I}}$ and N be right R-modules. Then the following implications hold:*

    (i) *If N has the SIP, then N is $\bigoplus_{i \in \mathcal{I}} M_i$-Rickart if and only if N is M_i-Rickart for all $i \in \mathcal{I}$, $\mathcal{I} = \{1, 2, \ldots, n\}$.*

    (ii) *If N has the SSIP, then N is $\bigoplus_{i \in \mathcal{I}} M_i$-Rickart if and only if N is M_i-Rickart for all $i \in \mathcal{I}$, $\mathcal{I}$ is an arbitrary index set.*

(iii) *If N has the SSIP, then N is $\prod_{i \in \mathcal{I}} M_i$-Rickart if and only if N is M_i-Rickart for all $i \in \mathcal{I}$, $\mathcal{I}$ is an arbitrary index set.*

Corollary 5.3. *(Corollary 2.10, [34]) For each $i \in \mathcal{I} = \{1, 2, \ldots, n\}$, M_i is $\bigoplus_{j \in \mathcal{I}} M_j$-Rickart if and only if M_i is M_j-Rickart for all $j \in \mathcal{I}$.*

While from Corollary 5.3 M_i is $\bigoplus_{j=1}^{n} M_j$-Rickart if M_i is M_j-Rickart for all $1 \le j \le n$, the next example shows that $\bigoplus_{i=1}^{n} M_i$ may not be M_j-Rickart even though M_i is M_j-Rickart for all $1 \le i \le n$.

Example 5.4. *Let $R = \mathbb{Z}[x]$ and let $M_1 = M_2 = N = \mathbb{Z}[x]$ be right R-modules. While M_i is N-Rickart for all $i = 1, 2$, $M_1 \oplus M_2$ is not N-Rickart: Consider $\varphi = (2, x) \in Hom_R(M_1 \oplus M_2, N)$. Then $Ker\varphi = (x, -2)R$ is not a direct summand of $M_1 \oplus M_2$.*

In the next result, we present conditions under which $\bigoplus_{i=1}^{n} M_i$ is M_j-Rickart.

Theorem 5.5. *(Theorem 2.12, [34]) Assume that there exists an ordered index set $\mathcal{I} = \{1, 2, \ldots, n\}$ for a class of R-modules $\{M_i\}_{i \in \mathcal{I}}$ such that M_i is M_j-injective for all $i < j \in \mathcal{I}$. Then $\bigoplus_{i=1}^{n} M_i$ is N-Rickart if and only if M_i is N-Rickart for all $i \in \mathcal{I}$, for any right R-module N.*

Example 5.4 also exhibits that the one-sided relative injective condition in Theorem 5.5 is not superfluous. (See that M_1 is not M_2-injective in that example.)

Corollary 5.6. *(Corollary 2.13, [34]) Assume that there exists an ordering $\mathcal{I} = \{1, 2, \ldots, n\}$ for a class of R-modules $\{M_i\}_{i \in \mathcal{I}}$ such that M_i is M_j-injective for all $i < j \in \mathcal{I}$. Then $\bigoplus_{i=1}^{n} M_i$ is a Rickart module if and only if M_i is M_j-Rickart for all $i, j \in \mathcal{I}$.*

Definition 5.7. *([34]) A module M is called N-C_2 (or relatively C_2 to N) if any submodule $N' \le N$ with $N' \cong M' \le^{\oplus} M$ implies $N' \le^{\oplus} N$. Hence, M has C_2 condition iff M is M-C_2.*

We now provide another instance when $\bigoplus_{i=1}^{n} M_i$ is M_j-Rickart if M_i is M_j-Rickart for all $1 \le i \le n$ (cf. Corollary 5.6).

Theorem 5.8. *(Theorem 2.29, [34]) Let $\{M_i\}_{i \in \mathcal{I}}$ be a class of right R-modules where $\mathcal{I} = \{1, 2, \ldots, n\}$. Assume that M_i is M_j-C_2 for all $i, j \in \mathcal{I}$. Then $\bigoplus_{i=1}^{n} M_i$ is a Rickart module if and only if M_i is M_j-Rickart for all $i, j \in \mathcal{I}$.*

In the next example, we show that the relative C_2 condition in Theorem 5.8 and the one-sided relative injective condition in Corollary 5.6 are not superfluous.

Example 5.9. *It is easy to see that $\mathbb{Z}[\frac{1}{2}]$ and $\mathbb{Z}$ are Rickart $\mathbb{Z}$-modules as each is a domain. $\mathbb{Z}[\frac{1}{2}]$ and $\mathbb{Z}$ are relatively Rickart to each other by Theorem 5.1 and because every $0 \neq \varphi \in Hom_{\mathbb{Z}}(\mathbb{Z}, \mathbb{Z}[\frac{1}{2}])$ is a monomorphism. Further, $\mathbb{Z}[\frac{1}{2}]$ is $\mathbb{Z}$-C_2, but $\mathbb{Z}$ is not $\mathbb{Z}[\frac{1}{2}]$-C_2 and $\mathbb{Z}[\frac{1}{2}]$ is not $\mathbb{Z}$-injective. For $\psi \in End_{\mathbb{Z}}(\mathbb{Z}[\frac{1}{2}] \oplus \mathbb{Z})$ defined by $\psi : (a, m) \mapsto (3a - m, 0)$, $Ker\psi = \{(m, 3m) \mid m \in \mathbb{Z}\} \not\leq^{\oplus} \mathbb{Z}[\frac{1}{2}] \oplus \mathbb{Z}$, hence $\mathbb{Z}[\frac{1}{2}] \oplus \mathbb{Z}$ is not a Rickart $\mathbb{Z}$-module.*

Corollary 5.10. *(Corollary 2.31, [34]) Let M be a Rickart module with C_2 condition. Then any finite direct sum of copies of M is a Rickart module.*

Next example follows from Corollary 5.10.

Example 5.11. *Consider $R = \prod_{n=1}^{\infty} \mathbb{Z}_2$ and $M = \bigoplus_{n=1}^{\infty} \mathbb{Z}_2$ as a right R-module. Since M is a nonsingular quasi-injective R-module, M is a Rickart module with C_2 condition. Thus by Corollary 5.10, $M^{(n)}$ is a Rickart module for any $n \in \mathbb{N}$.*

Theorem 5.12. *(Theorem 2.33, [34]) A module $M = M_1 \oplus M_2$ is Rickart if and only if M_1 and M_2 are Rickart modules, M_1 is M_2-Rickart, and for any $\varphi \in End_R(M)$ with $Ker\varphi \cap M_1 = 0$, $Ker\varphi \leq^{\oplus} M$.*

We remark that in Example 3.29 while $M_1 = \mathbb{Z}[x]$ is M_1-Rickart, it does not satisfy the last part of the statement of Theorem 5.12. Thus $\mathbb{Z}[x] \oplus \mathbb{Z}[x]$ is not a Rickart $\mathbb{Z}[x]$-module.

We conclude this section with providing a complete characterization for an arbitrary direct sum of Rickart modules to be Rickart, provided that each module is fully invariant in the direct sum.

Proposition 5.13. *(Proposition 2.34, [34]) Let $M_i \trianglelefteq \bigoplus_{j \in \mathcal{I}} M_j, \forall i \in \mathcal{I}$, $\mathcal{I}$ is an arbitrary index set. Then $\bigoplus_{j \in \mathcal{I}} M_j$ is a Rickart module if and only if M_i is a Rickart module for all $i \in \mathcal{I}$.*

6. Free Rickart and free Baer modules

The last section of our paper is devoted to the case of a special case of direct sums, namely that of free modules over the base ring. We will obtain conditions for the base ring such that free (and projective) modules over the base ring are Rickart modules. To obtain our first main result of this section (Theorem 6.2), we begin with the following well-known result of L. Small.

Theorem 6.1. *(Proposition 7.63, [30]) R is a right semihereditary ring iff $\mathrm{Mat}_n(R)$ is a right Rickart ring for all $n \in \mathbb{N}$.*

Theorem 6.2. *(Theorem 3.6, [34]) The following are equivalent for a ring R:*

(a) *every finitely generated free (projective) right R-module is a Rickart module;*

(b) $\mathrm{Mat}_n(R)$ *is a right Rickart ring for all $n \in \mathbb{N}$;*

(c) *every finite direct sum of copies of $R^{(k)}$ is a Rickart R-module for some $k \in \mathbb{N}$;*

(d) $\mathrm{Mat}_k(R)$ *is a right semihereditary ring for some $k \in \mathbb{N}$;*

(e) R *is a right semihereditary ring.*

Note that (d)⇔(e) in Theorem 6.2 was also proved by Small in a conceptual manner, using different arguments.

We recall that a module is said to be *torsionless* if it can be embedded in a direct product of copies of the base ring. In our next result we provide a characterization of rings R for which every finitely generated free right R-module is Baer.

Theorem 6.3. *(Theorem 3.5, [47]) The following are equivalent for a ring R:*

(a) *every finitely generated free (projective) right R-module is a Baer module;*

(b) *every finitely generated torsionless right R-module is projective;*

(c) *every finitely generated torsionless left R-module is projective;*

(d) R *is left semihereditary and right Π-coherent (i.e. every finitely generated torsionless right R-module is finitely presented);*

(e) R *is right semihereditary and left Π-coherent;*

(f) $M_n(R)$ *is Baer ring for all $n \in \mathbb{N}$.*

In particular, a ring R satisfying these equivalent conditions is right and left semihereditary.

Remark 6.4. *Note that Theorem 6.3 generalizes Theorem 2.2 in [16], which states that, for a von Neumann regular ring R, every finitely generated torsionless right R-module embeds in a free right R-module (FGTF property) iff $M_n(R)$ is a Baer ring for every $n \in \mathbb{N}$. Our result in fact establishes that every finitely generated torsionless right module is projective iff $M_n(R)$ is Baer for all $n \in \mathbb{N}$, even in the absence of von Neumann regularity of R.*

As a consequence of Theorems 6.2 and 6.3, we have the next three corollaries.

Corollary 6.5. *(Corollary 3.10, [34]) A ring R is left Π-coherent and every finitely generated free R-module is Rickart iff every finitely generated free R-module is Baer.*

We obtain a characterization of Prüfer domains in terms of the Rickart or Baer property for finitely generated free (projective) right R-modules.

Corollary 6.6. *(Corollary 3.7, [34] and Corollary 15, [54]) Let R be a commutative integral domain. Then the following conditions are equivalent:*

(a) *every finitely generated free (projective) right R-module is a Baer module;*

(b) *every finitely generated free (projective) right R-module is a Rickart module;*

(c) *the free R-module $R^{(k)}$ is a Rickart module for some $k \geq 2$;*

(d) *the free R-module $R^{(2)}$ is a Rickart module;*

(e) $\mathsf{Mat}_2(R)$ *is a right Rickart ring;*

(f) R *is a Prüfer domain.*

Note that in Part (c) of Corollary 6.6, $k \geq 2$ is required. For $k = 1$ we have the example of the commutative domain $\mathbb{Z}[x]$ (obviously a Rickart $\mathbb{Z}$-module), which is not a Prüfer domain.

We also obtain the following characterization of a Prüfer domain R in terms of the summand intersection property for finitely generated free (projective) right R-modules.

Corollary 6.7. *(Corollary 3.8, [34]) Let R be a commutative integral domain. Then the following conditions are equivalent:*

(a) *every finitely generated free (projective) right R-module has the SIP;*

(b) *the free R-module $R^{(k)}$ has the SIP for some $k \geq 3$;*

(c) *the free R-module $R^{(3)}$ has the SIP;*

(d) R *is a Prüfer domain.*

The next example shows a commutative integral domain R for which the free module $R^{(2)}$ has the SIP yet R is not a Prüfer domain. Thus, by Corollary 6.7 $R^{(3)}$ does not have the SIP. In this case, $R^{(2)}$ is not a Rickart R-module as well.

Example 6.8. *Consider $R = \mathbb{Z}[x]$, which is not a Prüfer domain. Let $M = (R \oplus R)_R$. If $(g, h)R$ and $(g', h')R$ are two proper direct summands of $R \oplus R$ for $g, g', h, h' \in R$, then by simple calculations we can show that either $(g, h)R \cap (g', h')R = (0, 0)$ or $(g, h)R = (g', h')R$. Thus M has the SIP but $R \oplus R \oplus R$ can not satisfy the SIP as a $\mathbb{Z}[x]$-module by Corollary 6.7. Furthermore, let $N = R_R$. By Example 5.4 we know that M is not N-Rickart. Thus, by Theorem 5.1 M is not a Rickart $\mathbb{Z}[x]$-module.*

Note that $\mathbb{Z}[x]$ and $\mathbb{Z}[x] \oplus \mathbb{Z}[x]$ are Rickart $\mathbb{Z}$-modules because $\mathbb{Z}[x] \oplus \mathbb{Z}[x] \cong_{\mathbb{Z}} \mathbb{Z}^{(\mathbb{N})} \oplus \mathbb{Z}^{(\mathbb{N})} \cong_{\mathbb{Z}} \mathbb{Z}^{(\mathbb{N})}$ (Remark 2.28, [32]).

Definition 6.9. *A ring R is said to be* right n-hereditary *if every n-generated right ideal of R is projective.*

Theorem 6.10. *(Proposition 3.13, [34]) The following conditions are equivalent for a ring R and a fixed $n \in \mathbb{N}$:*

    (a) *every n-generated free (projective) right R-module is a Rickart module;*

    (b) *the free R-module $R^{(n)}$ is a Rickart module;*

    (c) $\mathrm{Mat}_n(R)$ *is a right Rickart ring;*

    (d) R *is a right n-hereditary ring.*

For a fixed $n \in \mathbb{N}$, we obtain the following characterization for every n-generated free R-module to be Baer.

Theorem 6.11. *(Theorem 3.12, [47]) The following conditions are equivalent for a ring R and a fixed $n \in \mathbb{N}$:*

    (a) *every n-generated free (projective) right R-module is a Baer module;*

    (b) *every n-generated torsionless right R-module is projective.*

It is interesting to note that, as opposed to related notions (such as injectivity, quasi-injectivity, continuity and quasi-continuity), having $M \oplus M$ Baer does not imply that $M \oplus M \oplus M$ is also Baer.

We start with a lemma and by recalling the concept of an n-fir.

Definition 6.12. *A ring R is said to be a* right n-fir *if any right ideal that can be generated with $\leq n$ elements is free of unique rank (i.e., for every $I \leq R_R$, $I \cong R^{(k)}$ for some $k \leq n$, and if $I \cong R^{(l)} \Rightarrow k = l$) (for alternate definitions, see Theorem 1.1, [10]).*

The definition of (right) n-firs is left-right symmetric, thus we will call such rings simply n-firs. For more information on n-firs, see [10].

Theorem 6.13. *(Theorem 3.16, [47]) Let R be a n-fir. Then $R^{(n)}$ is a Baer R-module. Consequently, $M_n(R)$ is a Baer ring.*

In general, a module $M^{(n)}$ may be Rickart but $M^{(n+1)}$ may not be Rickart as the next example shows. Also, we remark that a right n-hereditary ring may always not be a right $(n + 1)$-hereditary ring. In Example 6.8, while $\mathbb{Z}[x]$ is a right 1-hereditary ring, it is not a right 2-hereditary ring. The following example is due to Jøndrup (see Theorem 2.3, [24] and [47]).

Example 6.14. *([24]) Let n be any natural number, K be any commutative field, and let R be the K-algebra on the $2(n + 1)$ generators X_i, Y_i*

$(i = 1, \ldots, n + 1)$ *with the defining relation*

$$\sum_{i=1}^{n+1} X_i Y_i = 0.$$

Since R is an n-fir (Theorem 2.3, [24]), $R^{(n)}$ is a Baer R-module by Theorem 6.13. In particular, since R is not $(n + 1)$-hereditary, $R^{(n+1)}$ is not a Rickart R-module.

Next, we provide an alternate proof of an earlier result of Small using the theory of Rickart modules (see Theorem 7.62, [30]).

Theorem 6.15. *For any $k \in \mathbb{N}$, R is a right hereditary ring if and only if $\mathrm{Mat}_k(R)$ is a right hereditary ring.*

Theorem 6.16. *(Theorem 2.26, [32] and Proposition 3.20, [34]) The following conditions are equivalent for a ring R:*
  (a) *every free (projective) right R-module is a Rickart module;*
  (b) *every direct sum of copies of $R^{(k)}$ is a Rickart R-module for some $k \in \mathbb{N}$;*
  (c) *every column finite matrix ring over R, $CFM(R)$, is a right Rickart ring;*
  (d) *the free R-module $R^{(R)}$ is a Rickart module;*
  (e) *$CFM_{\Gamma_0}(R)$ is a right Rickart ring for $|\Gamma_0| = |R|$;*
  (f) *$\mathrm{Mat}_k(R)$ is a right hereditary ring for some $k \in \mathbb{N}$;*
  (g) *R is a right hereditary ring.*

In the following we characterize the class of rings R for which every projective R-module is a Baer module. A ring R is said to be a *semiprimary ring* if the Jacobson radical, $\mathrm{Rad}(R)$, is nilpotent and $R/\mathrm{Rad}(R)$ is semisimple.

Theorem 6.17. *(Theorem 3.3, [47] and Corollary 3.23, [34]) The following conditions are equivalent for a ring R:*
  (a) *every free (projective) right R-module is a Rickart module and R is a semiprimary ring;*
  (b) *every free (projective) right R-module is a Baer module;*
  (c) *every free (projective) right R-module has the SSIP;*
  (d) *R is a right hereditary, semiprimary ring.*

We remark that in the preceding result, 'projective' can be replaced by 'flat'. The semiprimary condition in Theorem 6.17(d) is not superfluous as next example shows.

Example 6.18. *$\mathbb{Z}$ is a non-semiprimary right hereditary ring. $\mathbb{Z}^{(\mathbb{R})}$ is a Rickart $\mathbb{Z}$-module which is not a Baer $\mathbb{Z}$-module (Remark 2.28, [32]).*

From Theorem 2.20 in [46] (see Theorem 6.21) we showed that a ring is semisimple artinian if and only if every R-module is Baer. For the commutative rings one can restrict the requirement of "every R-module" to "every free R-module" to obtain the same conclusion.

Proposition 6.19. *(Theorem 6, [55]) Let R be a commutative ring. Every free R-module is Baer if and only if R is semisimple artinian. In particular, every R-module is Baer if every free R-module is so.*

Theorem 6.20. *(Theorem 3.18, [34]) The following are equivalent for a ring R:*

(a) *every finitely generated free (projective) right R-module is a Rickart module with C_2 condition;*

(b) *every finitely generated free (projective) right R-module is a Rickart module with C_3 condition;*

(c) *the free module $R^{(k)}$ is a Rickart module with C_2 condition for some $k \in \mathbb{N}$;*

(d) *the free module $R^{(k)}$ is a Rickart module with C_3 condition for some $k \geq 2$;*

(e) *the free module $R^{(2)}$ is a Rickart module with C_3 condition;*

(f) *R is a von Neumann regular ring.*

We remark that in Part(d) of Theorem 6.20, $k \geq 2$ is required. For $k = 1$, even though R_R may be a Rickart module with C_3 condition, R may not be a von Neumann regular ring. In Example 6.8, $\mathbb{Z}[x]$ is a Rickart $\mathbb{Z}[x]$-module with C_3 condition while $\mathbb{Z}[x]$ is not a von Neumann regular ring.

We now characterize the semisimple artinian rings in terms of Rickart and Baer modules.

Theorem 6.21. *(Theorem 2.20, [46] and Theorem 2.25, [32]) The following conditions are equivalent for a ring R:*

(a) *every right R-module is a Baer module;*

(b) *every right R-module is a Rickart module;*

(c) *every extending right R-module is a Rickart module;*

(d) *every injective right R-module is a Rickart module;*

(e) *every injective right R-module is a Baer module;*

(f) *R is a semisimple artinian ring.*

We extend Theorem 6.1 to a module theoretic setting using Lemma 3.6 (every direct sum of copies of an arbitrary retractable module is retractable).

Proposition 6.22. *(Proposition 3.2, [34]) Let M be a right R-module. If every finite direct sum of copies of M is a Rickart module then $End_R(M)$*

is a right semihereditary ring. Conversely, if M is a retractable module and if $End_R(M)$ is a right semihereditary ring, then every finite direct sum of copies of M is a Rickart module.

The next example illustrates the necessary direction in Proposition 6.22.

Example 6.23. *Consider $M = \mathbb{Q} \oplus \mathbb{Z}$ as a $\mathbb{Z}$-module. Then $M^{(n)} = \mathbb{Q}^{(n)} \oplus \mathbb{Z}^{(n)}$ is a Baer, hence, Rickart $\mathbb{Z}$-module for any $n \in \mathbb{N}$ by Corollary 4.10 (see also Example 4.12). Thus, from Proposition 6.22, $S = End_{\mathbb{Z}}(M) = \left(\begin{smallmatrix} \mathbb{Q} & \mathbb{Q} \\ 0 & \mathbb{Z} \end{smallmatrix}\right)$ is a right semihereditary ring. Note that $\left(\begin{smallmatrix} \mathbb{Q} & \mathbb{Q} \\ 0 & \mathbb{Z} \end{smallmatrix}\right)$ is a left hereditary ring but is not a right hereditary ring (see Example 2.33, [30]).*

The following example shows that the condition "*M is a retractable module*" in the hypothesis of the converse in Proposition 6.22, is not superfluous.

Example 6.24. *Consider $M = \mathbb{Z}_{p^\infty}$ as a right $\mathbb{Z}$-module. Then it is well-known that M is not retractable. Note that $End_{\mathbb{Z}}(M)$ is the ring of p-adic integers which is a Dedekind domain and hence is a (semi)hereditary ring. However, M is not a Rickart $\mathbb{Z}$-module, (and neither are direct sums of copies of M).*

As a consequence of Theorem 6.3, we can obtain the following result for finite direct sums of copies of an arbitrary retractable Baer module.

Corollary 6.25. *(Corollary 3.7, [47]) Let M be a retractable module. Then every finite direct sum of copies of M is a Baer module iff $End(M)$ is left semihereditary and right Π-coherent.*

Corollary 6.26. *(Corollary 3.14, [34]) Let M be a retractable module. Then $M^{(n)}$ is a Rickart module iff $End_R(M)$ is a right n-hereditary ring for a fixed $n \in \mathbb{N}$.*

Our next result provides a rich source of more examples of when the concepts of Rickart and Baer modules differ.

Proposition 6.27. *(Proposition 3.11, [34]) Let R be a right semihereditary ring which is not a Baer ring. Then every finitely generated free R-module is Rickart, but is not Baer.*

In Example 4.6, the ring R exhibits right semihereditary which is not Baer.

Proposition 6.28. *(Proposition 3.19, [34]) Let M be an indecomposable artinian Rickart module. Then any finite direct sum of copies of M is a Rickart module and satisfies C_2 condition.*

The next proposition extends Theorem 6.16 to endomorphism rings of finitely generated retractable modules.

Proposition 6.29. *(Corollary 3.21, [34]) Let M be a finitely generated retractable module. Then every direct sum of copies of M is a Rickart module iff $End_R(M)$ is a right hereditary ring.*

Proposition 6.30. *(Proposition 2.29, [32]) Let R be a right hereditary ring which is not a Baer ring. Then every free right R-module is Rickart, but is not Baer.*

Proposition 6.31. *Let R be a right hereditary ring which is not a semiprimary ring. Then there exists an index set $\mathcal{I}$ such that $M^{(\mathcal{I})}$ is a Rickart R-module, but is not a Baer R-module.*

Example 6.32. *From Example 6.18, since $\mathbb{Z}$ is a non-semiprimary right hereditary ring, there exists an index set $\mathbb{R}$ such that $\mathbb{Z}^{(\mathbb{R})}$ is a Rickart $\mathbb{Z}$-module which is not a Baer $\mathbb{Z}$-module.*

In the next result we provide a characterization for an arbitrary direct sum of copies of a Baer module to be Baer, for the case when M is finitely generated and retractable. In contrast to Corollary 6.25, we require the modules to be finitely generated.

Theorem 6.33. *(Theorem 3.4, [47]) Let M be a finitely generated retractable module. Then every direct sum of copies of M is a Baer module iff $End_R(M)$ is semiprimary and (right) hereditary.*

Given the connection provided by Theorem 3.4 between extending modules and Baer modules, we obtain the following result concerning Σ-extending (respectively, n-Σ-extending) modules, i.e., modules M with the property that direct sums of arbitrary (respectively, n) copies of M are extending. We generalize in this the results of Lemma 2.4 on polyform modules in [11] (recall that every polyform module is $\mathcal{K}$-nonsingular).

Theorem 6.34. *(Theorem 3.18, [47]) Let M be a $\mathcal{K}$-nonsingular module, with $S = End_R(M)$.*

(1) If $M^{(n)}$ is extending, then every n-generated right torsionless S-module is projective; it follows that S is a right n-hereditary ring.

(2) If $M^{(n)}$ is extending for every $n \in \mathbb{N}$, then S is right a semi-hereditary and left Π-coherent ring.

(3) If $M^{(\mathcal{I})}$ is extending for every index set $\mathcal{I}$, and M is finitely generated, then S is a semiprimary hereditary ring.

A more detailed discussion on these and related conditions for a module to be Σ-extending will appear in a sequel to this paper.

Proposition 6.35. *(Proposition 2.14, [47]) If a Baer module M can be decomposed into a finite direct sum of indecomposable summands, then every arbitrary direct sum decomposition of M is finite.*

If the endomorphism ring of a module is a PID (principal ideal domain), we obtain the following result, due to Wilson, which has been reformulated to our setting (Lemma 4, [52]).

Proposition 6.36. *(Proposition 3.11, [47]) Let M be a finite direct sum of copies of some finite rank, torsion-free module whose endomorphism ring is a PID. Then M is Baer module.*

We conclude this paper with information on some references for further results on the topics we have discussed. The list of these references is only suggestive and is not complete by any means. For results on Baer, quasi-Baer, and Rickart rings, see for example, [1], [2], [3], [4], [5], [6], [12], [14], [15], [20], [23], [24], [26], [35], [42], [54]. Results on Baer, quasi-Baer, and Rickart modules and related notions can be found in [21], [27], [28], [29], [31], [32], [33], [34], [44], [45], [46], [47], [49], [50], [51], [52], [55]. For results on (FI-)extending and (quasi-)continuous modules, see for example, [7], [9], [11], [13], [19], [22], [25], [36], [37], [38], [39], [43], [48].

OPEN PROBLEMS:
1. Obtain a characterization for a finite (an infinite) direct sum of Baer modules to be Baer.
2. Obtain a characterization for a finite (an infinite) direct sum of quasi-Baer modules to be quasi-Baer.
3. Obtain a characterization for a finite (an infinite) direct sum of Rickart modules to be Rickart.

Acknowledgments. The authors are thankful to the Ohio State University, Columbus and Lima, and Math Research Institute, Columbus, for the support of this research work. We also thank X. Zhang for his help in proof-reading the manuscript.

References

[1] S.K. Berberian, Baer ∗-Rings, Springer-Verlag: Berlin-Heidelberg-New York, 1972
[2] G.F. Birkenmeier, H.E. Heatherly, J.Y. Kim and J.K. Park, *Triangular matrix representations*, J. Algebra, **230**(2) (2000), 558–595
[3] G.F. Birkenmeier, J.Y. Kim and J.K. Park, *Quasi-Baer ring extensions and biregular rings*, Bull. Austral. Math. Soc., **61** (2000), 39–52

[4] G.F. Birkenmeier, J.Y. Kim and J.K. Park, *On quasi-Baer rings*, Contemp. Math., Amer. Math. Soc., Providence, RI, **259** (2000), 67–92

[5] G.F. Birkenmeier, J.Y. Kim and J.K. Park, *A sheaf representation of quasi-Baer rings*, J. Pure Appl. Algebra, **146** (2000), 209–223

[6] G.F. Birkenmeier, J.Y. Kim and J.K. Park, *Polynomial extensions of Baer and quasi-Baer rings*, J. Pure Appl. Algebra, **159** (2001), 25–42

[7] G.F. Birkenmeier, B.J. Müller and S.T. Rizvi, *Modules in which every fully invariant submodule is essential in a direct summand*, Comm. Algebra, **30**(3) (2002), 1395–1415

[8] S.U. Chase, *Direct products of modules*, Trans. Amer. Math. Soc., **97**(3) (1960), 457–473

[9] A.W. Chatters and S.M. Khuri, *Endomorphism rings of modules over nonsingular CS rings*, J. London Math. Soc., **21**(2) (1980), 434–444

[10] P.M. Cohn, *Free Rings And Their Relations*, Academic Press: London & New York (1971)

[11] J. Clark and R. Wisbauer, *Polyform and projective Σ-extending modules*, Alg. Colloq., **5**(4) (1998), 391–408

[12] W.E. Clark, *Twisted matrix units semigroup algebras*, Duke Math. J., **34** (1967), 417–423

[13] N.V. Dung, D.V. Huynh, P.F. Smith and R. Wisbauer, *Extending Modules*, Pitman Research Notes in Mathematics Series, 313; Longman Scientific & Technical: Harlow (1994)

[14] S. Endo, *Note on p.p. rings*, Nagoya Math. J., **17** (1960), 167–170

[15] M.W. Evans, *On commutative P.P. rings*, Pacific J. Math., **41**(3) (1972), 687–697

[16] C. Faith, *Embedding torsionless modules in projectives*, Publ. Mat., **34**(2) (1990), 379–387

[17] C. Faith, *Algebra: Rings, Modules And Categories I*, Springer-Verlag, New York-Heidelberg (1973)

[18] K.R. Goodearl, *Von Neumann Regular Rings*, Pitman, London (1979); 2nd edn, Krieger (1991)

[19] K. Hanada, Y. Kuratomi and K. Oshiro, *On direct sums of extending modules and internal exchange property*, J. Algebra, **250**(1) (2002), 115–133

[20] A. Hattori, *A foundation of torsion theory for modules over general rings*, Nagoya Math. J., **17** (1960), 147–158

[21] J. Hausen, *Modules with the summand intersection property*, Comm. Algebra, **17**(1) (1989), 135–148

[22] A. Harmancı and P.F. Smith, *Finite direct sums of CS-modules*, Houston J. Math., **19**(4) (1993), 523–532

[23] H.L. Jin, J. Doh and J.K. Park, *Group actions on quasi-Baer rings*, Canad. Math. Bull., **52**(4) (2009), 564–582

[24] S. Jøndrup, *p.p. rings and finitely generated flat ideals*, Proc. Amer. Math. Soc., **28**(2) (1971), 431–435

[25] J. Kado, Y. Kuratomi and K. Oshiro, *CS-property of direct sums of uniform modules*, International Symposium on Ring Theory, Trends in Math., Birkäuser (2001), 149–159

[26] I. Kaplansky, *Rings Of Operators*, W. A. Benjamin Inc., New York-Amsterdam (1968)

[27] S.M. Khuri, *Endomorphism rings and lattice isomorphisms*, J. Algebra, **56**(2) (1979), 401–408

[28] S.M. Khuri, *Endomorphism rings of nonsingular modules*, Ann. S. Math. Québec, **IV**(2) (1980), 145–152

[29] S.M. Khuri, *Nonsingular retractable modules and their endomorphism rings*, Bull. Austral. Math. Soc., **43**(1) (1991), 63–71

[30] T.Y. Lam, Lectures On Modules And Rings, GTM 189, Springer-Verlag, Berlin-Heidelberg-New York (1999)

[31] Q. Liu and B.Y. Ouyang, *Rickart modules*, Nanjing Daxue Xuebao Shuxue Bannian Kan, **23**(1) (2003), 157–166

[32] G. Lee, S.T. Rizvi and C.S. Roman, *Rickart modules*, Comm. Algebra, **38**(11) (2010), 4005–4027

[33] G. Lee, S.T. Rizvi and C.S. Roman, *Dual Rickart modules*, Comm. Algebra, **39**(12) (2011), 4036–4058

[34] G. Lee, S.T. Rizvi and C.S. Roman, *Direct sums of Rickart modules*, J. Algebra, to appear (DOI: 10.1016/j.jalgebra.2011.12.003)

[35] S. Maeda, *On a ring whose principal right ideals generated by idempotents form a lattice*, J. Sci. Hiroshima Univ. Ser. A, **24** (1960), 509–525

[36] S.H. Mohamed and B.J. Müller, Continuous And Discrete Modules, Lecture Note Ser. No. 147, London Math. Soc., Cambridge Univ. Press (1990)

[37] S.H. Mohamed and B.J. Müller, *Ojective modules*, Comm. Algebra, **30**(4) (2002), 1817–1827

[38] B.J. Müller and S.T. Rizvi, *On injective and quasi-continuous modules*, J. Pure Appl. Algebra, **28**(2) (1983), 197–210

[39] B.J. Müller and S.T. Rizvi, *Direct sums of indecomposable modules*, Osaka J. Math., **21**(2) (1984), 365–374

[40] C. Mucke, Zerlegungseigenschaften von stetigen und quasi-stetigen Moduln, Algebra Berichte 57, Verlag Reinhard Fischer, Munich (1988)

[41] J. von Neumann, Continuous Geometry, (with a foreword by Israel Halperin), Princeton Landmarks in Mathematics, Princeton University Press (1960)

[42] W.K. Nicholson, *On p.p.-endomorphism rings*, Canad. Math. Bull., **36**(2) (1993), 227–230

[43] S.T. Rizvi, *Commutative rings for which every continuous module is quasi-injective*, Arch. Math. (Basel), **50**(5) (1988), 435–442

[44] S.T. Rizvi and C.S. Roman, *Baer and quasi-Baer modules*, Comm. Algebra, **32**(1) (2004), 103–123

[45] S.T. Rizvi and C.S. Roman, *Baer property of modules and applications*, Advances in Ring Theory, (2005), 225–241

[46] S.T. Rizvi and C.S. Roman, *On K-nonsingular modules and applications*, Comm. Algebra, **35**(9) (2007), 2960–2982

[47] S.T. Rizvi and C.S. Roman, *On direct sums of Baer modules*, J. Algebra, **321**(2) (2009), 682–696

[48] C.S. Roman, *A generalized relative injectivity for extending modules*, Proceedings of the International Conference on Modules and Representation Theory, Presa Univ. Clujeana, Cluj-Napoca, (2009), 171–181

[49] C.S. Roman, *Baer and quasi-Baer modules*, Doctoral Dissertation, The Ohio State University (2004)

[50] V. Stephenson and G.M. Tsukerman, *Rings of endomorphisms of projective modules*, Sibirsk. Mat. Ž., **11** (1970), 228–232

[51] R. Wiegand, *Endomorphism rings of ideals in a commutative regular ring*, Proc. Amer. Math. Soc., **23** (1969), 442–449

[52] G.V. Wilson, *Modules with the summand intersection property*, Comm. Algebra, **14**(1) (1986), 21–38

[53] R. Wisbauer, Modules And Algebras: Bimodule Structure And Group Actions On Algebras, Pitman Monographs and Surveys in Pure and Applied Mathematics 81, Longman, Harlow (1996)

[54] K.G. Wolfson, *Baer rings of endomorphisms*, Math. Ann., **143** (1961), 19–28

[55] G.M. Tsukerman, *Rings of endomorphisms of free modules*, Sibirsk. Mat. Ž., **7** (1966), 1161–1167

[56] J. Zelmanowitz, *Regular modules*, Trans. Amer. Math. Soc., **163** (1972), 340–355

DEPARTMENT OF MATHEMATICS
THE OHIO STATE UNIVERSITY
COLUMBUS, OHIO, 43210, U.S.A.
E-mail address: lgy999@math.osu.edu

DEPARTMENT OF MATHEMATICS
THE OHIO STATE UNIVERSITY
LIMA, OHIO, 45804, U.S.A.
E-mail address: rizvi.1@osu.edu

DEPARTMENT OF MATHEMATICS
THE OHIO STATE UNIVERSITY
LIMA, OHIO, 45804, U.S.A.
E-mail address: cosmin@math.osu.edu

Proceedings of the Sixth China-Japan-Korea
International Conference on Ring Theory
June 27-July 2, 2011   Suwon, Korea

NOTES ON SIMPLE-BAER MODULES AND RINGS

Lixin Mao

ABSTRACT. In this paper, we first define and study simple-Baer
modules. Then we investigate when the endomorphism ring of a
module is a simple-Baer or universally mininjective ring.

1. Introduction

Following [11], a right R-module M_R with $S = \mathrm{End}M_R$ is called a
Baer module if the right annihilator in M of any nonempty subset of S is
generated by an idempotent element of S. In [7], the concept of Rickart
modules, which is a generalization of Baer modules, was introduced. A
right R-module M_R with $S = \mathrm{End}M_R$ is called a *Rickart module* if the
right annihilator in M of any principal left ideal of S is generated by an
idempotent element of S. In this paper, we further extend the concept
of Rickart modules to simple-Baer modules by replacing "principal left
ideal" with "simple left ideal". Some properties of simple-Baer modules
are obtained.

Let M_R be a right R-module with $S = \mathrm{End}M_R$. We first prove that:
(1) M_R is a simple-Baer module if and only if S is a right simple-Baer
ring and $_SM$ is simple-flat; (2) S is a left universally mininjective ring if
and only if M_R is a simple-Baer module and S is a left minannihilator
ring; (3) S is a right universally mininjective ring if and only if αM is a
direct summand of M_R for any simple right ideal αS of S and S is a right
minannihilator ring. Next we consider the properties of those R-modules
with simple-Baer endomorphism rings in terms of covers and envelopes.
For example, we prove that: (1) If S is a right simple-Baer ring, then
$r_M(I)$ has a monic addM_R-cover for any simple left ideal I of S; (2) If
S is a left simple-Baer ring, then M/IM has an epic addM_R-envelope
for any simple right ideal I of S.

Throughout this paper, R is an associative ring with identity and
all modules are unitary. We denote by M_R (resp. $_RM$) a right (resp.
left) R-module and by $S = \mathrm{End}M_R$ the endomorphism ring of M_R and

---

2010 Mathematics Subject Classification : 16D20, 16D40, 16D50.

Keywords : Simple-Baer module, simple-Baer ring, universally mininjective ring,
(pre)cover, (pre)envelope.

by addM_R the class consisting of all modules isomorphic to direct summands of finite direct sums of copies of M_R. For a right R-module M_R, $X \subseteq R$ and $Y \subseteq S$, $l_M(X) = \{u \in M : ux = 0 \text{ for all } x \in X\}$ and $r_M(Y) = \{m \in M : ym = 0 \text{ for all } y \in Y\}$. If $X = \{x\}$, we abbreviate $l_M(X)$ to $l_M(x)$. Similarly, if $Y = \{y\}$, we abbreviate $r_M(Y)$ to $r_M(y)$. For unexplained concepts and notations, we refer the reader to [1, 4, 6, 12, 13].

2. Main results

Recall that R is a *right simple-Baer ring* [9] if the right annihilator of every simple left ideal of R is a direct summand of R_R. We generalize the concept of simple-Baer rings to the general setting of modules as follows.

Definition 2.1. A right R-module M_R with $S = \text{End}M_R$ is called a *simple-Baer module* if, for any simple left ideal I of S, $r_M(I) = eM$ with $e^2 = e \in S$.

Obviously, R is a right simple-Baer ring if and only if R_R is a simple-Baer module.

Example 2.2. (1) Let M_R be a right R-module with $S = \text{End}M_R$. If the left socle of S $\text{Soc}_l(S) = 0$, then M_R is clearly a simple-Baer module.

(2) Let M_R be a right R-module with $S = \text{End}M_R$. If the Jacobson radical of S $J(S) = 0$, or more general, S is a semiprime ring, then M_R is a simple-Baer module.

In fact, if I is any simple left ideal of S, then $I^2 \neq 0$, and so $I = Se$ for some $e^2 = e \in S$ by [13, 2.7]. Thus $r_M(I) = (1 - e)M$, and so M_R is a simple-Baer module.

As a consequence, if R is a semiprime ring, then any finitely generated free right R-module R^n is a simple-Baer module since $S = \text{End}R^n \cong M_n(R)$ is a semiprime ring.

(3) Any Rickart module (e.g. semisimple module, nonsingular injective module, Baer module) is obviously simple-Baer. However, the converse is not true in general.

For example, let R be the ring of 2×2 matrices over the integers, then $R[X]$ is not a Rickart left $R[X]$-module since $R[X] \begin{pmatrix} 2 & 0 \\ X & 0 \end{pmatrix}$ is not a projective left ideal of $R[X]$ (see [5, Example p.432]). But $R[X]$ is a simple-Baer left $R[X]$-module by [10, Remark 1.1].

Recall that a right R-module M_R is said to *have the strong summand intersection property* [7] if the intersection of any family of direct summands of M_R is a direct summand of M_R.

Proposition 2.3. *Let M_R be a right R-module with the strong summand intersection property and $S = \mathrm{End}M_R$. Then M_R is a simple-Baer module if and only if $r_M(I)$ is a direct summand of M_R for any semisimple left ideal I of S.*

Proof. " $\Rightarrow$ " Let I be any semisimple left ideal of S. Then $I = \oplus_{i \in \Lambda} S\alpha_i$ with each $S\alpha_i$ simple. By hypothesis, each $r_M(\alpha_i)$ is a direct summand of M_R, and so $r_M(\oplus_{i \in \Lambda} S\alpha_i) = \cap_{i \in \Lambda} r_M(\alpha_i)$ is also a direct summand of M_R.

" $\Leftarrow$ " is trivial. $\qquad\qquad\square$

The next result shows that the simple-Baer property of modules is inherited by direct summands under some condition.

Theorem 2.4. *Let M_R be a simple-Baer R-module with $S = \mathrm{End}M_R$. Then eM is a simple-Baer R-module for any $e^2 = e \in S$ satisfying $SeS = S$.*

Proof. Write $T = \mathrm{End}(eM)_R$, $\pi : M \to eM$ to be the canonical projection and $\lambda : eM \to M$ to be the canonical injection. Let $T\alpha$ be any simple left ideal of T and $\beta = \lambda\alpha\pi \in S$. We first show that $S\beta$ is a simple left ideal of S.

If $s\beta \neq 0$ for some $s \in S$, then $SeSs\beta \neq 0$, and so $eSs\beta \neq 0$. Hence there exists $b \in S$ such that $ebs\beta \neq 0$. So there exists $x \in M$ such that $ebs\beta(x) \neq 0$, whence

$$\pi(ebs)\lambda\alpha(ex) = (ebs)(\alpha(ex)) = ebs\beta(x) \neq 0.$$

Thus $0 \neq (\pi(ebs)\lambda)\alpha \in T\alpha$. Hence $T\pi(ebs)\lambda\alpha = T\alpha$. Thus there exists $t \in T$ such that $\alpha = t\pi(ebs)\lambda\alpha$, and so

$$\beta = \lambda\alpha\pi = (\lambda t\pi)(ebs)(\lambda\alpha\pi) = (\lambda t\pi)ebs\beta \in Ss\beta.$$

Hence $S\beta = Ss\beta$. Thus $S\beta$ is a simple left ideal of S.

By hypothesis, $r_M(\beta)$ is a direct summand of M_R. Note that

$$r_M(\beta) = r_{eM}(\alpha) \oplus (1 - e)M.$$

So $r_{eM}(\alpha)$ is a direct summand of M_R. Since

$$r_{eM}(\alpha) \leq (eM)_R \leq M_R,$$

we have that $r_{eM}(\alpha)$ is a direct summand of $(eM)_R$. Consequently eM is a simple-Baer R-module. $\qquad\qquad\square$

According to [9], a left R-module $_RN$ is said to be *simple-flat* if $r_R(I)N = r_N(I)$ for any simple left ideal I of R.

The following result exhibits the connection between a simple-Baer module and its endomorphism ring.

Theorem 2.5. *Let M_R be a right R-module with $S = \text{End}M_R$. The following are equivalent:*

(1) *M_R is a simple-Baer module.*

(2) *S is a right simple-Baer ring and $_SM$ is simple-flat.*

Proof. (1) $\Rightarrow$ (2) Let $S\alpha$ be a simple left ideal of S. Then $r_M(\alpha) = eM$ for some $e^2 = e \in S$ by (1). Thus $e \in r_S(\alpha)$, and so $eS \subseteq r_S(\alpha)$.

Conversely, let $\beta \in r_S(\alpha)$, then $\alpha\beta = 0$. For any $y \in M$, $\beta y \in r_M(\alpha) = eM$. Hence $\beta y = e\beta y$, and so $\beta = e\beta \in eS$. Thus $r_S(\alpha) \subseteq eS$. Consequently $r_S(\alpha) = eS$. It follows that S is a right simple-Baer ring.

Now if $x \in r_M(\alpha)$, then $x \in eM$ and so $x \in r_S(\alpha)M$. Therefore $r_M(\alpha) \subseteq r_S(\alpha)M$.

In addition, it is clear that $r_S(\alpha)M \subseteq r_M(\alpha)$. Thus $r_S(\alpha)M = r_M(\alpha)$ and hence $_SM$ is simple-flat.

(2) $\Rightarrow$ (1) Let $S\alpha$ be a simple left ideal of S. Then $r_S(\alpha) = eS$ for some $e^2 = e \in S$ since S is left simple-Baer. But $_SM$ is min-flat, and so $r_M(\alpha) = r_S(\alpha)M = eM$. Thus M_R is a simple-Baer module. $\square$

Recall that a right R-module M_R is a *minannihilator module* [9] if $l_Mr_R(I) = MI$ for any simple left ideal I of R.

Following [10], R is called a *left minannihilator ring* if R_R is a minannihilator module and R is called a *left universally mininjective ring* if every simple left ideal is a direct summand of $_RR$.

It is known that R is a left universally mininjective ring if and only if every right R-module is a minannihilator module if and only if every right R-module is simple-flat if and only if R is a left minannihilator right simple-Baer ring (see [9, Theorem 3.9]).

We next characterize the universally mininjectivity of the endomorphism ring of a right R-module. The following lemma is needed.

Lemma 2.6. *Let M_R be a right R-module with $S = \text{End}M_R$.*

(1) *If M_R is quasi-injective and $_SM$ is simple-flat, then S is a left minannihilator ring.*

(2) *If M_R is quasi-projective and $_SM$ is a minannihilator module, then S is a right minannihilator ring.*

Proof. (1) Let $S\alpha$ be a simple left ideal of S, $\beta \in S$ and $r_S(\alpha) \subseteq r_S(\beta)$. Write $\iota : \text{im}(\alpha) \to M$ to be the inclusion and define $\gamma : \text{im}(\alpha) \to M$ by $\gamma(\alpha(x)) = \beta(x)$ for $x \in M$. If $\alpha(x) = 0$, then

$$x \in r_M(\alpha) = r_S(\alpha)M \subseteq r_S(\beta)M.$$

So $\beta(x) = 0$. Thus γ is well-defined.

Since M_R is quasi-injective, there exists $\theta \in S$ such that $\gamma = \theta\iota$. Thus for $t \in M$,

$$\beta(t) = \gamma(\alpha(t)) = (\theta\alpha)(t).$$

Therefore $\beta = \theta\alpha \in S\alpha$. By [9, Theorem 2.3], S is a left minannihilator ring.

(2) Let αS be a simple right ideal of S, $\beta \in S$ and $l_S(\alpha) \subseteq l_S(\beta)$.

Define $\varphi : M \to \alpha M$ by $\varphi(x) = \alpha x$ for $x \in M$ and define $\psi : M \to \alpha M$ by

$$\psi(x) = \beta x$$

for $x \in M$. Since $l_S(\alpha) \subseteq l_S(\beta) \subseteq l_S(\beta x)$, $\beta x \in \alpha M$ by [9, Theorem 2.3], and whence ψ is well-defined.

Since M_R is quasi-projective and φ is an epimorphism, there exists $\theta \in S$ such that $\psi = \varphi\theta$. Thus for any $x \in M$,

$$\beta x = \psi(x) = (\varphi\theta)(x) = (\alpha\theta)(x).$$

Hence $\beta = \alpha\theta \in \alpha S$. So S is a right minannihilator ring by [9, Theorem 2.3]. $\square$

Theorem 2.7. *Let M_R be a right R-module with $S = \mathrm{End}M_R$. The following are equivalent:*

(1) S is a left universally mininjective ring.

(2) M_R is a simple-Baer module and S is a left minannihilator ring.

(3) M_R is a simple-Baer module and if $S\alpha$ is a simple left ideal of S, then $\alpha M = eM$ with $e^2 = e \in S$.

     Moreover, if M_R is quasi-injective, then the above conditions are also equivalent to

(4) M_R is a simple-Baer module.

Proof. (1) $\Rightarrow$ (2) By [9, Theorem 3.9], S is a left minannihilator right simple-Baer ring and $_S M$ simple-flat. By Theorem 2.5, M_R is a simple-Baer module.

(2) $\Rightarrow$ (1) Since S is a right simple-Baer ring by Theorem 2.5, S is a left universally mininjective ring by [9, Theorem 3.9] and (2).

(1) $\Rightarrow$ (3) M_R is a simple-Baer module since (1) is equivalent to (2).

Let $S\alpha$ be a simple left ideal of S. Then $S\alpha = St$ with $t^2 = t \in S$ by (1). So there exists $\gamma \in S$ such that $\gamma\alpha = t$. Thus $\alpha = \alpha t = \alpha\gamma\alpha$, and so $\alpha M = (\alpha\gamma)M$ with $(\alpha\gamma)^2 = \alpha\gamma \in S$.

(3) $\Rightarrow$ (1) Let $S\alpha$ be a simple left ideal of S. Then there exists $e^2 = e \in S$ such that $r_M(\alpha) = eM$ since M_R is a simple-Baer module. In addition, $\alpha M = tM$ with $t^2 = t \in S$ by (3).

Define $\beta : M = \alpha M \oplus (1-t)M \to M$ by

$$\beta(\alpha x + y) = (1-e)x$$

for $x \in M$ and $y \in (1-t)M$. We claim that β is well-defined.

In fact, let $\alpha x + y = 0$, then $\alpha x = 0$. Hence $x \in r_M(\alpha) = eM$, and so $x = ex, (1-e)x = 0$.

For any $u \in M$, we have

$$(\alpha - \alpha\beta\alpha)u = \alpha u - \alpha\beta(\alpha u) = \alpha u - \alpha(1 - e)u = 0.$$

So $\alpha - \alpha\beta\alpha = 0$. Thus $S\alpha = S(\beta\alpha)$ with $(\beta\alpha)^2 = \beta\alpha \in S$, and so (1) follows.

$(2) \Rightarrow (4)$ is trivial.

$(4) \Rightarrow (2)$ follows from Theorem 2.5 and Lemma 2.6.   □

Theorem 2.8. *Let M_R be a right R-module with $S = \mathrm{End}M_R$. The following are equivalent:*

(1) If αS is a simple right ideal of S, then $\alpha M = hM$ for some $h^2 = h \in S$.

(2) S is a left simple-Baer ring and $_SM$ is a minannihilator module.

Proof. $(1) \Rightarrow (2)$ Let αS be a simple right ideal of S. Then $\alpha M = hM$ for some $h^2 = h \in S$ by (1). It is easy to see that $(1 - h)\alpha = 0$. Hence $S(1 - h) \subseteq l_S(\alpha)$.

Conversely, let $\beta \in l_S(\alpha)$, then $\beta\alpha = 0$, and so $\beta h = 0$. Thus $\beta = \beta(1 - h) \in S(1 - h)$, and hence $l_S(\alpha) \subseteq S(1 - h)$. So $l_S(\alpha) = S(1 - h)$. Therefore S is a left simple-Baer ring.

Now let $x \in r_M l_S(\alpha)$, then $(1 - h)x = 0$. Thus $x = hx \in hM = \alpha M$. Hence $r_M l_S(\alpha) = \alpha M$, and so $_SM$ is a minannihilator module.

$(2) \Rightarrow (1)$ Let αS be a simple right ideal of S. Then $l_S(\alpha) = Se$ for some $e^2 = e \in S$ since S is a left simple-Baer ring. So $\alpha M = r_M l_S(\alpha) = r_M(e) = (1 - e)M$.   □

Theorem 2.9. *Let M_R be a right R-module with $S = \mathrm{End}M_R$. The following are equivalent:*

(1) S is a right universally mininjective ring.

(2) $\alpha M = hM$ for some $h^2 = h \in S$ for any simple right ideal αS of S and S is a right minannihilator ring.

(3) For any simple right ideal αS of S, $\alpha M = hM$ for some $h^2 = h \in S$ and $r_M(\alpha) = eM$ for some $e^2 = e \in S$.

Moreover, if M_R is quasi-projective, then the above conditions are also equivalent to

(4) $\alpha M = hM$ for some $h^2 = h \in S$ for any simple right ideal αS of S.

Proof. $(1) \Rightarrow (2)$ By [9, Theorem 3.9] and (1), S is a right minannihilator left simple-Baer ring and $_SM$ is a minannihilator module. So (2) holds by Theorem 2.8.

$(2) \Rightarrow (1)$ Since S is a left simple-Baer ring by Theorem 2.8, S is a right universally mininjective ring by [9, Theorem 3.9] and (2).

(1) $\Rightarrow$ (3) Let αS be a simple right ideal of S. Then $\alpha S = tS$ with $t^2 = t \in S$ by (1). So there exists $\gamma \in S$ such that $\alpha\gamma = t$. Thus $\alpha = t\alpha = \alpha\gamma\alpha$, and so $r_M(\alpha) = (1 - \gamma\alpha)M$.

The proof of (3) $\Rightarrow$ (1) is similar to that of (3) $\Rightarrow$ (1) in Theorem 2.7.

(2) $\Rightarrow$ (4) is trivial.

(4) $\Rightarrow$ (2) follows from Theorem 2.8 and Lemma 2.6. $\square$

Let $\mathcal{C}$ be a class of right R-modules and M a right R-module. Following [3, 4], we say that a homomorphism $\phi : C \to M$ is a $\mathcal{C}$-precover of M if $C \in \mathcal{C}$ and the abelian group homomorphism $\operatorname{Hom}_R(C', \phi)$: $\operatorname{Hom}_R(C', C) \to \operatorname{Hom}_R(C', M)$ is surjective for every $C' \in \mathcal{C}$. A $\mathcal{C}$-precover $\phi : C \to M$ is said to be a $\mathcal{C}$-cover of M if every endomorphism $g : C \to C$ such that $\phi g = \phi$ is an isomorphism. Dually we have the definitions of a $\mathcal{C}$-preenvelope and a $\mathcal{C}$-envelope. $\mathcal{C}$-covers ($\mathcal{C}$-envelopes) may not exist in general, but if they exist, they are unique up to isomorphism.

Finally, we investigate the properties of the endomorphism ring of a right R-module in terms of (pre)covers and (pre)envelopes of some special right R-modules.

Lemma 2.10. *Let M_R be a right R-module with $S = \operatorname{End}M_R$ and $I \subseteq S$. Then*

$$\operatorname{Hom}_R(M/IM, M) \cong l_S(I).$$

Proof. Let $\pi : M \to M/IM$ be the canonical map.

Define $\varphi : \operatorname{Hom}_R(M/IM, M) \to l_S(I)$ via

$$\varphi(f) = f\pi$$

for any $f \in \operatorname{Hom}_R(M/IM, M)$ and define $\phi : l_S(I) \to \operatorname{Hom}_R(M/IM, M)$ via

$$\phi(g)(\overline{x}) = g(x)$$

for any $g \in l_S(I)$ and $x \in M$.

It is easy to check that φ and ϕ are well-defined, $\varphi\phi = 1$ and $\phi\varphi = 1$. Thus $\operatorname{Hom}_R(M/IM, M) \cong l_S(I)$. $\square$

Recall that R a *left simple-coherent ring* [9] if the right annihilator of every simple left ideal of R is finitely generated. A right simple-coherent ring is defined similarly.

Obviously, R is a right simple-Baer ring if and only if R is a right simple-coherent ring and aR is flat for every simple left ideal Ra of R.

Proposition 2.11. *Let M_R be a right R-module with $S = \operatorname{End}M_R$.*

(1) S is a right simple-coherent ring if and only if $r_M(I)$ has an $\operatorname{add}M_R$-precover, where I is any simple left ideal of S.

(2) S is a left simple-coherent ring if and only if M/IM has an $\operatorname{add}M_R$-preenvelope, where I is any simple right ideal of S.

Proof. (1) Let I be any simple left ideal of S. By [8, Lemma 3.16], we have the isomorphism

$$\operatorname{Hom}_R(M, r_M(I)) \cong r_S(I).$$

In addition, $\operatorname{Hom}_R(M, r_M(I))$ is a finitely generated right S-module if and only if $r_M(I)$ has an addM_R-precover by [2, Lemma 3(2)]. So S is a right simple-coherent ring if and only if $r_M(I)$ has an addM_R-precover.

(2) Let I be any simple right ideal of S. Then, by [2, Lemma 3(1)], $\operatorname{Hom}_R(M/IM, M)$ is a finitely generated left S-module if and only if M/IM has an addM_R-preenvelope. By Lemma 2.10, S is a left simple-coherent ring if and only if M/IM has an addM_R-preenvelope. $\square$

Theorem 2.12. *Let M_R be a right R-module with $S = \operatorname{End}M_R$. If S is a right simple-Baer ring, then $r_M(I)$ has a monic addM_R-cover for any simple left ideal I of S.*

Proof. Let $S\alpha$ be any simple left ideal of S. Then $r_S(\alpha) = eS$ for some $e^2 = e \in S$. So $eM \subseteq r_M(\alpha)$. We claim that the inclusion $\iota : eM \to r_M(\alpha)$ is an addM_R-cover of $r_M(\alpha)$.

In fact, let $\varphi : M^m \to r_M(\alpha)$ with m a positive integer be any right R-homomorphism, $\lambda_i : M \to M^m$ the ith canonical injection, $i = 1, 2, \cdots, m$ and $\mu : r_M(\alpha) \to M$ the inclusion, then

$$\mu\varphi\lambda_i \in r_S(\alpha) = eS.$$

So there exists $t_i \in S$ such that $\mu\varphi\lambda_i = et_i$.

Define $\psi : M^m \to eM$ by

$$\psi(x_1, x_2, \cdots, x_m) = e\sum_{i=1}^m t_i x_i$$

for any $x_i \in M$. Then we have

$$\varphi(x_1, x_2, \cdots, x_m) = \mu\varphi\sum_{i=1}^m \lambda_i(x_i) = e\sum_{i=1}^m t_i x_i = \iota\psi(x_1, x_2, \cdots, x_m),$$

whence $\varphi = \iota\psi$. So ι is an addM_R-precover of $r_M(\alpha)$. Since ι is monic, ι is also an addM_R-cover of $r_M(\alpha)$. $\square$

Theorem 2.13. *Let M_R be a right R-module with $S = \operatorname{End}M_R$. If S is a left simple-Baer ring, then M/IM has an epic addM_R-envelope for any simple right ideal I of S. The converse holds if M_R is finitely generated projective and $_SM$ is flat.*

Proof. Let αS be any simple right ideal of S. Then $l_S(\alpha) = Se$ for some $e^2 = e \in S$.

Define $\rho : M/\alpha M \to eM$ by

$$\rho(\overline{x}) = ex.$$

Then ρ is well-defined and is an epimorphism. We will show that $\rho :$ $M/\alpha M \to eM$ is an addM_R-envelope of $M/\alpha M$.

Let $\varphi : M/\alpha M \to M^m$ be any R-homomorphism with m a positive integer and $\pi_i : M^m \to M$ be the ith projection, $1 \le i \le m$.

Note that $\pi_i \varphi \tau \in l_S(\alpha) = Se$, where $\tau : M \to M/\alpha M$ is the canonical map. So there exist $s_i \in S$ such that $\pi_i \varphi \tau = s_i e$.

Define $h : eM \to M^m$ via

$$h(ex) = (s_1 ex, s_2 ex, \cdots, s_m ex)$$

for any $x \in M$. Then

$$\pi_i h \rho(\overline{x}) = \pi_i h(ex) = s_i ex = \pi_i \varphi(\overline{x}).$$

Thus $\pi_i h \rho = \pi_i \varphi$ and hence $h\rho = \varphi$. So ρ is an addM_R-preenvelope. Since ρ is epic, ρ is also an addM_R-envelope.

Conversely, if αS is any simple right ideal of S and $M/\alpha M$ has an epic addM_R-envelope $f : M/\alpha M \to N$, then by Proposition 2.11, $l_S(\alpha)$ is finitely generated, and so $S\alpha$ is finitely presented.

Let I be any right ideal of S. We claim that I is simple-flat.

In fact, if $\gamma : S/\alpha S \to I$ is any S-homomorphism, since $_S M$ is flat, we get the following commutative diagram:

$$
\begin{array}{ccc}
(S/\alpha S) \otimes_S M & \xrightarrow{\gamma \otimes_S 1} & I \otimes_S M \\
\Big\downarrow{\cong} & & \Big\downarrow{\cong} \\
M/\alpha M & \xrightarrow{\theta} & IM.
\end{array}
$$

Let $\iota : IM \to M$ be the inclusion. Then there exists $\xi : N \to M$ such that the following diagram is commutative:

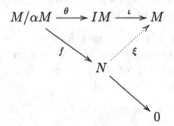

So $\ker(f) \subseteq \ker(\theta)$. Thus there is $\phi : N \to IM$ such that $\phi f = \theta$.

By [1, Proposition 20.10], we have the following commutative diagram:

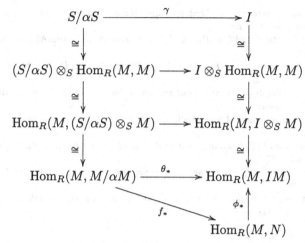

Thus γ factors through the finitely generated projective right S-module $\mathrm{Hom}_R(M,N)$. Hence I is simple-flat by [9, Theorem 2.6].

The S-module exact sequence $0 \to I \to S \to S/I \to 0$ induces the exact sequence

$$0 = \mathrm{Tor}_2^S(S, S/S\alpha) \to \mathrm{Tor}_2^S(S/I, S/S\alpha) \to \mathrm{Tor}_1^S(I, S/S\alpha) \to \mathrm{Tor}_1^S(S, S/S\alpha) = 0.$$

Thus $\mathrm{Tor}_2^S(S/I, S/S\alpha) \cong \mathrm{Tor}_1^S(I, S/S\alpha) = 0$ by [9, Theorem 2.6].

On the other hand, the S-module exact sequence $0 \to S\alpha \to S \to S/S\alpha \to 0$ induces the exact sequence

$$0 = \mathrm{Tor}_2^S(S/I, S) \to \mathrm{Tor}_2^S(S/I, S/S\alpha) \to \mathrm{Tor}_1^S(S/I, S\alpha) \to \mathrm{Tor}_1^S(S/I, S) = 0.$$

Thus $\mathrm{Tor}_1^S(S/I, S\alpha) \cong \mathrm{Tor}_2^S(S/I, S/S\alpha) = 0$. Hence $S\alpha$ is flat, and so is projective. It follows that S is a left simple-Baer ring.   □

Acknowledgments. This research was supported by NSFC (No. 11071111, 11171149), NSF of Jiangsu Province of China (No. BK2011068), Jiangsu 333 Project, Jiangsu Six Major Talents Peak Project.

References

[1] F.W. Anderson and K.R. Fuller, *Rings and Categories of Modules*, Springer-Verlag, New York, 1974.

[2] L. Angeleri-Hügel, *Endocoherent modules*, Pacific J. Math. **212** (2003), 1-11.

[3] E.E. Enochs, *Injective and flat covers, envelopes and resolvents*, Israel J. Math. **39** (1981), 189-209.

[4] E.E. Enochs and O.M.G. Jenda, *Relative Homological Algebra*, Walter de Gruyter, Berlin-New York, 2000.

[5] J. Jøndrup, *p.p. rings and finitely generated flat ideals*, Proc. Amer. Math. Soc. **28** (1971), 431-435.

[6] T.Y. Lam, *Lectures on Modules and Rings*, Springer-Verlag, New York-Heidelberg-Berlin, 1999.

[7] G. Lee, S.T. Rizvi and C.S. Roman, *Rickart modules*, Comm. Algebra **38** (2010), 4005-4027.

[8] L.X. Mao, *Properties of P-coherent and Baer modules*, Period. Math. Hungar. **60** (2010), 97-114.

[9] L.X. Mao, *Simple-Baer rings and minannihilator modules*, Comm. Algebra **40** (2012), to appear.

[10] W.K. Nicholson and M.F. Yousif, *Mininjective rings*, J. Algebra **187** (1997), 548-578.

[11] S.T. Rizvi and C.S. Roman, *Baer and quasi-Baer modules*, Comm. Algebra **32** (2004), 103-123.

[12] J.J. Rotman, *An Introduction to Homological Algebra*, Academic Press, New York, 1979.

[13] R. Wisbauer, *Foundations of Module and Ring Theory*, Philadelphia, Gordon and Breach, 1991.

INSTITUTE OF MATHEMATICS
NANJING INSTITUTE OF TECHNOLOGY
NANJING 211167, P.R. CHINA
E-mail address: maolx2@hotmail.com

Proceedings of the Sixth China-Japan-Korea
International Conference on Ring Theory
June 27-July 2, 2011 Suwon, Korea

A NOTE ON QUASI-JOHNS RINGS

LIANG SHEN

ABSTRACT. A ring R is called right quasi-Johns if every essential
right ideal of R is a right annihilator and R satisfies ACC on essential right ideals. Several characterizations of quasi-Johns rings
are given in this short article. It is also shown that a left Kasch
ring with ACC on right annihilators is semilocal.

1. Introduction

Throughout this paper rings are associative with identity. Write
J, S_l, S_r, Z_l and Z_r for the Jacobson radical, the left socle, the right
socle, the left singular ideal and the right singular ideal of a ring R,
respectively. Let R be a ring and M an R-module. We use $N \subseteq^{ess} M$ to
mean that N is an essential submodule of M. And Soc(M) denotes the
socle of M. For a subset X of a ring R, the left annihilator of X in R is
$l(X) = \{r \in R : rx = 0 \text{ for all } x \in X\}$. Right annihilators are defined
analogously. $M_n(R)$ (R_n) denotes the set of $n \times n$ ($n \times 1$) matrices over
R.

Recall that a ring R is *quasi-Frobenius* (*QF*) if R is right noetherian
and every one-sided ideal is an annihilator. There are three unsolved
conjectures on *QF* rings (see [11]). One of them is Faith-Menal conjecture, which came out of the research on Johns rings. Recall that a
ring R is called *right Johns* if R is right noetherian and every right ideal
of R is a right annihilator. The left side can be defined similarly. In
[6], Johns used a false result of Kurshan [9, Theorem 3.3] to show that
right Johns rings are right artinian. Later in [5], Faith and Menal gave
a counter example to inform that right Johns rings may not be right
artinian. Then they asked whether a *strongly right Johns* ring (any matrix ring $M_n(R)$ over R is right Johns, $n \geq 1$) is right artinian. This is
named Faith-Menal conjecture by Nicholson and Yousif. An important
key to this conjecture is to show that such rings are semilocal.

In this short note, we discuss on quasi-Johns rings which are real
generalizations of Johns rings. A ring R is called a *right quasi-Johns* if

---

2010 Mathematics Subject Classification : 16L60, 16P70.
Keywords : Quasi-Johns ring, Johns ring, Kasch ring, semilocal ring.

every essential right ideal of R is a right annihilator and R satisfies ACC on essential right ideals. Several characterizations of quasi-Johns rings are explored. We also show some connections between quasi-Johns rings and QF rings in cases of some annihilator conditions. Some questions are presented. At last, a result on semilocal rings is given. It is shown that a left Kasch ring with ACC on right annihilators is semilocal (see Theorem 2.23). Recall that a ring R is called *right Kasch* if every maximal right ideal of R is a right annihilator, or equivalently, $l(I) \neq 0$ for each proper right ideal I of R. The left side can be defined similarly.

2. Results

Recall that a ring R is called *right dual* (*right quasi-dual*) if every right ideal (essential right ideal) of R is a right annihilator. The left sides can be defined similarly.

Definition 2.1. A ring R is called a right quasi-Johns ring if R is right quasi-dual and with ACC on essential right ideals of R. Left quasi-Johns rings can be defined similarly. R is a quasi-Johns ring if it is both left and right quasi-Johns.

It is obvious that a right Johns ring is right quasi-Johns. By the following example, the converse is not true. Let R be a ring and $_R W_R$ a bimodule. Recall that the *trivial extension* of W by R is the additive group $T(R, W) = R \oplus W$ endowed with the multiplication $(a, w)(a', w') = (aa', aw' + wa')$.

Example 2.2. ([12, Example 2.3]) Let $M_1 = M_2 = \mathbb{Z}_2$ and $R = T(\mathbb{Z}_2, M)$ be the trivial extension of ring $\mathbb{Z}_2$ and the $\mathbb{Z}_2$-module $M = M_1 \oplus M_2$. Then R is quasi-Johns but not Johns.

Proof. It is proved in [12, Example 2.3] that R is a commutative finite quasi-dual ring which is not dual. So R is a quasi-Johns ring which is not Johns. □

The next example show that a right quasi-Johns ring may not be left quasi-Johns.

Example 2.3. ([11, Example 8.16]) Let D be any countable, existentially closed division ring over a field F, and let $R = D \otimes_F F(x)$. Then $R = T(R, D)$ is a right quasi-Johns ring but not a left quasi-Johns ring.

Proof. It is proved in [11, Example 8.16] that R is right Johns but not right artinian. If R is left quasi-Johns, R is left Kasch by the following Proposition 2.10. Thus, according to [11, Theorem 8.9], R is right artinian. This is impossible. □

Now we look at some basic properties of quasi-Johns rings.

Lemma 2.4. ([4, Lemma 2]) *Let M be an R-module. Then M satisfies ACC on essential submodules if and only if $M/Soc(M)$ is noetherian.*

Theorem 2.5. *If R is right quasi-Johns, then eRe is also right quasi-Johns, where $e^2 = e \in R$ and $ReR=R$.*

Proof. By [12, Theorem 3.8], eRe is right quasi-dual. Now we show that eRe satisfies ACC on essential right ideals of eRe. Set $S = eRe$. By the above lemma, we only need to prove that $S/Soc(S_S)$ is right noetherian. The result is obtained by the next two steps.

(i) $Soc(S_S)= eSoc(R_R)e$. Firstly, we prove $Soc(S_S) \supseteq eSoc(R_R)e$. It is only need to show that xR being a minimal right ideal of R with $exe \neq 0$ implies that $exeS$ is a minimal right ideal of S. If not, there is an element $y \in S$ such that $0 \neq exeyS \subsetneq exeS$. Since xR is minimal and $exe \neq 0$, $exeR$ is also a minimal right ideals of R. But $0 \neq exeySR$, so $exeySR = exeR$, which implies that $exeyS = exeS$. This is a contradiction. Secondly, we show $Soc(S_S) \subseteq eSoc(R_R)e$. It is only need to show that, for $y \in S$, yS being a minimal right ideal of S implies that yR is a minimal right ideal of R. If not, there is an element $r \in R$ such that $0 \subsetneq yrR \subsetneq yR$. Since $yrR \neq 0$ and $ReR=R$, $0 \neq yrRe$ which is a right ideal of S. As yS is minimal and $yrRe = yerRe \subseteq yS$, $yrRe = yS$. $ReR = R$ shows that $yrR = yR$. It is a contradiction.

(ii) $\frac{S}{eSoc(R_R)e}$ is right noetherian. Since R is right quasi-Johns, $\frac{R}{Soc(R_R)}$ is right noetherian by the above lemma. Set $\frac{S}{eSoc(R_R)e} = \hat{S}$ and $\frac{R}{Soc(R_R)} = \overline{R}$. Let $\hat{I}_1 \subseteq \hat{I}_2 \subseteq \cdots \subseteq \hat{I}_n \subseteq \cdots$ be an ascending chain of right ideals of $\hat{S}$, where $I_k \subseteq S$, $\forall k \geq 1$. We might as well assume $I_1 \subseteq I_2 \subseteq \cdots \subseteq I_n \subseteq \cdots$ is an ascending chain of right ideals of S, where $eSoc(R_R)e \subseteq I_k$, $\forall k \geq 1$. Now we show that there exists an integer $n_0 \geq 1$ such that for any $n \geq n_0$, $\hat{I}_n = \hat{I}_{n_0}$. Since $\overline{R}$ is right noetherian, there exists $n_0 \geq 1$ such that for any $n \geq n_0$, $\overline{I_n R} = \overline{I_{n_0} R}$. Thus, $I_n R + Soc(R_R) = I_{n_0} R + Soc(R_R)$, $\forall n \geq n_0$. So $eI_n Re + eSoc(R_R)e = eI_{n_0} Re + eSoc(R_R)e$. Since $I_k \subseteq S$, $\forall k \geq 1$, it is clear that $I_n + eSoc(R_R)e = I_{n_0} + eSoc(R_R)e$. That is $\hat{I}_n = \hat{I}_{n_0}$, $\forall n \geq n_0$. $\square$

Remark 2.6. The assumption that $ReR = R$ in the above theorem is necessary. By [12, Remark 3.9], there exists a QF ring R and an idempotent $e \in R$ such that $ReR \neq R$ and $eRe \cong S$ where S is not right quasi-dual. So R is right quasi-Johns but eRe is not right quasi-Johns.

Kerr [7] constructed a ring R with ACC on right annihilators such that $M_2(R)$ does not have ACC on right annihilators. The next lemma shows that the chain condition "ACC on essential right ideals" can be inherited to any $M_n(R)$, $n > 1$.

Lemma 2.7. *Let n be a positive integer. R satisfies ACC on essential right ideals of R if and only if $M_n(R)$ satisfies ACC on essential right ideals of $M_n(R)$.*

Proof. We only need to prove the case of $n = 2$. The others are similarly. Consider any right ideal T of $M_2(R)$. Then T has the form $[I \ I] = \{[\alpha \ \beta] | \alpha, \beta \in I\}$, where I is a submodule of the free right R-module R_2. It is not difficult to see that T is essential in $M_2(R)$ if and only if I is essential in R_2. Thus $M_2(R)$ satisfies ACC on essential right ideals if and only if R_2 satisfies ACC on essential submodules. According to Lemma 2.4, R_2 satisfies ACC on essential submodules if and only if $R_2/Soc(R_2)$ is right noetherian. Since $\frac{R_2}{Soc(R_2)} \simeq \frac{R}{S_r} \oplus \frac{R}{S_r}$, $\frac{R_2}{Soc(R_2)}$ is right noetherian if and only if $\frac{R}{S_r}$ is right noetherian if and only if R satisfies ACC on essential right ideals. $\square$

Theorem 2.8. *Let n be a positive integer. $M_n(R)$ is right quasi-Johns if and only if R is right quasi-Johns and every n-generated singular right R-module is cogenerated by R.*

Proof. By [12, Theorem 3.2], $M_n(R)$ is right quasi-dual if and only if every n-generated singular right R-module is cogenerated by R. Then by Lemma 2.7, $M_n(R)$ is right quasi-Johns if and only if R is right quasi-Johns and every n-generated singular right R-module is cogenerated by R. $\square$

Question 2.9. Let R be a right quasi-Johns ring, is $M_n(R)$ $(n > 1)$ also a right quasi-Johns ring? By Lemma 2.7, it is equivalent to ask whether $M_n(R)$ over a right quasi-dual ring R is also right quasi-dual.

Proposition 2.10. *If R is a right quasi-Johns ring, then:*
(1) *R is right Kasch;*
(2) *$J=\mathbf{r}(S_r) = Z_l$ is nilpotent;*
(3) *$\mathbf{l}(J) \subseteq^{ess} R_R$, $\mathbf{l}(J) \subseteq^{ess} {}_R R$, $S_r = \mathbf{r}(Z_r)$;*
(4) *$Z_r \subseteq J$, $S_l \subseteq S_r$.*

Proof. For (1), since R is right quasi-dual, by [12, Lemma 2.5], R is right Kasch.

For (2), it is clear $J \subseteq \mathbf{r}(S_r)$. Since R is right Kasch, for any maximal right ideal M of R, R/M has an image xR in S_r. It is easy to see $M = \mathbf{r}(x)$. As J is the intersection of all maximal right ideals of R, $\mathbf{r}(S_r) \subseteq J$. Thus $J = \mathbf{r}(S_r)$. And by [12, Lemma 2.6] and [12, Theorem 2.9], $J = Z_l$ is nilpotent.

For (3), by [15, Theorem 3.2], $\mathbf{l}(J) \subseteq^{ess} {}_R R$ and $S_r = \mathbf{r}(Z_r)$. Now we show $\mathbf{l}(J) \subseteq^{ess} R_R$. For each x that is not in $\mathbf{l}(J)$, as J is nilpotent, there exists $n \geq 1$ such that $xJ^n \neq 0$ and $xJ^{n+1} = 0$. Then $0 \neq xJ^n \in \mathbf{l}(J)$.

Thus $l(J) \subseteq^{ess} R_R$.

For (4), by [11, Lemma 4.20 (2)], Z_r is nilpotent. Therefore $Z_r \subseteq J$. Thus $S_l \subseteq r(J) \subseteq r(Z_r) = S_r$. $\qquad\square$

Corollary 2.11. *If R is a right Johns ring, then:*
(1) *R is right Kasch;*
(2) *$J=r(S_r) = Z_l$ is nilpotent;*
(3) *$l(J) \subseteq^{ess} R_R$, $l(J) \subseteq^{ess}{}_R R$, $S_r = r(Z_r)$;*
(4) *$Z_r \subseteq J$, $S_l \subseteq S_r$.*

Corollary 2.12. *If R is a quasi-Johns (or Johns) ring, then:*
(1) *R is left and right Kasch;*
(2) *$J=Z_l = Z_r = r(S_r) = l(S_l)$ is nilpotent;*
(3) *$S_r = S_l = l(J) = r(J)$ which is essential both in R_R and ${}_R R$.*

It is known by [11, Theorem 8.11 (6)] that any two-sided Johns ring is QF. Example 2.2 shows that a finite two-sided quasi-Johns ring may not be QF. Next we show several relations between quasi-Johns rings and QF rings in cases of some annihilator conditions.

Recall that a ring R is called a *right minannihilator* ring if every minimal right ideal of R is an annihilator. A ring R is called *right mininjective* if every homomorphism from a minimal right ideal I of R to R_R can be extended to one from R_R to R_R. And a ring R is called *right minsymmetric* if tR is a minimal right ideal of R, $t \in R$, implies that Rt is a minimal left ideal of R. The left sides of these definitions can be defined similarly.

Lemma 2.13. ([10, Proposition 2.4] *The following are equivalent for a left minannihilator ring R:*
(1) *R is right mininjective.*
(2) *R is right minsymmetric.*
(3) *$S_r \subseteq S_l$.*

Lemma 2.14. ([13, Theorem 2.5]) *If R is a left and right mininjective ring with ACC on right annihilators in which $S_r \subseteq^{ess} R_R$, then R is QF.*

Proposition 2.15. *If R is a right quasi-Johns and right minannihilator ring, then R is left mininjective.*

Proof. By Proposition 2.10, $S_l \subseteq S_r$. According to the above lemma, R is left mininjective. $\qquad\square$

Remark 2.16. The "right minannihilator condition" in the above proposition is necessary. Example 2.2 is a right quasi-Johns ring R which is not left mininjective. If it is left mininjective, as it is finite and commutative, R is left and right mininjective and with ACC on right annihilators.

According to Corollary 2.12, $S_r \subseteq^{ess} R_R$. Thus, the ring R is QF by Lemma 2.14. This is impossible.

Theorem 2.17. *Let R be a quasi-Johns and semilocal ring. The following are equivalent:*

(1) *R is QF;*

(2) *Every one-sided small ideal of R is an annihilator;*

(3) *Every one-sided singular ideal of R is an annihilator;*

(4) *R is a two-sided minannihilator ring;*

(5) *Every one-sided principal ideal of R is an annihilator.*

Proof. It is obvious that $(1) \Rightarrow (2)$, (3), (4), (5) and $(5) \Rightarrow (4)$. For $(2) \Rightarrow (4)$ and $(3) \Rightarrow (4)$. By Corollary 2.12, $J = Z_l = Z_r$. Since every one-sided minimal ideal of R is either a direct summand of R or nilpotent (see [3, Lemma 10-3.3]), every one-sided minimal ideal of R is either a direct summand or contained in J, which is the biggest one-sided small and singular ideal of R, respectively. Now we only need to show $(4) \Rightarrow (1)$. By Proposition 2.15 and Proposition 2.10, R is two-sided mininjective and J is nilpotent. Thus, R is a semiprimary ring. Then by [11, Theorem 3.24], S_l is finite dimensional. So R is left noetherian because R/S_l is left noetherian. By Lemma 2.14, R is QF.   □

Remark 2.18. The conditions of (2), (3), (4), (5) are necessary because Example 2.2 is a quasi-Johns and semilocal ring which is not QF.

Question 2.19. Can the "semilocal" condition in the above theorem be removed?

Question 2.20. Can the two-sided quasi-Johns conditions in the above theorem be replaced by one-sided?

Question 2.21. Can the two-sided annihilator conditions in (2), (3), (4), (5) be replaced by one-sided?

It is known that an important key to Faith-Menal conjecture is to prove the relative rings are semilocal. At last we give a new result on semilocal rings. First we need some preparations.

Recall that a ring R is called *right C2* if every right ideal which is isomorphic to a direct summand of R_R is also a direct summand of R_R. R is called *strongly right C2* if every matrix ring $M_n(R)$ over R is right C2, $n \geq 1$, or equivalently, $\mathbf{r}(I) \neq 0$ for each finitely generated proper left ideal I of R (see [2, Lemma 2.3]). Since left Kasch rings satisfy $\mathbf{r}(I) \neq 0$ for each proper left ideal I of R (see [11, Proposition 1.44]), we have the following relations:

left Kasch ring $\Longrightarrow$ strongly right C2 ring $\Longrightarrow$ right C2 ring.

And neither of the converses is true. Given a von Neumann regular ring R which is not semisimple, then it is easy to see that R is strongly right C2 but not Kasch. And [8, Question] show that a right C2 ring may not be strongly right C2. It is known by [11, Lemma 8.8] that a right C2 ring with right finite uniform dimension is semilocal. Thus, a left Kasch ring with right finite uniform dimension is semilocal. Next we will show that a left Kasch ring with ACC on right annihilators is also a semilocal ring.

Let M be an R-module and $\{K_\lambda\}_\wedge$ be a family of proper submodules of M. $\{K_\lambda\}_\wedge$ is called *coindependent* if for every $\lambda \in \wedge$ and finite subset $I \subseteq \wedge \setminus \{\lambda\}$ $K_\lambda + \cap_{i \in I} K_i = M$ (if I is the empty set, then set $\cap_{i \in I} K_i = M$).

Lemma 2.22. ([14, Lemma 4.2]) *Let R be a ring. Then R is semilocal if and only if R_R ($_R R$) has no infinite coindependent family of submodules.*

Theorem 2.23. *If R is a left Kasch ring and with ACC on right annihilators, then R is semilocal.*

Proof. If R is not semilocal, by Lemma 2.22, $_R R$ has infinite coindependent family of proper submodules $\{K_i \subset {}_R R\}_{i=1}^\infty$. By the definition of coindependent family of submodules,

$$\cap_{i=1}^m K_i + K_{m+1} = R, \ m = 1, 2, \cdots.$$

Thus $\mathbf{r}(\cap_{i=1}^m K_i) \cap \mathbf{r}(K_{m+1}) = 0, \ m = 1, 2, \cdots$. Since R is left Kasch and K_i are proper left ideals of R, $\mathbf{r}(K_m) \neq 0, \ m = 1, 2, \cdots$. As $\mathbf{r}(\cap_{i=1}^m K_i) \oplus \mathbf{r}(K_{m+1}) \subseteq \mathbf{r}(\cap_{i=1}^{m+1} K_i), \ m = 1, 2, \cdots$, we have an infinite chain $\mathbf{r}(K_1) \subset \mathbf{r}(K_1 \cap K_2) \subset \mathbf{r}(K_1 \cap K_2 \cap K_3) \subset \cdots$ of right annihilators. This is a contradiction. □

Question 2.24. Is a strongly right C2 ring with ACC on right annihilators a semilocal ring?

Question 2.25. Is a right C2 ring with ACC on right annihilators a semilocal ring?

Acknowledgments. The research is supported by China Scholarship Council. The author is also partially supported by the National Natural Science Foundation of China (No. 10971024), the Natural Science Foundation of Jiangsu Province (No. BK2010393), and Specialized Research Fund for the Doctoral Program of Higher Education (No. 200802860024).

References

[1] F.W. Anderson and K.R. Fuller, Rings and Categories of Modules, 2nd ed, New York: Springer-Verlag, 1992.

[2] J.L. Chen and W.X. Li, *On Artiness of right CF rings*, Comm. Algebra 32 (2004), 4485-4494.

[3] J. Dauns, Modules and Rings, Cambridge University Press, 1994.

[4] N.V. Dung, D.V. Huynh and R. Wisbauer, *Quasi-injective modules with acc or dcc on essential submodules*, Arch. Math. 53 (1989), 252-255.

[5] C. Faith, and P. Menal, *A counter example to a conjecture of Johns*, Proc. Amer. Math. Soc. 116 (1992), 21-26.

[6] B. Johns, *Annihilator conditions in noetherian rings*, J. Algetra 49 (1977), 222-224.

[7] J.W. Kerr, *An example of a Goldie ring whose matrix ring is not Goldie*, J. Algebra 61 (1979), 590-592.

[8] F. Kourki, *When maximal linearly independent subsets of a free module have the same cardinality?*, Modules and comodules, Trends in Mathematics (2008), 281-293.

[9] R.P. Kurshan, *Rings whose cyclic modules have finitely generated socle*, J. Algetra 15 (1970), 376-386.

[10] W.K. Nicholson and M.F. Yousif, *Mininjective rings*, J. Algebra 187 (1997), 548-578.

[11] W.K. Nicholson and M.F. Yousif, Quasi-Frobenius Rings, Cambridge Tracts in Mathematics 158, Cambridge University Press, 2003.

[12] S.S. Page and Y.Q. Zhou, *Quasi-dual rings*, Comm. Algebra 28 (2000), 489-504.

[13] L. Shen and J.L. Chen, *New characterizations of quasi-Frobenius rings*, Comm. Algebra 34 (2006), 2157-2165.

[14] L. Shen and J.L. Chen, *On strong Goldie dimension*, Comm. Algebra 35 (2007), 3018-3025.

[15] Y.Q. Zhou, *Rings in which certain right ideals are direct summands of annihilators rings*, J. Aust. Math. Soc. 73 (2002), 335-346.

DEPARTMENT OF MATHEMATICS
SOUTHEAST UNIVERSITY, NANJING 210096, P.R. CHINA
E-mail address: lshen@seu.edu.cn

Proceedings of the Sixth China-Japan-Korea
International Conference on Ring Theory
June 27-July 2, 2011 Suwon, Korea

VON NEUMANN REGULAR RINGS SATISFYING
GENERALIZED ALMOST COMPARABILITY

MAMORU KUTAMI

ABSTRACT. We treat some interesting properties (Morita invariantness, the strict unperforation, separativity, and etc.) for
von Neumann regular rings satisfying generalized almost comparability.

1. Introduction

In the paper, we investigate some interesting properties for (von
Neumann) regular rings satisfying generalized almost comparability by
connecting with regular rings satisfying almost comparability. The notion of almost comparability for regular rings was first introduced by
Ara and Goodearl [1], for giving an alternative proof of the outstanding
O'Meara's Theorem: directly finite simple regular rings with weak comparability are unit-regular ([11, Theorem 1]). And, the study of regular
rings satisfying almost comparability was continued in [4], [5], [9] and
etc.. After that, the notion of generalized almost comparability was defined in [10] as an extension of almost comparability, and the forms of
regular rings satisfying generalized almost comparability was determined
by using regular rings satisfying almost comparability.

In Section 2, we recall various comparabilities connecting with generalized almost comparability. In Section 3, we show the following results
by using the forms of regular rings satisfying generalized almost comparability which was given in [10]. Let R be a regular ring satisfying
generalized almost comparability. Then: (1) Any matrix rings $M_n(R)$
of R satisfy generalized almost comparability if and only if R has the
zero right socle or satisfies the comparability axiom with the only simple
right ideal up to isomorphism (Theorem 3.7). (2) For any finitely generated projective R-modules A, B and any positive integer n, $nA \prec nB$
implies $A \prec B$ (Theorem 3.17). (3) For any finitely generated projective

2010 Mathematics Subject Classification : 16E50, 16D70.

Keywords : von Neumann regular rings, generalized almost comparability, almost
comparability.

R-modules A and B, $A \oplus A \cong A \oplus B \cong B \oplus B$ implies $A \cong B$ (Theorem 3.23).

2. Notations and definitions

Throughout this paper, R is a ring with identity and R-modules are unitary right R-modules. We recall some notations and definitions.

Notation 2.1. For two R-modules M and N, we use $M \lesssim N$ (resp. $M \lesssim_\oplus N$, $M \prec N$, $M \prec_\oplus N$) to mean that there exists an isomorphism from M to a submodule of N (resp. a direct summand of N, a proper submodule of N, a proper direct summand of N). For a submodule M of an R-module N, $M \leq_\oplus N$ (resp. $M < N$, $M <_\oplus N$) means that M is a direct summand of N (resp. a proper submodule of N, a proper direct summand of N). For a cardinal number k and an R-module M, kM denotes the direct sum of k-copies of M. For a ring R, $L(R_R)$ means the family of all principal right ideals of R. For a set S, we use $|S|$ to denote the cardinal number of S.

Definition 2.2. An R-module M is *directly finite* provided that M is not isomorphic to a proper direct summand of itself. If M is not directly finite, then M is said to be *directly infinite*. A ring R is *directly finite* (resp. *directly infinite*) if the R-module R_R is directly finite (resp. directly infinite), and R is *stably finite* if the ring $M_n(R)$ of $n \times n$ matrices over R is directly finite for all positive integers n. It is well-known that every direct summand of a directly finite module is directly finite and a ring R is stably finite if and only if every finitely generated projective R-module is directly finite. An idempotent e in a ring R is *infinite* if eR is a directly infinite module. A simple ring R is said to be *purely infinite* if every nonzero right ideal of R contains an infinite idempotent (see [3], for details). Note that a simple regular ring R is purely infinite if and only if xR is directly infinite for any nonzero element $x \in R$, equivalently both $xR \prec yR$ and $yR \prec xR$ for any nonzero elements $x, y \in R$. Here, a ring R is said to be (von Neumann) *regular* if for each $x \in R$ there exists an element y of R such that $xyx = x$, and a ring R is said to be *unit-regular* if for each $x \in R$ there exists a unit element (i.e. an invertible element) u of R such that $xux = x$. A ring R is said to be *an exchange ring* if the R-module R_R satisfies the exchange property, where an R-module M *satisfies the exchange property* if, for every R-module A and any decompositions $A = M' \oplus N = \oplus_{i \in I} A_i$ with $M' \cong M$, there exist submodules $A_i' \leq A_i$ $(i \in I)$ such that $A = M' \oplus (\oplus_{i \in I} A_i')$.

We shall give well-known elementary properties for regular rings and unit-regular rings: (1) Let R be a regular ring. Then any projective

R-module P is written by a direct sum of cyclic submodules of P, and any finitely generated submodule of P is a direct summand of P ([7, Theorem 1.11]). Also any projective R-module satisfies the exchange property, whence R is an exchange ring. (2) Let R be a unit-regular ring. Then if $A \oplus C \cong B \oplus C$ for any R-modules A, B, C such that C is finitely generated projective, then $A \cong B$ ([7, Theorem 4.14]). Therefore any finitely generated projective R-module is directly finite, whence a unit-regular ring R is stably finite.

Now, we recall definitions of various comparability containing almost comparability and generalized almost comparability for regular rings which will be used in Section 3 below.

Definition 2.3. A regular ring R *satisfies s-comparability* provided that, for each $x, y \in R$, either $xR \lesssim s(yR)$ or $yR \lesssim s(xR)$, where s is a positive integer. In particular, R is said to *satisfy the comparability axiom* if it satisfies 1-comparability. We note that a regular ring satisfying s-comparability for some positive integer $s > 1$ satisfies automatically 2-comparability ([5, Theorem 2.8]). A regular ring R satisfies *almost comparability* if, for each $x, y \in R$, either $xR \prec yR \oplus zR$ for all nonzero principal right ideals zR of R or $yR \prec xR \oplus zR$ for all nonzero principal right ideals zR of R. It is known that, for simple regular rings, almost comparability is equivalent to 2-comparability ([4, Corollary 4.5]), and that almost comparability is a strict middle condition between 1-comparability and 2-comparability. Hence any regular ring satisfying almost comparability is prime. Also, it is known from [9, Theorem 1.11] that almost comparability for regular rings is Morita invariant. A regular ring R satisfies *generalized almost comparability* if, for each $x, y \in R$ and each nonzero element $z \in R$, either $xR \prec yR \oplus zR$ or $yR \prec xR \oplus zR$, which is equivalent to the condition that for each $x, y \in R$ and each nonzero element $z \in R$, either $xR \lesssim yR \oplus zR$ or $yR \lesssim xR \oplus zR$. Obviously regular rings satisfying almost comparability have generalized almost comparability, and hence the notion of generalized almost comparability is an extension for one of almost comparability.

All basic results concerning regular rings can be found in Goodearl's book [7].

3. Generalized almost comparability

We begin with giving the forms of regular rings satisfying generalized almost comparability. To do this, the following definition will be useful in the sequel.

Definition 3.1. Let R be a regular ring. Any non-empty subfamily X of $L(R_R)$ satisfies *the comparability axiom* if, for each $xR, yR \in X$, either $xR \lesssim yR$ or $yR \lesssim xR$, where $L(R_R)$ is the family of all principal right ideals of R. Note that $L(R_R)$ satisfies the comparability axiom if and only if so does R.

Theorem 3.2. ([10, Theorem 1.13]) *Regular rings satisfying generalized almost comparability are completely determined as follows.*
(I) *Any directly finite regular ring satisfying generalized almost comparability is one of the following* (a), (b) *and* (c):
   (a) *A directly finite regular ring with zero right socle satisfying almost comparability.*
   (b) *A directly finite regular ring with the only simple right ideal up to isomorphism satisfying the comparability axiom.*
   (c) *A ring which is isomorphic to a ring $D_1 \times D_2$, where D_1 and D_2 are division rings.*
(II) *Any directly infinite regular ring satisfying generalized almost comparability is one of the following* (d) *and* (e):
   (d) (i) *A directly infinite regular ring with zero right socle satisfying almost comparability, or* (ii) *A directly infinite regular ring R with zero right socle which has nonzero idempotents e, f such that $eRf = 0$, $L(R_R) = \{0\} \cup X \cup Y \cup Z$ (disjoint union) and both $X \cup Z$ and $Y \cup Z$ satisfy the comparability axiom, where $X = \{xR \in L(R_R) \mid eR \lesssim xR$ and $fR \not\lesssim xR\}$, $Y = \{yR \in L(R_R) \mid fR \lesssim yR$ and $eR \not\lesssim yR\}$ and $Z = \{zR \in L(R_R) \mid eR \oplus fR \lesssim zR\}$.*
   (e) (i) *A directly infinite regular ring with the only simple right ideal up to isomorphism satisfying the comparability axiom, or* (ii) *A ring which is isomorphic to a ring $D \times T$, where D is a division ring and T is a purely infinite simple regular ring.*

Remark 3.3. For a regular ring R satisfying generalized almost comparability with (d)(ii) in Theorem 3.2 above, we have the following results (1)–(3) from [10, Proposition 1.5, its proof and Remark 1]: (1) $eR \oplus eR \prec eR$ and $fR \oplus fR \prec fR$; (2) For each $xR \in X$ and $yR \in Y$, we have $eR \prec xR \prec eR$ and $fR \prec yR \prec fR$; (3) Any nonzero projective R-module is directly infinite.

Now, we first ask Morita invariantness for generalized almost comparability over regular rings. To see this, we newly give a definition of generalized almost comparability for finitely generated projective modules, as follows.

Definition 3.4. Let R be a ring. A finitely generated projective R-module P satisfies *generalized almost comparability* if, for each direct summands Q, Q' of P and each nonzero direct summand T of P, either

$Q \prec Q' \oplus T$ or $Q' \prec Q \oplus T$. Note that generalized almost comparability for finitely generated projective R-modules is inherited by direct summands, and that a regular ring R satisfies generalized almost comparability if and only if so does the R-module R_R.

From the definition above, we have the following obviously.

Lemma 3.5. *Let R be a regular ring. Then the following conditions are equivalent:*

(a) *Any nonzero finitely generated projective R-module satisfies generalized almost comparability.*

(b) *For any positive integer n, nR_R satisfies generalized almost comparability.*

For an R-module M_R, we put $add(M_R) = \{$an R-module $N \mid N \lesssim_\oplus nM$ for some positive integer $n\}$. The following lemma follows from equivalences of the Hom and Tensor functors between the categories $add(M_R)$ and $add(S_S)$, where $S = End_R(M)$ (see [13, 46.7]).

Lemma 3.6. *Let M be a finitely generated projective R-module over a ring R, and set $S = End_R(M)$. Hence M is an (S, R)-bimodule. If M is flat as a left S-module, then M satisfies generalized almost comparability if and only if so does S as an S-module. In particular, a finitely generated projective R-module P over a regular ring R satisfies generalized almost comparability if and only if so does the ring $End_R(P)$.*

Generalized almost comparability for regular rings is not inherited by matrix rings in general. In fact, using Lemmas 3.5, 3.6 and Theorem 3.2, we can give a criterion for Morita invariantness of generalized almost comparability over regular rings, as follows.

Theorem 3.7. *Let R be a regular ring satisfying generalized almost comparability. Then the following conditions are equivalent:*

(a) *R has the zero right socle, or R satisfies the comparability axiom with the only simple right ideal up to isomorphism.*

(b) *R satisfies either (a), (b), (d)(i), (d)(ii), or (e)(i) in Theorem 3.2.*

(c) *For any positive integer n, $M_n(R)$ satisfies generalized almost comparability.*

(d) *For any nonzero finitely generated projective R-module P, the ring $End_R(P)$ satisfies generalized almost comparability.*

(e) *Every ring S which is Morita equivalent to R satisfies generalized almost comparability.*

Proof. (c) $\iff$ (d) $\iff$ (e) follows from Lemmas 3.5 and 3.6, and (a) $\iff$ (b) follows from Theorem 3.2 and [10, Proposition 1.3 and Theorem 1.4]. Thus we prove that (b) $\iff$ (c) holds. If R is a regular ring satisfying generalized almost comparability with either (a), (b), (d)(i) or (e)(i) in Theorem 3.2, then we see that $M_n(R)$ satisfies generalized almost comparability for all positive integers n, because almost comparability for regular rings inherits in matrix rings from [9, Theorem 1.11]. On the other hand, if R is a regular ring satisfying generalized almost comparability with (c) in Theorem 3.2, then R is isomorphic to a ring $D_1 \times D_2$, where D_1 and D_2 are division rings. Then $M_n(R) \cong M_n(D_1) \times M_n(D_2)$, and so $M_n(R)$ is non-prime and directly finite, whence we see from Theorem 3.2(I) that $M_n(R)$, for each $n(\geq 2)$, does not satisfy generalized almost comparability. Also, if R is a regular ring satisfying generalized almost comparability with (e)(ii) in Theorem 3.2, then R is isomorphic to a ring $D \times T$, where D is a division ring and T is a purely infinite simple regular ring. Then $M_n(R) \cong M_n(D) \times M_n(T)$, and so $M_n(R)$ is non-prime and directly infinite, whence $M_n(R)$, for each $n(\geq 2)$, does not satisfy generalized almost comparability from Theorem 3.2(II) and Remark 3.3(3). Now, let R be a regular ring satisfying generalized almost comparability with (d)(ii) in Theorem 3.2, whence R is a directly infinite regular ring with zero right socle which has nonzero idempotents e, f such that $eRf = 0$, $L(R_R) = \{0\} \cup X \cup Y \cup Z$ (disjoint union) and both $X \cup Z$ and $Y \cup Z$ satisfy the comparability axiom, where $X = \{xR \in L(R_R) \mid eR \lesssim xR$ and $fR \not\lesssim xR\}$, $Y = \{yR \in L(R_R) \mid fR \lesssim yR$ and $eR \not\lesssim yR\}$ and $Z = \{zR \in L(R_R) \mid eR \oplus fR \lesssim zR\}$. We first claim that nR_R satisfies generalized almost comparability for any positive integer n. Let P, Q be nonzero direct summands of nR_R, whence we have finite direct sum decompositions of nonzero principal right ideals of R for P, Q, as follows:

$$P \cong (\oplus_{i \in I} x_i R) \oplus (\oplus_{j \in J} y_j R) \oplus (\oplus_{k \in K} z_k R),$$
$$Q \cong (\oplus_{i \in I'} x_i' R) \oplus (\oplus_{j \in J'} y_j' R) \oplus (\oplus_{k \in K'} z_k' R),$$

where each $x_i R, x_i' R \in X$, $y_j R, y_j' R \in Y$ and $z_k R, z_k' R \in Z$. Note that if $I \neq \phi$ then $eR \prec \oplus_{i \in I} x_i R \prec \oplus_{i \in I} eR \prec eR$ from Remark 3.3, whence $eR \prec \oplus_{i \in I} x_i R \prec eR$. Similarly, if $I' \neq \phi$ (resp. $J \neq \phi$, $J' \neq \phi$), then $eR \prec \oplus_{i \in I'} x_i' R \prec eR$ (resp. $fR \prec \oplus_{j \in J} y_j R \prec fR$, $fR \prec \oplus_{j \in J'} y_j' R \prec fR$). Now we divide into three cases as follows.

Case 1. Assume that $|K| = |K'| = \phi$. If $|I| \neq \phi, |J'| \neq \phi$ and $|I'| = |J| = \phi$, then either $\oplus_{i \in I} x_i R \prec (\oplus_{j \in J'} y_j' R) \oplus T$ or $\oplus_{j \in J'} y_j' R \prec (\oplus_{i \in I} x_i R) \oplus T$ for each nonzero direct summand T of nR_R, because T has a nonzero submodule which is isomorphic to some principal right ideal of R containing either eR or fR. Similarly, if $|I'| \neq \phi, |J| \neq \phi$ and $|I| = |J'| = \phi$, then either $\oplus_{j \in J} y_j R \prec (\oplus_{i \in I'} x_i' R) \oplus T$ or $\oplus_{i \in I'} x_i' R \prec (\oplus_{j \in J} y_j R) \oplus T$ for each nonzero direct summand T of nR_R. Thus, for the above cases, either $P \prec Q \oplus T$ or $Q \prec P \oplus T$ for each nonzero direct summand T of nR_R. For the other cases, we see that either $P \prec Q$ or $Q \prec P$ obviously.

Case 2. Assume that $|K| \neq \phi$ and $|K'| = \phi$. Since $z_k R \in Z$, we have $eR \oplus fR \prec z_k R$, so $Q \prec P$. Similaly, if $|K'| \neq \phi$ and $|K| = \phi$, then $P \prec Q$.

Case 3. Assume that $|K| \neq \phi$ and $|K'| \neq \phi$. Since $z_k R \in Z$, we see that $z_k R \cong eR \oplus wR$ for some nonzero right ideal wR of R, so $eR \oplus z_k R \cong eR \oplus (eR \oplus wR) \prec eR \oplus wR \cong z_k R$. Thus $eR \oplus z_k R \prec z_k R$. Similarly $fR \oplus z_k R \prec z_k R$. Hence $\oplus_{k \in K} z_k R \prec P \prec \oplus_{k \in K} z_k R$. Also, $\oplus_{k \in K'} z'_k R \prec Q \prec \oplus_{k \in K'} z'_k R$. Now, we put $K = \{1, 2, \cdots, m\}$ and $K' = \{1, 2, \cdots, n\}$. Since $z_1 R, z'_1 R \in Z$, using the comparability axiom for Z, we may assume that $z_1 R \lesssim z'_1 R$, so $z_1 R \oplus w_1 R \cong z'_1 R$ for some right ideal $w_1 R$ of R. Next we can compare $w_1 R$ and $z_2 R$, because both $X \cup Z$ and $Y \cup Z$ satisfy the comparability axiom. Continuing the procedure, we see that $\oplus_{k \in K} z_k R \lesssim \oplus_{k \in K'} z'_k R$ or $\oplus_{k \in K'} z'_k R \lesssim \oplus_{k \in K} z_k R$, whence either $P \prec Q$ or $Q \prec P$.

From Cases 1–3 above, we see that nR_R satisfies generalized almost comparability as desired. Thus so does $M_n(R)$ for all n from Lemma 3.6, whence (b) $\iff$ (c) holds. The proof is complete. $\square$

For a regular ring R satisfying generalized almost comparability, if R is directly finite, then so is $M_n(R)$ for any positive integer n, from Lemma 3.8(1) below. Connecting with the direct finiteness, it is well-known from [7, Corollary 5.6] that if an R-module M contains no infinite direct sums of nonzero pairwise isomorphic submodules, then M is directly finite. But the converse does not hold in general, even in case of regular rings with the comparability axiom. Now we ask that if a regular ring R satisfying generalized almost comparability contains no infinite direct sum of nonzero pairwise isomorphic submodules then so does $M_n(R)$ for any positive integer n. To see this, we need some results below.

Lemma 3.8. ([10, Theorem 2.7]) *Let R be a regular ring satisfying generalized almost comparability. Then we have the following.*

(1) *R has the property* (DF), *which means that any finite direct sums of directly finite projective R-modules are directly finite. In particular, if R is directly finite, then R is stably finite.*

(2) *For any projective R-modules A, B, C such that C is finitely generated and directly finite, $A \oplus C \prec_\oplus B \oplus C$ implies $A \prec_\oplus B$.*

Notation 3.9. We use $SubP(R)$ to denote the family of all R-modules M such that M is a submodule of some projective R-module.

Theorem 3.10. *Let R be a regular ring satisfying generalized almost comparability. For each $M, N \in SubP(R)$, if both M and N contain no infinite direct sums of nonzero pairwise isomorphic submodules, then so does $M \oplus N$.*

Proof. We prove Theorem 3.10 by modifying the proof of [8, Proposition 2.5]. Assume that M and N are nonzero, and that these contain no infinite direct sums of nonzero pairwise isomorphic submodules, whence M and N are directly finite. Now, we assume that there exists a nonzero R-module X such that $\aleph_0 X \lesssim M \oplus N$, and hence $tX \lesssim M \oplus N$ for all positive integers t. Then we may consider that X is a nonzero cyclic projective R-module, because $M, N \in SubP(R)$. Since $X \lesssim M \oplus N$ ($\in SubP(R)$), we have that $X \lesssim_\oplus M \oplus N$. Using the exchange property for X, we have decompositions $M = M_1 \oplus M_1^*$ and $N = N_1 \oplus N_1^*$ such that $X \cong M_1 \oplus N_1$. Note that both M_1 and N_1 are cyclic projective and directly finite. From Lemma 3.8(1), we see that X is directly finite. For each positive integer $k (\geq 2)$, we have that $kX \prec_\oplus M \oplus N = (M_1 \oplus M_1^*) \oplus (N_1 \oplus N_1^*)$, hence $X \oplus (k-1)X \prec_\oplus (M_1 \oplus M_1^{**}) \oplus (N_1 \oplus N_1^{**}) \cong X \oplus (M_1^{**} \oplus N_1^{**})$ for some finitely generated direct summands M_1^{**}, N_1^{**} of M_1^*, N_1^* respectively. Since X is finitely generated projective and directly finite, we have $(k-1)X \prec_\oplus M_1^{**} \oplus N_1^{**} \lesssim_\oplus M_1^* \oplus N_1^*$ from Lemma 3.8(2). Therefore we see that $tX \lesssim M_1^* \oplus N_1^*$ for all positive integers t. Since $M_1^*, N_1^* \in SubP(R)$, using the argument above, we have decompositions $M_1^* = M_2 \oplus M_2^*$ and $N_1^* = N_2 \oplus N_2^*$ such that $X \cong M_2 \oplus N_2$ and $tX \lesssim_\oplus M_2^* \oplus N_2^*$ for all positive integers t. Continuing the procedure, we see that $\aleph_0 X \cong (M_1 \oplus N_1) \oplus (M_2 \oplus N_2) \oplus \cdots \cong (M_1 \oplus M_2 \oplus \cdots) \oplus (N_1 \oplus N_2 \oplus \cdots)$, which is directly infinite. From Lemma 3.8(1), we have that either $M_1 \oplus M_2 \oplus \cdots$ or $N_1 \oplus N_2 \oplus \cdots$ is directly infinite. Thus, either $M_1 \oplus M_2 \oplus \cdots$ or $N_1 \oplus N_2 \oplus \cdots$ contains an infinite direct sum of nonzero pairwise isomorphic submodules, which contradicts the assumption. The proof is complete. $\square$

Corollary 3.11. *Let R be a regular ring satisfying generalized almost comparability. For each $M \in SubP(R)$, the following conditions are equivalent:*

(a) *M contains no infinite direct sum of nonzero pairwise isomorphic submodules.*

(b) *For any positive integer n, nM contains no infinite direct sums of nonzero pairwise isomorphic submodules.*

Lemma 3.12. ([8, Lemma 2.7]) *Let P be a finitely generated projective R-module over a regular ring R, and let $S = End_R(P)$. Then the following conditions are equivalent:*

(a) *There exists a nonzero R-submodule Q of P such that $\aleph_0 Q \lesssim P$.*

(b) *There exists a nonzero S-submodule T of S_S such that $\aleph_0 T \lesssim S_S$.*

From Corollary 3.11 and Lemma 3.12 above, we have the following result.

Theorem 3.13. *Let R be a regular ring satisfying generalized almost comparability. If R contains no infinite direct sums of nonzero pairwise isomorphic submodules, then so does $M_n(R)$ for any positive integer n.*

Next, we assure the strict unperforation property for finitely generated projectives over regular rings satisfying generalized almost comparability in Theorem 3.17 below. To do this, we need some following results.

Lemma 3.14. ([5, Lemma 1.2]) *Let R be a regular ring, and let A, B, C be finitely generated projective R-modules. If $A \oplus B \cong kC$ for some positive integer k, then there exists a decomposition $C = C_0 \oplus C_1 \oplus \cdots \oplus C_k$ such that $A \cong C_1 \oplus 2C_2 \oplus \cdots \oplus kC_k$ and $B \cong kC_0 \oplus (k-1)C_1 \oplus \cdots \oplus C_{k-1}$.*

Lemma 3.15. ([10, Proposition 2.4]) *Let R be a regular ring satisfying almost comparability, and let A, B be projective R-modules such that A is finitely generated. If $nA \prec nB$ for some positive integer n, then $A \prec B$.*

Lemma 3.16. *Let R be a regular ring satisfying generalized almost comparability with (d)(ii) in Theorem 3.2, and let Z be one in (d)(ii) above.*

(1) *If $aR \cong bR \oplus cR$ for some nonzero elements $a, b, c \in R$ such that $aR \in Z$, then there exist principal right ideals $b^*R, c^*R \in Z$ such that $b^*R \oplus c^*R \prec aR \prec b^*R \oplus c^*R$. Thus, we may consider bR, cR as $bR, cR \in Z$ for treating the comparability.*

(2) *For any nonzero elements $a, b \in R$ such that $aR \in Z$, there exist a principal right ideal $b^*R \in Z$ such that $aR \oplus b^*R \prec aR \oplus bR \prec aR \oplus b^*R$. Thus, we may consider bR as $bR \in Z$ for treating the comparability.*

Proof. Since $aR \in Z$, we note that $aR \cong eR \oplus fR \oplus a'R$ for some principal right ideal $a'R$ of R, where e, f are one in Theorem 3.2(d)(ii). Noting that $eR \oplus eR \prec eR$ and $fR \oplus fR \prec fR$, we see that $aR \oplus eR \oplus fR \prec aR$.
(1) Put $B := bR \oplus eR \oplus fR$ and $C := cR \oplus eR \oplus fR$, whence $B \oplus C \prec aR$, because $aR \cong bR \oplus cR$. Thus $B \cong b^*R$ and $C \cong c^*R$ for some nonzero elements $b^*, c^* \in R$. Then $b^*R, c^*R \in Z$ and $b^*R \oplus c^*R \prec aR \prec b^*R \oplus c^*R$.
(2) Note again that $aR \oplus eR \oplus fR \prec aR$, whence $aR \oplus bR \oplus eR \oplus fR \prec aR \oplus bR \prec aR \oplus bR \oplus eR \oplus fR$. Assume $bR \in X$, so $eR \prec bR \prec eR$. Put $B := bR \oplus eR \oplus fR$. Then we have $B \prec eR \oplus fR \prec aR$ since $eR \oplus eR \prec eR$, whence $B \cong b^*R$ for some nonzero $b^* \in R$. Thus $b^*R \in Z$ and $aR \oplus b^*R \prec aR \oplus bR \prec aR \oplus b^*R$. Similarly, we have the same result in case that $bR \in Y \cup Z$. □

Theorem 3.17. *Let R be a regular ring satisfying generalized almost comparability, and let A, B be finitely generated projective R-modules. If $nA \prec nB$ for some positive integer n, then $A \prec B$.*

Proof. By Lemma 3.15 above, Theorem 3.17 holds for a regular ring satisfying generalized almost comparability with either (a), (b), (d)(i) or (e)(i) in Theorem 3.2. Also, it holds for a regular ring satisfying generalized almost comparability with (c) in Theorem 3.2, using [7, Corollary 3.11] and the unit-regularity for R. Now, let R be a regular ring satisfying generalized almost comparability with (d)(ii) in Theorem 3.2. By the mathematical induction, we claim that $nA \prec nB$ implies $A \prec B$ for any finitely generated projective R-modules A, B and any positive integer n. We assume that $nA \prec nB$. Then we may assume that A is directly infinite from [10, Corollary 2.8]. Let $A \cong \oplus_{i=1}^{l} a_i R$ and $B \cong \oplus_{j=1}^{m} b_j R$, where all $a_i, b_j \ (\neq 0) \in R$. If some $a_i R \in X$, then there exists some $b_j R \notin Y$ since $nA \prec nB$, whence the $b_j R \in X \cup Z$, where X, Y, Z are one of Theorem 3.2(d)(ii). Similarly, if some $a_i R \in Y$, then we have some $b_j R \in Y \cup Z$. Thus if all $a_i R \in X \cup Y$, then we have $A \prec B$ from Remark 3.3. Thus we treat the case that some $a_i R \in Z$. Since $nA \prec nB$, using Lemma 3.16, we may consider that all $a_i R, b_j R \in Z$. Noting that $nA \prec_\oplus nB$, we have a decomposition $nB \cong C \oplus D$ such that $A \leq C$ and $(n-1)A \leq D$. From Lemma 3.14, there exists a decomposition $B = B_0 \oplus B_1 \oplus \cdots \oplus B_n$ such that $C \cong B_1 \oplus 2B_2 \oplus \cdots \oplus nB_n$ and $D \cong nB_0 \oplus (n-1)B_1 \oplus \cdots \oplus B_{n-1}$. Then, from Lemma 3.16 again, we may consider that each B_i (where $0 \leq i \leq n$) is isomorphic to a direct sum of principal right ideals of R all which are in Z. Using the comparability axiom for $X \cup Z$ and $Y \cup Z$, we see, from the similar proof of Case 3 in Theorem 3.7, that either $B_0 \lesssim B_2 \oplus 2B_3 \oplus \cdots \oplus (n-1)B_n$ or $B_2 \oplus 2B_3 \oplus \cdots \oplus (n-1)B_n \lesssim B_0$. If $B_0 \lesssim B_2 \oplus 2B_3 \oplus \cdots \oplus (n-1)B_n$, then we see, from the direct infiniteness for A, that $(n-1)A \prec (n-1)A \leq D \lesssim (n-1)B_0 \oplus (n-1)B_1 \oplus \cdots \oplus (n-1)B_n \lesssim (n-1)B$, whence $(n-1)A \prec (n-1)B$. Using the hypothesis for the mathematical induction, we have $A \prec B$. On the other hand, if $B_2 \oplus 2B_3 \oplus \cdots \oplus (n-1)B_n \lesssim B_0$, then $A \prec A \leq C \lesssim B_0 \oplus B_1 \oplus \cdots \oplus B_n = B$, whence $A \prec B$. Therefore Theorem 3.17 holds. Finally, let R be a regular ring satisfying generalized almost comparability with (e)(ii) in Theorem 3.2, whence R is isomorphic to a ring $D \times T$, where D is a division ring and T is a purely infinite simple regular ring. Then we have $m(t_1 T) \prec t_2 T$ for any nonzero $t_1, t_2 \in T$ and any positive integer m, because $t_1, t_2 (\neq 0) \in T$ implies $t_1 T \prec t_2 T$ and so $t_1 T \oplus wT \cong t_2 T$ for some nonzero $w \in T$. Noting that $t_1 T \prec wT$ from the property of T again, we have $2(t_1 T) \prec t_2 T$, whence $m(t_1 T) \prec t_2 T$ for any positive integer m, as desired. Thus we see that Theorem 3.17 holds. The proof is complete. □

Finally, we can show that any regular ring satisfying generalized almost comparability is separative, in Theorem 3.23 below. To see this, we recall the definition and well-known results for separative rings.

Definition 3.18. ([2]) A ring R is said to be *separative* (resp. *strongly separative*) if, for any finitely generated projective R-modules A, B, we have that $A \oplus A \cong A \oplus B \cong B \oplus B$ (resp. $A \oplus A \cong A \oplus B$) implies $A \cong B$. More generally, an ideal I of R is said to be *separative* (resp. *strongly separative*) if the above implication holds for any $A, B \in FP(I)$, where $FP(I)$ means the family of all finitely generated projective R-modules P such that $P = PI$. Obviously, any unit-regular ring is strongly separative.

Lemma 3.19. ([2, Sections 2 and 5]) *For a ring R, the following conditions are equivalent:*

(a) *R is separative (resp. strongly separative).*

(b) *For any finitely generated projective R-modules A, B, C, if $A \oplus C \cong B \oplus C$ and C is isomorphic to direct summands of both mA and nB for some positive integers m and n (resp. if $A \oplus C \cong B \oplus C$ and C is isomorphic to a direct summand of mA for some positive integer m), then $A \cong B$.*

Lemma 3.20. ([2, Lemma 4.1 and Theorem 4.2]) *Let R be an exchange ring, and let I be an ideal of R. Then we have the following.*

(1) *R is separative if and only if so are I and R/I.*

(2) *I is separative if and only if all corner rings gRg, for idempotents $g \in I$, are separative.*

Now, we recall the definition of generalized s-comparability, given by E. Pardo [12].

Definition 3.21. Let R be a regular ring, and let s be a positive integer. Then R is said to satisfy *generalized s-comparability* provided that for each finitely generated projective R-modules P, Q, there exists a central idempotent e of R such that both $Pe \lesssim_{\oplus} s(Qe)$ and $Q(1-e) \lesssim_{\oplus} s(P(1-e))$, which is equivalent that, for each $x, y \in R$, there exists a central idempotent e of R such that both $e(xR) \lesssim s(e(yR))$ and $(1-e)yR \lesssim s((1-e)(xR))$ ([6, Theorem 2]). In particular, if $s = 1$, then R is said to satisfy *general comparability*. Thus generalized s-comparability is an extension for both general comparability and s-comparability.

Lemma 3.22. ([12, Theorems 2.3 and 3.9]) *Any exchange ring satisfying generalized s-comparability is separative. Therefore any regular ring satisfying almost comparability or general comparability is separative. In particular, any directly finite exchange ring satisfying s-comparability is strongly separative.*

Using the above, we can show the interesting result, as follows.

Theorem 3.23. *Any regular ring R satisfying generalized almost comparability is separative. In addition, if R is directly finite, then it is strongly separative.*

Proof. Let R be a regular ring satisfying generalized almost comparability, and we claim that R is separative. From Theorem 3.2 and Lemma 3.22, it is sufficient to prove that R with (d)(ii) in Theorem 3.2 is separative. Then R is a directly infinite regular ring with zero right socle which has nonzero idempotents e, f such that $eRf = 0$, $L(R_R) = \{0\} \cup X \cup Y \cup Z$ (disjoint union) and both $X \cup Z$ and $Y \cup Z$ satisfy the comparability axiom, where $X = \{xR \in L(R_R) \mid eR \lesssim xR$ and $fR \not\lesssim xR\}$, $Y = \{yR \in L(R_R) \mid fR \lesssim yR$ and $eR \not\lesssim yR\}$ and $Z = \{zR \in L(R_R) \mid eR \oplus fR \lesssim zR\}$. Now, we put $I = RfR$. Then R/I satisfies the comparability axiom, hence it is separative by Lemma 3.22, because the family of all principal right ideals of R/I is induced by $X \cup Z \cup \{0\}$ by [7, Proposition 2.18 and Corollary 2.23], and $X \cup Z \cup \{0\}$ satisfies the comparability axiom. Also, for any nonzero idempotent $g \in I$, there exists a positive integer n such that $gR \lesssim n(fR)$ from [7, Corollary 2.23], whence $gR \prec fR$ from Remark 3.3, so $gR \in Y$. Hence the family of all finitely generated submodules of gR equals to $Y \cup \{0\}$. Noting that $Y \cup \{0\}$ satisfies the comparability axiom, we see that gRg satisfies the comparability axiom from [7, Proposition 2.4], whence I is separative by Lemmas 3.20(2) and 3.22. Thus R is separative by Lemma 3.20(1), as desired. In addition, if R is directly finite, then R is strongly separative from Theorem 3.2 and Lemma 3.22.　　　　□

Acknowledgments. This work was supported by JSPS KAKENHI (21540041).

References

[1] P. Ara and K.R. Goodearl, *The almost isomorphism relation for simple regular rings*, Publ. Mat. UAB **36** (1992), 369–388.

[2] P. Ara, K.R. Goodearl, K.C. O'Meara and E. Pardo, *Separative cancellation for projective modules over exchange rings*, Israel J. Math. **105** (1998), 105–137.

[3] P. Ara, K.R. Goodearl and E. Pardo, K_0 *of purely infinite simple regular rings*, K-Theory **26** (2002), 69–100.

[4] P. Ara, K.R. Goodearl, E. Pardo and D.V. Tyukavkin, K-*theoretically simple von Neumann regular rings*, J. Algebra **174** (1995), 659–677.

[5] P. Ara, K.C. O'Meara and D.V. Tyukavkin, *Cancellation of projective modules over regular rings with comparability*, J. Pure Appl. Algebra **107** (1996), 19–38.

[6] H. Chen, *Regular rings with generalized s-comparability*, SEAMS Bull. Math. **25** (2001), 15–21.

[7] K.R. Goodearl, Von Neumann Regular Rings, 2nd ed. Malabar, Florida: Krieger, 1991.

[8] M. Kutami, *Regular rings with comparability and some related properties*, Comm. Algebra **30**(7) (2002), 3337–3349.

[9] M. Kutami, *On regular rings satisfying almost comparability*, Comm. Algebra **35**(7) (2007), 2171–2182.

[10] M. Kutami, *Regular rings satisfying generalized almost comparability*, Comm. Algebra **37**(5) (2009), 1579–1593.

[11] K.C. O'Meara, *Simple regular rings satisfying weak comparability*, J. Algebra **141** (1991), 162–186.

[12] E. Pardo, *Comparability, separability, and exchange rings*, Comm. Algebra **24**(9) (1996), 2915–2929.

[13] R. Wisbauer, Foundations of Module and Ring Theory, Gordon and Breach Science Publishers, Amsterdam, 1991.

DEPARTMENT OF MATHEMATICS
FACULTY OF SCIENCE, YAMAGUCHI UNIVERSITY
YAMAGUCHI 753-8512, JAPAN
E-mail address: kutami@yamaguchi-u.ac.jp

Proceedings of the Sixth China-Japan-Korea
International Conference on Ring Theory
June 27-July 2, 2011   Suwon, Korea

A NEW PSEUDORANDOM NUMBER
GENERATOR AST

HUILING SONG

ABSTRACT. In this paper, we give a method to construct a finite
field using an Artin-Schreier tower, such as $\mathbb{F}_{3^{3^r}}$, and a multiplication algorithm using the recursive basis. Using the Artin-Schreier
tower, we propose a new pseudorandom number generator AST
for the ternary case.

1. Introduction

The theory of finite fields is increasingly important not only in several
areas of mathematics, including linear and abstract algebra, number theory and algebraic geometry, but also in computer science, information
theory, and various field in engineering. For example, there is a method
of pseudorandom number generators as an application of the theory of
finite fields. It is widely used in simulations in various area and crypto-
graphic systems. A pseudorandom number generator is an algorithm for
generating a sequence of numbers with components 0 and 1 of size w,
and is produced by a generator from initial seeds. TGFSR and MT are
pseudorandom number generators suggested by Matsumoto, which are
widely used. Getting a hint from these, in [9], we have proposed a pseudorandom number generator AST for the binary case, which is a modified
version of the twisted generalized feedback shift register (TGFSR). Similarly, this method can be also used in the ternary case, what is that
we can proposed a new pseudorandom number generator AST for the
ternary case. It is base on finite fields using the specific Artin-Schreier
tower. In this paper, we give the construction of finite fields for $p = 3$,
and give a new pseudorandom number generator.

Here is the plan of this paper. In section 2, we will describe a construction of finite fields using the specific Artin-Schreier tower starting
from the ternary field $\mathbb{F}_3$ and a multiplication algorithm. In sections 3
and 4, we will define a linear recurrence equation and introduce several

2010 Mathematics Subject Classification : 12Y05, 11K45.
Keywords : Artin-Schreier tower, pseudorandom number generator.

examples of pseudorandom number generators for the binary case. Finally, we will define certain matrices as an application of section 2 and propose a new pseudorandom number generator using them.

2. Construction using Artin-Schreier towers

2.1. Recursive structures for $p = 3$ using an Artin-Schreier tower

Definition 1 ([2]). Let K_0 be the prime field $\mathbb{F}_3 = \{0, 1, 2\}$ and $f_1(x) := x^3 - x - 1$ be a polynomial in $\mathbb{F}_3[x]$, we define

$$K_1 := K_0[x]/(f_1(x)) = K_0(\alpha_1) = \mathbb{F}_3(\alpha_1) = \mathbb{F}_{3^3},$$

where $\alpha_1 := \bar{x} \in K_1$ be the image of x in K_1. Suppose that α_{r-1} and $f_{r-1}(x)$ are defined for $r \geq 2$. Define $f_r(x)$, K_r and α_r as follows:

$$f_r(x) := x^3 - x - (\alpha_1 \cdots \alpha_{r-1})^2, \quad K_r := K_{r-1}[x]/(f_r(x)),$$

$$\alpha_r := \bar{x} \in K_r = K_{r-1}(\alpha_r).$$

Then we have the tower of finite fields inductively:

$$K_0 \subset K_1 = K_0(\alpha_1) \subset K_2 = K_1(\alpha_2) \subset \cdots \subset K_r = K_{r-1}(\alpha_r) \subset \cdots .$$

We call this sequence of extensions the *Artin-Schreier tower*.

The polynomial f_r in the definition is known to be irreducible over K_{r-1} by analyzing the Artin-Schreier extensions (see [2]). Because of its natural definition of the tower, this Artin-Schreier tower has a beautiful recursive structure which is a key structure of our current work.

Let us explain the recursive structure. Since the basis of K_1 over K_0 is 1, α_1 and α_1^2, we have an expression

$$K_1 = \{s_0 1 + t_0\alpha_1 + u_0\alpha_1^2 \mid s_0, t_0, u_0 \in K_0\}.$$

The basis of K_2 over K_1 is 1, α_2 and α_2^2, then the basis of K_2 over K_0 is

$$1, \alpha_1, \alpha_1^2, \alpha_2, \alpha_1\alpha_2, \alpha_1^2\alpha_2, \alpha_2^2, \alpha_1\alpha_2^2, \alpha_1^2\alpha_2^2,$$

thus we have an expression

$$
\begin{aligned}
K_2 &= \{s_1 1 + t_1\alpha_2 + u_1\alpha_2^2 \mid s_1, t_1, u_1 \in K_1\} \\
&= \{s_{01} 1 + t_{01}\alpha_1 + u_{01}\alpha_1^2 + s_{02}\alpha_2 + t_{02}\alpha_1\alpha_2 + u_{02}\alpha_1^2\alpha_2 + s_{03}\alpha_2^2 + \\
&\quad t_{03}\alpha_1\alpha_2^2 + u_{03}\alpha_1^2\alpha_2^2 \mid s_{0i}, t_{0i}, u_{0i} \in K_0 \ 1 \leq i \leq 3\}.
\end{aligned}
$$

Similarly, the basis of K_r over K_{r-1} is 1, α_r and α_r^2, so that we have the basis of K_r over K_0 as

$$\underbrace{\overbrace{\overbrace{(1,\alpha_1,\alpha_1^2}^{3}|\alpha_2,\alpha_1\alpha_2,\alpha_1^2\alpha_2|\alpha_2^2,\alpha_1\alpha_2^2,\alpha_1^2\alpha_2^2}^{3^2}|\cdots|\underbrace{\alpha_{r-1},\ldots,(\alpha_1\cdots\alpha_{r-1})^2}_{3^{r-1}}}^{3^{r-2}}\Big|\alpha_r,\ldots,(\alpha_1\cdots\alpha_r)^2)}_{3^r}.$$

Note that the last $\frac{2}{3}$ part of this basis is given by multiplying α_r and α_r^2 with the first $\frac{1}{3}$ part of this basis which is the recursive structure of the basis of this extensions.

2.2. Multiplication algorithm for $p = 3$

Using the recursive structures of the basis exhibited above, we can make an algorithm of multiplication on the Artin-Schreier extensions without the power expression of each element. We write $s_1 + s_2\alpha_r + s_3\alpha_r^2 \in K_r$ as (s_1, s_2, s_3) with $s_1, s_2, s_3 \in K_{r-1}$. And we also write the multiplication of two elements of K_r, $(s_1 + s_2\alpha_r + s_3\alpha_r^2)(t_1 + t_2\alpha_r + t_3\alpha_r^2)$ as $(s_1, s_2, s_3)(t_1, t_2, t_3)$. Taking the multiplication of two elements $s_1 + s_2\alpha_r + s_3\alpha_r^2, t_1 + t_2\alpha_r + t_3\alpha_r^2 \in K_r$ inside the field K_r, we have

$$
\begin{aligned}
&(s_1 + s_2\alpha_r + s_3\alpha_r^2)(t_1 + t_2\alpha_r + t_3\alpha_r^2)\\
=\ & s_1t_1 + s_1t_2\alpha_r + s_1t_3\alpha_r^2 + s_2t_1\alpha_r\\
&+s_2t_2\alpha_r^2 + s_2t_3\alpha_r^3 + s_3t_1\alpha_r^2 + s_3t_2\alpha_r^3 + s_3t_3\alpha_r^4\\
=\ & s_1t_1 + (s_1t_2 + s_2t_1)\alpha_r + (s_1t_3 + s_2t_2 + s_3t_1)\alpha_r^2\\
&+(s_2t_3 + s_3t_2)\alpha_r^3 + s_3t_3\alpha_r^4.
\end{aligned}
$$

Since α_r is a root of $f_r(x) = x^3 - x - (\alpha_1 \cdots \alpha_{r-1})^2$, we have

$$\alpha_r^3 = \alpha_r + (\alpha_1 \cdots \alpha_{r-1})^2,$$

thus we can write

$$
\begin{aligned}
&(s_1 + s_2\alpha_r + s_3\alpha_r^2)(t_1 + t_2\alpha_r + t_3\alpha_r^2)\\
=\ & s_1t_1 + (s_2t_3 + s_3t_2)(\alpha_1, \cdots, \alpha_{r-1})^2\\
&+(s_1t_2 + s_2t_1 + (s_2t_3 + s_3t_2) + s_3t_3(\alpha_1, \cdots, \alpha_{r-1})^2)\alpha_r\\
&+(s_1t_3 + s_2t_2 + s_3t_1 + s_3t_3)\alpha_r^2.
\end{aligned}
$$

Since $(s_1t_1 + (s_2t_3 + s_3t_2)(\alpha_1, \cdots, \alpha_{r-1})^2)$, $(s_1t_2 + s_2t_1 + (s_2t_3 + s_3t_2) + s_3t_3(\alpha_1, \cdots, \alpha_{r-1})^2)$ and $(s_1t_3 + s_2t_2 + s_3t_1 + s_3t_3)$, marking as S_1, S_2 and S_3, are the elements of K_{r-1} we have

$$(s_1 + s_2\alpha_r + s_3\alpha_r^2)(t_1 + t_2\alpha_r + t_3\alpha_r^2) = (S_1, S_2, S_3).$$

Doing the above operation recursively, we can express an element of K_r as a vector of length 3^r over K_0 with the basis shown in the end of the last subsection. Note that $(\alpha_1 \cdots \alpha_{r-1})^2$ can be regarded as the vector $(0, \cdots, 0, 1)$ over K_0. By the argument above, we have the matrix which expresses the multiplication of two elements.

Theorem 1. *For an element* $(t_1, t_2, \ldots, t_{3r}) \in K_r$, *we define the* $3^r \times 3^r$ *matrix* $T^{(r)}(t_1, t_2, \ldots, t_{3r})$ *inductively as follows:*

$$T^{(1)}(t_1, t_2, t_3) := \begin{pmatrix} t_1 & t_2 & t_3 \\ t_3 & t_1 + t_3 & t_2 \\ t_2 & t_2 + t_3 & t_1 + t_3 \end{pmatrix},$$

$$T^{(r)}(t_1, t_2, \ldots, t_{3r}) := \begin{pmatrix} M_1 & M_2 & M_3 \\ M_3 \cdot M_0 & M_1 + M_3 & M_2 \\ M_2 \cdot M_0 & M_2 + M_3 \cdot M_0 & M_1 + M_3 \end{pmatrix} \quad (r \geq 2),$$

$$M_1 := T^{(r-1)}(s_1, s_2, \ldots, s_{3^{r-1}}),$$

$$M_2 := T^{(r-1)}(s_{3^{r-1}+1}, s_{3^{r-1}+2}, \ldots, s_{2 \cdot 3^{r-1}}),$$

$$M_3 := T^{(r-1)}(s_{2 \cdot 3^{r-1}+1}, s_{2 \cdot 3^{r-1}+2}, \ldots, s_{3^{r-1}}),$$

$$M_0 := T^{(r-1)}(0, 0, \ldots, 0, 1).$$

Then we can multiply two elements of K_r *expressed as* 3^r-*dimensional vectors.*

For two elements $(s_1, s_2, \ldots, s_{3r}), (t_1, t_2, \ldots, t_{3r}) \in K_r$, *the multipication of these elements is given by the matrix multiplication:*

$$(s_1, s_2, \ldots, s_{3r}) \cdot (t_1, t_2, \ldots, t_{3r}) = (s_1, s_2, \ldots, s_{3r}) \cdot T^{(r)}(t_1, t_2, \ldots, t_{3r}).$$

Proof. The case for K_1 is clear from the argument above the theorem. When $r = 2$, let $(s_1, s_2, \cdots, s_9), (t_1, t_2, \cdots, t_9)$ be two elements of K_2, then

$$(s_1, s_2, s_3, s_4, s_5, s_6, s_7, s_8, s_9)(t_1, t_2, t_3, t_4, t_5, t_6, t_7, t_8, t_9)$$

$$= ((s_1, s_2, s_3) + (s_4, s_5, s_6)\alpha_2 + (s_7, s_8, s_9)\alpha_2{}^2)((t_1, t_2, t_3) + (t_4, t_5, t_6)\alpha_2$$

$$+ (t_7, t_8, t_9)\alpha_2{}^2)$$

$$= (s_1, s_2, s_3)(t_1, t_2, t_3)$$

$$\quad + ((s_1, s_2, s_3)(t_4, t_5, t_6) + (s_4, s_5, s_6)(t_1, t_2, t_3))\alpha_2$$

$$\quad + ((s_1, s_2, s_3)(t_7, t_8, t_9) + (s_4, s_5, s_6)(t_4, t_5, t_6) + (s_7, s_8, s_9)(t_1, t_2, t_3))\alpha_2{}^2$$

$$\quad + ((s_4, s_5, s_6)(t_7, t_8, t_9) + (s_7, s_8, s_9)(t_4, t_5, t_6))\alpha_2{}^3$$

$$\quad + (s_7, s_8, s_9)(t_7, t_8, t_9)\alpha_2{}^4$$

$$=(s_1, s_2, s_3)(t_1, t_2, t_3)$$
$$+ ((s_1, s_2, s_3)(t_4, t_5, t_6) + (s_4, s_5, s_6)(t_1, t_2, t_3))\alpha_2$$
$$+ ((s_1, s_2, s_3)(t_7, t_8, t_9) + (s_4, s_5, s_6)(t_4, t_5, t_6)$$
$$+ (s_7, s_8, s_9)(t_1, t_2, t_3))\alpha_2{}^2$$
$$+ ((s_4, s_5, s_6)(t_7, t_8, t_9) + (s_7, s_8, s_9)(t_4, t_5, t_6))(\alpha_2 + (0, 0, 1))$$
$$+ ((s_7, s_8, s_9)(t_7, t_8, t_9))(\alpha_2{}^2 + (0, 0, 1)\alpha_2)$$
$$=(s_1, s_2, s_3)(t_1, t_2, t_3) + (s_4, s_5, s_6)(t_7, t_8, t_9)(0, 0, 1)$$
$$+ (s_7, s_8, s_9)(t_4, t_5, t_6)(0, 0, 1)$$
$$+ ((s_1, s_2, s_3)(t_4, t_5, t_6) + (s_4, s_5, s_6)((t_1, t_2, t_3) + (t_7, t_8, t_9))$$
$$+ (s_7, s_8, s_9)((t_4, t_5, t_6) + (t_7, t_8, t_9)(0, 0, 1)))\alpha_2$$
$$+ ((s_1, s_2, s_3)(t_7, t_8, t_9) + (s_4, s_5, s_6)(t_4, t_5, t_6)$$
$$+ (s_7, s_8, s_9)((t_1, t_2, t_3) + (t_7, t_8, t_9)))\alpha_2{}^2$$
$$=(s_1, s_2, s_3)T^{(1)}(t_1, t_2, t_3) + (s_4, s_5, s_6)T^{(1)}(t_7, t_8, t_9)T^{(1)}(0, 0, 1)$$
$$+ (s_7, s_8, s_9)T^{(1)}(t_4, t_5, t_6)T^{(1)}(0, 0, 1)$$
$$+ ((s_1, s_2, s_3)T^{(1)}(t_4, t_5, t_6)$$
$$+ (s_4, s_5, s_6)(T^{(1)}(t_1, t_2, t_3) + T^{(1)}(t_7, t_8, t_9))$$
$$+ (s_7, s_8, s_9)(T^{(1)}(t_4, t_5, t_6) + T^{(1)}(t_7, t_8, t_9)T^{(1)}(0, 0, 1)))\alpha_2$$
$$+ ((s_1, s_2, s_3)T^{(1)}(t_7, t_8, t_9) + (s_4, s_5, s_6)T^{(1)}(t_4, t_5, t_6)$$
$$+ (s_7, s_8, s_9)(T^{(1)}(t_1, t_2, t_3) + T^{(1)}(t_7, t_8, t_9)))\alpha_2{}^2$$
$$=(s_1, s_2, s_3, s_4, s_5, s_6, s_7, s_8, s_9)\begin{pmatrix} T_1 & T_2 & T_3 \\ T_3T_0 & T_1 + T_3 & T_2 \\ T_2T_0 & T_2 + T_3T_0 & T_1 + T_3 \end{pmatrix}$$
$$=(s_1, s_2, s_3, s_4, s_5, s_6, s_7, s_8, s_9)T^{(2)}(t_1, t_2, t_3, t_4, t_5, t_6, t_7, t_8, t_9).$$

Where $T_1 = T^{(1)}(t_1, t_2, t_3)$, $T_2 = T^{(1)}(t_4, t_5, t_6)$, $T_3 = T^{(1)}(t_7, t_8, t_9)$, and $T_0 = T^{(1)}(0, 0, 1)$. For the case K_r, let $(s_1, \ldots, s_{3r})$ and $(t_1, \ldots, t_{3r})$ be two elements of K_r. We obtain the following by induction:

$$(s_1, \ldots, s_{3r})(t_1, \ldots, t_{3r})$$
$$=((s_1, \ldots, s_{3r-1}) + (s_{3r-1+1}, \ldots, s_{2\cdot3r-1})\alpha_{r-1} + (s_{2\cdot3r-1+1}, \ldots, s_{3r})\alpha^2_{r-1})$$
$$((t_1, \ldots, t_{3r-1}) + (t_{3r-1+1}, \ldots, t_{2\cdot3r-1})\alpha_{r-1} + (t_{2\cdot3r-1+1}, \ldots, t_{3r})\alpha^2_{r-1}),$$
$$=(s_1, \ldots, s_{3r})\begin{pmatrix} M_1 & M_2 & M_3 \\ M_3 \cdot M_0 & M_1 + M_3 & M_2 \\ M_2 \cdot M_0 & M_2 + M_3 \cdot M_0 & M_1 + M_3 \end{pmatrix}$$

$$=(s_1,\ldots,s_{3^r})T^{(r)}(t_1,\ldots,t_{3^r}).$$

$\square$

Along the way, we get an algorithm for multiplication of two elements of K_r as below:

Algorithm
Input: $r, (s_1,\ldots,s_{3^r}), (t_1,\ldots,t_{3^r})$
Output: $(u_1,\ldots,u_{3^r})$
Procedure:
1. $T_i^0 \leftarrow t_i \ (1 \le i \le 3^r), U^0 \leftarrow 1;$
2. for $(j=1, j \le r, j = j+1);$
　　for $(i = 1, i \le 3^{r-j}, i = i+1);$

$$T_i^j \leftarrow \begin{pmatrix} T_{3i-2}^{(j-1)} & T_{3i-1}^{(j-1)} & T_{3i}^{(j-1)} \\ T_{3i}^{(j-1)}U^{(j-1)} & T_{3i-2}^{(j-1)} + T_{3i}^{(j-1)} & T_{3i-1}^{(j-1)} \\ T_{3i-1}^{(j-1)}U^{(j-1)} & T_{3i-1}^{(j-1)} + T_{3i}^{(j-1)}U^{(j-1)} & T_{3i-2}^{(j-1)} + T_{3i}^{(j-1)} \end{pmatrix}$$

$$U^j \leftarrow \begin{pmatrix} 0 & 0 & U^{(j-1)} \\ (U^{(j-1)})^2 & U^{(j-1)} & 0 \\ 0 & (U^{(j-1)})^2 & U^{(j-1)} \end{pmatrix}$$

3. $(u_1,\ldots,u_{3^r}) \leftarrow (s_1,\ldots,s_{3^r})T_1^r$
4. return $(u_1,\ldots,u_{3^r})$

---

3. Linear recurrence equations on finite fields

There is a method of pseudorandom number generators, which produces a sequence of numbers, as an application of the theory of finite fields.

Definition 2. ([1]) The most useful type of generator of pseudorandom processes updates a current sequence of numbers in a manner that appears to be random. Such a deterministic generator, g , yields numbers recursively, in a fxed sequence. The previousk numbers (often just the single previous number) determine(s) the next number:

$$x_i = g(x_{i-1},\ldots,x_{i-k}).$$

Because the set of numbers directly representable in the computer is fnite, the sequence will repeat. The set of values at the start of the recursion is called the **seed**. Each time the recursion is begun with the same seed, the same sequence is generated. The length of the sequence prior to beginning to repeat is called the **period or cycle length**.

Definition 3. Let W be a w-dimensional vector space over $\mathbb{F}_2$ which we regard as the state space of the generator. Let $g : W^n \to W$ be a linear state map. Let $x_{n-1},\ldots,x_1, x_0$ be initial nw-arrays with $x_0,\ldots,x_{n-1} \in W$. We define the linear recurrence equation

$$x_{j+n} = g(x_{j+n-1},\ldots,x_j)$$

as n-th order linear recurrence.

Let $S := W^n$, and $f : S \to S$ be a linear state transition map. Then n-th order linear recurrence equation can be transformed into the first order linear recurrence equation as follow:

$$f(x_{j+n-1}, \ldots, x_j) = (g(x_{j+n-1}, \ldots, x_j), x_{j+n-1}, \ldots, x_{j+1}) \ (j = 0, 1, \cdots).$$

This definition explain how to translate n-th order linear recurrence equations into first order linear recurrence equations.

4. Pseudorandom number generators for $p = 2$

4.1. TGFSR

Definition 4. ([5], [6]) Let n, m and w be positive integers with $n > m$. The twisted GFSR generator (TGFSR) is the same as the GFSR generator except that it is based on the linear recurrence equation

$$(1) \qquad x_{j+n} := x_{j+m} + x_j A \qquad (j = 0, 1, \ldots),$$

where each x_j is a word regarded as a row vector over $\mathbb{F}_2$ of size w, A is a $w \times w$ matrix with entries in $\mathbb{F}_2$ and $+$ means the addition as $\mathbb{F}_2$-vectors.

It can be transformed into the first order linear recurrence map as:

$$f : (x_{j+n-1}, \ldots, x_{j+1}, x_j) \mapsto (x_{j+m} + x_j A, x_{j+n-1}, \ldots, x_{j+2}, x_{j+1})$$

$$(j = 0, 1, \cdots),$$

where f is a linear state transition map, which multiply nw-bit vector by matrix B,

$$B = \begin{pmatrix} & I_w & & & \\ & & I_w & & \\ I_w & & & \ddots & \\ & & & & I_w \\ A & & & & \end{pmatrix}.$$

Since we have the equation

$$(x_{j+n}, x_{j+n-1}, \ldots, x_{j+1}) = (x_{j+n-1}, x_{j+n-2}, \ldots, x_j)B$$

for $j = 0, 1, \ldots$, it is clear that the period of the sequence of numbers is equal to just the order of the matrix B.

With a suitable choice of n, m, and A, the TGFSR generator attains the maximal period $2^{nw} - 1$, that is, it produces all possible states except the zerostate in a period.

4.2. MT

Definition 5 ([7]). Mersenne Twister is based on the following linear recurrence equation

$$(2) \qquad x_{j+n} = x_{j+m} + (x_j^u | x_{j+1}^l) A \qquad (j = 0, 1, \dots),$$

where x_k^l stands for the extraction of the lower r bits of x_k, x_k^u stands for the extraction of the upper $w - r$ bits of x_k, and $(x_j^u | x_{j+1}^l)$ stands for the concatenation of x_j^u and x_{j+1}^l. It requires several constants, an integer n which is the degree of the recurrence equation, an integer r with $0 \le r \le w - 1$, an integer m with $1 \le m \le n$, and a $w \times w$ matrix A with entries in $\mathbb{F}_2$. Let $x_{n-1}, \cdots, x_1, x_0$ be initial seeds. Then, the generator produces x_n by the above recurrence equation with $j = 0$. By putting $j = 1, 2, \dots$, the generator determines $x_{n+1}, x_{n+2}, \cdots$.

If one eliminates the lower r bits from the $(n \times w)$-array $x_{j+n-1}, \dots \dots$, x_{j+1}, x_j, then the dimension of the state space is $nw - r$, which can be taken any number. This is the great advantage of MT. See [7] for details.

Similarly, Mersenne Twister (2) can be transformed as:

$$f : (x_{j+n-1}, x_{j+n-2}, \dots, x_{j+1}, \{x_j^u\})$$

$$\mapsto (x_{j+m} + (x_j^u | x_{j+1}^l) A, x_{j+n-1}, \dots, x_{j+2}, \{x_{j+1}^u\}) \ (j = 0, 1, \cdots).$$

Now, let B be $(nw - r) \times (nw - r)$-array matrix as follows:

$$B = \begin{pmatrix} & I_w & & & \\ & & I_w & & \\ I_w & & & \ddots & \\ & & & & I_{w-r} \\ S & & & & \end{pmatrix},$$

where $S = \begin{pmatrix} 0 & I_r \\ I_{w-r} & 0 \end{pmatrix} A$, then, one gets

$$(x_{j+n}, x_{j+n-1}, \dots, x_{j+1}^u) = (x_{j+n-1}, x_{j+n-2}, \dots, x_j^u) B.$$

Note that MT can attain the maximal period (see [7]) with a suitable choice of B, but it is difficult to find such B with maximal period for TGFSR.

4.3. AST for $p = 2$

Definition 6 ([8]). Let $A^{(1)}(t_1, t_2)$ be the 2×2 matrix defined by
$$A^{(1)}(t_1, t_2) := \begin{pmatrix} t_1 & t_2 \\ t_2 & t_1 + t_2 \end{pmatrix}.$$

Define the $2^r \times 2^r$ matrix $A^{(r)}(t_1, \ldots, t_{2^r})$ as $\begin{pmatrix} S & T \\ U & V \end{pmatrix}$, where $2^{r-1} \times 2^{r-1}$ matrices S, T, U, V are defined recursively as follows:

$$S = A^{(r-1)}(t_1, \ldots, t_{2^{(r-1)}}),$$
$$T = A^{(r-1)}(t_{2^{(r-1)}+1}, \ldots, t_{2^r}),$$
$$U = A^{(r-1)}(t_{2^{(r-1)}+1}, \ldots, t_{2^r}) \cdot A^{(r-1)}(0, \ldots, 1),$$
$$V = A^{(r-1)}(t_1, \ldots, t_{2^{(r-1)}}) + A^{(r-1)}(t_{2^{(r-1)}+1}, \ldots, t_{2^r}).$$

Definition 7 ([9]). The multiplication of $x, (1 + \alpha_r) \in \mathbb{F}_{2^{2^r}}$ can be written as follows:

$$x(1 + \alpha_r) = x(\underbrace{1, 0, \ldots, 0}_{2^{r-1}} | 1, 0, \ldots, 0)$$

$$= x \cdot A^{(r)}(\underbrace{1, 0, \ldots, 0}_{2^{r-1}} | 1, 0, \ldots, 0)$$

$$= x \cdot \begin{pmatrix} I & I \\ A^{(r-1)}(\underbrace{0, \ldots, 0, 1}_{2^{r-1}}) & O \end{pmatrix}.$$

Then we define the $2^r \times 2^r$ matrix B_r as $A^{(r)}(\underbrace{1, 0, \ldots, 0}_{2^{r-1}} | 1, 0, \ldots, 0)$ defined in Definition 6, and the $2^{r-1} \times 2^{r-1}$ matrix A_{r-1} as $A^{(r-1)}(\underbrace{0, \ldots, 0, 1}_{2^{r-1}})$.

Here I is the identity matrix, and O is the zero matrix.

Definition 8 ([9]). Let W be a w-dimensional vector space over $\mathbb{F}_2$ which is the state space of the generator. Let n, w and r be positive integers with $n \geq 2$ and $r \geq 2$ so that $nw := 2^r$. Define a linear state map $g : W^n \to W^{\frac{n}{2}}$ as below.

Put $x_{n-1}, \ldots, x_1, x_0 \in W$ which we regard as an initial nw-array. We define the linear recurrence by

(3)
$$(x_{j+\frac{3}{2}n-1}, \ldots, x_{j+n}) := g(x_{j+n-1}, \ldots, x_j)$$
$$:= (x_{j+\frac{1}{2}n-1}, \ldots, x_j) \times A_{r-1} + (x_{j+n-1}, \ldots, x_{j+\frac{1}{2}n})$$
$$(j = 0, 1, \ldots),$$

where A_{r-1} is the matrix defined in Definition 7.

Put $S := W^n$, then the equation (3) can be transformed into the first order linear recurrence from S to S:

(4)
$$f(x_{j+n-1}, \ldots, x_j) = (g(x_{j+n-1}, \ldots, x_j), x_{j+n-1}, \ldots, x_{j+\frac{1}{2}n})$$
$$= (x_{j+\frac{3}{2}n-1}, \ldots, x_{j+n}, x_{j+n-1}, \ldots, x_{j+\frac{1}{2}n})$$
$$(j = 0, 1, \ldots).$$

We call this pseudorandom number generator the *Artin-Schreier Tower* *(AST)*.

This f is a linear state transition map. Since $2^r = nw$, the linear recurrence equation (4) is same as multiplying an nw-bit vector by B_r, where B_r is is already defined in Definition 7 as

$$B_r = \begin{pmatrix} I_{r-1} & I_{r-1} \\ A_{r-1} & O \end{pmatrix}.$$

Thus, for nonnegative integer j, we have

$$(x_{j+\frac{3}{2}n-1}, \ldots, x_{j+n}, x_{j+n-1}, \ldots, x_{j+\frac{1}{2}n})$$

$$= (x_{j+n-1}, \ldots, x_{j+\frac{1}{2}n}, x_{j+\frac{1}{2}n-1}, \ldots, x_j) \times B_r.$$

5. AST for $p = 3$

Definition 9. Define $3^{r+1} \times 3^{r+1}$ matrixes $\bar{A}_{r+1}$ and $\bar{B}_{r+1}$ as followings.

$$\bar{A}_{r+1} = \begin{pmatrix} O & O & \bar{A}_r \\ \bar{A}_r^2 & \bar{A}_r & O \\ O & \bar{A}_r^2 & \bar{A}_r \end{pmatrix}, \text{ and } \bar{B}_{r+1} = \begin{pmatrix} 2I & O & I \\ \bar{A}_r & O & O \\ O & \bar{A}_r & O \end{pmatrix},$$

where $\bar{A}_r = T^{(r)}(\underbrace{0, \cdots, 0, 1}_{3^r})$

and $\bar{A}_r = T^{(r)}(\underbrace{2, 0, \cdots, 0}_{3^r-1}, \underbrace{0, \cdots, 0}_{3^r-1}, \underbrace{1, 0, \cdots, 0}_{3^r-1})$, which is defined in Theorem 6.

Definition 10. Let W be a w-dimensional vector space over $\mathbb{F}_3$ which is the state space of the generator. Let n, w and r be positive integers with $n \geq 2$ and $r \geq 2$ so that $nw := 3^r$. Define a linear state map $g : W^n \to W^{\frac{2}{3}n}$ as below.

Put $x_{n-1}, \ldots, x_1, x_0 \in W$ which we regard as an initial nw-array. We define the linear recurrence by

$$(5) \quad (x_{j+\frac{5}{3}n-1}, \ldots, x_{j+\frac{4}{3}n}, x_{j+\frac{4}{3}n-1}, \ldots, x_{j+n}) := g(x_{j+n-1}, \ldots, x_j)$$

$$:= ((x_{j+\frac{2}{3}n-1}, \ldots, x_{j+\frac{1}{3}n}) \times \bar{A}_{r-1} + (x_{j+n-1}, \ldots, x_{j+\frac{2}{3}n}) \times 2I,$$

$$(x_{j+\frac{1}{3}n-1}, \ldots, x_j) \times \bar{A}_{r-1})$$

where $(j = 0, 1, \ldots)$ and $\bar{A}_{r-1}$ is the matrix defined in Definition 9.

Put $S := W^n$, then the equation (5) can be transformed into the first order linear recurrence from S to S:

$$(6) \quad f(x_{j+n-1}, \ldots, x_j) = (g(x_{j+n-1}, \ldots, x_j), x_{j+n-1}, \ldots, x_{j+\frac{2}{3}n})$$

$$= (x_{j+\frac{5}{3}n-1}, \ldots, x_{j+\frac{4}{3}n}, x_{j+\frac{4}{3}n-1}, \ldots, x_{j+n}, x_{j+n-1}, \ldots, x_{j+\frac{2}{3}n})$$

$$(j = 0, 1, \ldots).$$

This f is a linear state transition map. Since $3^r = nw$, the linear recurrence equation (6) is same as multiplying an nw-bit vector by $\bar{B}_r$, where $\bar{B}_r$ is is already defined in Definition 9 as

$$\bar{B}_{r+1} = \begin{pmatrix} 2I & O & I \\ \bar{A}_r & O & O \\ O & \bar{A}_r & O \end{pmatrix}.$$

Thus, for nonnegative integer j, we have

$$(x_{j+\frac{5}{3}n-1}, \ldots, x_{j+\frac{4}{3}n}, x_{j+\frac{4}{3}n-1}, \ldots, x_{j+n}, x_{j+n-1}, \ldots, x_{j+\frac{2}{3}n})$$

$$= (x_{j+n-1}, \ldots, x_j) \times \bar{B}_r.$$

Start with initial seeds $x_{n-1}, \ldots, x_1, x_0$ the state transition is given as follows:

$$(x_{n-1}, \cdots, x_{\frac{2}{3}n}, x_{\frac{2}{3}n-1}, \cdots, x_{\frac{1}{3}n}, x_{\frac{1}{3}n-1}, \cdots, x_0) \times \bar{B}_r$$

$$\downarrow$$

$$(x_{\frac{5}{3}n-1}, \cdots, x_{\frac{4}{3}n}, x_{\frac{4}{3}n-1}, \cdots, x_n, x_{n-1}, \cdots, x_{\frac{2}{3}n}) \times \bar{B}_r$$

$$\downarrow$$

$$\vdots$$

$$(x_{\frac{1}{3}n-1}, \cdots, x_0, \cdots\cdots\cdots\cdots) \times \bar{B}_r$$

$$\downarrow$$

$$(x_{n-1}, \cdots, x_{\frac{2}{3}n}, x_{\frac{2}{3}n-1}, \cdots, x_{\frac{1}{3}n}, x_{\frac{1}{3}n-1}, \cdots, x_0).$$

AST can generate $\frac{2}{3}n$ words by multiplying $\bar{B}_r$ for each time, thus the period of the sequence is $\frac{2}{3}n \times o(\bar{B}_r)$.

Example 1. *Let us consider the case $r = 4$. In this case, $\bar{B}_r$ is a $3^4 \times 3^4$ matrix and $nw = 3^4$ holds. Take the parameter $w = 2^5 = 32$, for example, then $n = 2^6 = 64$. Let $x_8, x_7, x_6, x_5, x_4, x_3, x_2, x_1, x_0$ be initial seeds. The transformation $f : \mathbb{F}_{3^{34}} \to \mathbb{F}_{3^{34}}$ produces a pseudorandom number sequence starting with the initial seeds $x_8, x_7, x_6, x_5, x_4, x_3, x_2, x_1, x_0$.*

By experiment, $o(\bar{B}_r)$ equals $\frac{3^{3^r}-1}{2}$ when $r \leq 4$. Then period of AST with $r = 4$ is $\frac{2}{3} \cdot 3^4 \cdot \frac{3^{3^4}-1}{2} \approx 2.394 \times 10^{40}$.

6. Concluding remarks

We defined matrices $\bar{B}_r$ and propose a new pseudorandom number generator using them. By experiment, we know that the order of $\bar{B}_r$ equals $\frac{3^{3^r}-1}{2}$ when $r \leq 4$. We conjectured that it is true for all the r. We are developing the proof of the order of the $\bar{B}_r$, and the test of the theoretical properties of the new pseudorandom number generator using the TestU01. We will treat these in another occasion.

References

[1] J. E. Gentle, Random Number Generation and Monte Carlo Methods, Second Edition, Springer-Verlag, 2005.

[2] H. Ito, and T. Kajiwara, and H. Song, A Tower of Artin-Schreier extensions of finite fields and its applications, to appear in JP J. of Algebra, Number Theory and Applications.

[3] D. E. Knuth, The Art of Computer Programming, Volume 2 Seminumerical algorithms, Third Edition, Addison-Wesley, 1997.

[4] H. Lidl, and H. Neiderreiter, Finite fields, Second Edition, Cambridge University Press, 1997.

[5] M. Matsumoto, and Y. Kurita, Twisted GFSR Generators, ACM Trans. on Modeling and Computer Simulation 2 (1992), 179-194.

[6] M. Matsumoto, and Y. Kurita, Twisted GFSR Generators II, ACM Trans. on Modeling and Computer Simulation 4 (1994), 254-266.

[7] M. Matsumoto, and T. Nishimura, Mersenne Twister: a 623-dimensionally equidistributed uniform pseudo-random number generator, ACM Trans. on Modeling and Computer Simulation 8 (1998), 3-30.

[8] H. Song, and H. Ito, On the construction of huge finite fields. AC2009 Proceedings (2009), 1-7.
    http://tnt.math.se.tmu.ac.jp/ac/2009/proceedings/ac2009-proceedings.pdf

[9] H. Song. H. Ito, and Y. Kitadai, A pseudorandom number generator using an Artin-Schreier tower, to appear in SUT Journal of Mathematics.

[10] P. L'Ecuyer and R. Simard. TestU01: A C Library for Empirical Testing of Random Number Generators ACM Trans. on Mathematical Software, Vol. 33, article 22, 2007.

DEPARTMENT OF MATHEMATICS, FACULTY OF FOUNDATION
HARBIN FINANCE UNIVERSITY, P.R. CHINA
AND
DEPARTMENT OF APPLIED MATHEMATICS, GRADUATE SCHOOL OF ENGINEERING
HIROSHIMA UNIVERSITY, JAPAN
E-mail address: huiling1978@hotmail.com

Proceedings of the Sixth China-Japan-Korea
International Conference on Ring Theory
June 27-July 2, 2011 Suwon, Korea

A NOTE ON PRIME RINGS WITH LEFT DERIVATIONS

NADEEM UR REHMAN

ABSTRACT. Let R be a prime rings and $I \neq 0$ an ideal of R. An additive mapping $D : R \to R$ is called a left derivation on R if $D(xy) = D(x)y + xD(y)$, holds for all $x, y \in R$. In the present paper, we have discussed the commutativity of a prime rings admitting a left derivation D satisfying several conditions.

1. Introduction

Throughout the present paper R will denote an associative ring with center $Z(R)$. For any $x, y \in R$, the symbol $[x, y]$ will represent the commutator $xy - yx$ and the symbol $(x \circ y)$ stands for the skew-commutator $xy + yx$. Recall that R is prime if $aRb = \{0\}$ implies that $a = 0$ or $b = 0$. An additive mapping $d : R \longrightarrow R$ is called a derivation if $d(xy) = d(x)y + xd(y)$, holds for all $x, y \in R$. In particular, for a fixed $a \in R$, the mapping $I_a : R \to R$ given by $I_a(x) = [a, x]$ is a derivation which is said to be an inner derivation.

An additive mapping $D : R \longrightarrow R$ is called a left derivation if $D(xy) = xD(y) + yd(x)$, holds for all $x, y \in R$. An additive mapping $D : R \longrightarrow R$ will be called a Jordan left derivations if $D(x^2) = 2xD(x)$ holds for all $x \in R$. It turns out that the notion of Jordan left derivations is in a close connection with so-called commuting mappings. A mapping F of a ring R into itself is said to be commuting on R if $F(x)x = xF(x)$ for all $x \in R$. There has been considerable interest for commuting and related mappings on prime rings. The fundamental result is due to Posner [10]. He proved that if a prime ring R admits a nonzero derivation that is commuting on R, then R is commutative. The analogous result was obtained for automorphisms [9]. It can be easily prove that in a non-commutative prime ring any left derivation is zero (for reference see

2010 Mathematics Subject Classification : 16N60, 16R50.
Keywords : Prime rings, derivations, left derivations.
This research is supported by UGC, India, Grant No. 36-8/2008(SR).

[3] and [11]), where further references can be found. Also in case of commutative ring, the notion of derivation and left derivation are coincide.

During the last few decade, several authors have proved commutativity theorems for prime rings or semiprime rings admitting automorphisms, derivations or generalized derivations which are centralizing or commuting on appropriate subset of R ([1, 2, 4, 6] and [7], for partial bibliography). In the year 2001, Ashraf and author [1] established that a prime ring R with a nonzero ideal I must be commutative if it admits a derivation d satisfying either of the properties: $d(xy) - xy \in Z(R)$, and $d(xy) + xy \in Z(R)$ for all $x, y \in R$. Now, it is natural to ask the question as to what we can say about the commutativity of R if the derivation d is replaced by a left derivation. In this paper, we investigate the commutativity of prime ring R admitting a left derivation satisfying any one of the following properties: (i) $D(xy) - xy \in Z(R)$, (ii) $D(xy) + xy \in Z(R)$ (iii) $[D(x), x] = 0$, (iv) $[D(x), y] = [x, y]$, (v) $D(x \circ y) = x \circ y$, (vi) $D(x) \circ y = x \circ y$ for all x, y in some appropriate subset of R.

2. Main results

We shall use without explicit mention the following basic commutator and anti-commutator identities:

(I) $[xy, z] = x[y, z] + [x, y]z$
(II) $[x, yz] = y[x, z] + [x, y]z$
(III) $x \circ yz = (x \circ y)z - y[x, z] = y(x \circ z) + [x, y]z$
(IV) $(xy) \circ z = x(y \circ z) - [x, z]y = (x \circ z)y + x[y, z]$.

We begin our discussion with the following lemma is essentially proved in [9].

Lemma 2.1. *If a prime ring R contains a commutative nonzero right ideal, then R is commutative.*

The next lemma is well known.

Lemma 2.2. *Let R be a prime ring, I a nonzero ideal of R. If $xIy = 0$ for $x, y \in R$, then $x = 0$ or $y = 0$.*

Now, we prove the following:

Theorem 2.3. *Let R be a prime ring and I a nonzero ideal of R. If R admits a left derivation D such that $D(xy) - xy \in Z(R)$, for all $x, y \in I$, then R is commutative.*

Proof. We have, $D(xy) - xy \in Z(R)$, for all $x, y \in I$. If $D = 0$, then $xy \in Z(R)$, for all $x, y \in I$. Using the same arguments as used in the beginning of the proof of Theorem 2.1 of [1], we get the required result. Hence, onward we assume that $D \neq 0$. For any $x, y \in I$, we have $D(xy) - xy \in Z(R)$. This can be rewritten as $xD(y) + yD(x) - xy \in Z(R)$, for all $x, y \in I$. Replacing y by yz, we obtain $x(yD(z) + zD(y) - yz) + yzD(x) \in Z(R)$, for all $x, y, z \in I$. Thus, in particular $[x(yD(z) + zD(y) - yz) + yzD(x), x] = 0$ for all $x, y, z \in I$. This yields that $[yzD(x), x] = 0$, for all $x, y, z \in I$ and hence

$$(1) \quad yz[D(x), x] + y[z, x]D(x) + [y, x]zD(x) = 0, \text{ for all } x, y, z \in I.$$

For any $w \in I$ replace y by wy in (1), we find that

$$w(yz[D(x), x] + y[z, x]D(x) + [y, x]zD(x)) + [w, x]yzD(x) = 0$$

for all $x, y, z, w \in I$. Now application of (1) gives that $[w, x]yzD(x) = 0$ for all $x, y, z, w \in I$ and hence $[w, x]yID(x) = \{0\}$, for all $x, y, w \in I$. Thus, Lemma 2.2 implies that for each $x \in I$, either $D(x) = \{0\}$ or $[w, x]y = 0$ for all $y, w \in I$, that is $[w, x]I = \{0\}$, and so $[w, x] = 0$ for all $w \in I$. The set of $x \in I$ for which these two properties hold are additive subgroups of I whose union is I. Therefore either $D(x) = (0)$ for all $x \in I$ or $[w, x] = 0$ for all $x, w \in I$. If $D(x) = \{0\}$, for all $x \in I$, then $0 = D(xr) = xD(r) + rD(x) = xD(r)$ for all $x \in I$ and $r \in R$. By Lemma 2.2, we get $D(r) = 0$ for all $r \in R$, a contradiction. On the other hand if $[w, x] = 0$ for all $x, w \in I$, then I is commutative and hence by Lemma 2.1, we get the required result. $\qquad \square$

Theorem 2.4. *Let R be a prime ring and I a non-zero ideal of R. If R admits a left derivation D such that $D(xy) + xy \in Z(R)$, for all $x, y \in I$, then R is commutative.*

Proof. Suppose D is a left derivation satisfying the property $D(xy) + xy \in Z(R)$, for all $x, y \in I$, then the left derivaton $(-D)$ satisfies the condition $(-D)(xy) - xy \in Z(R)$, for all $x, y \in I$. Hence by Theorem 2.3, R is commutative. $\qquad \square$

Theorem 2.5. *Let R be a prime ring and I a non-zero ideal of R. Then the following conditions are equivalent:*

    (i) R admits a left derivation D such that $D(xy) - xy \in Z(R)$ or $D(xy) + xy \in Z(R)$, for all $x, y \in I$.

    (ii) R is commutative.

Proof. Obviously, $(ii) \Rightarrow (i)$.

Now, we will prove that

$(i) \Rightarrow (ii)$. For each $x \in I$, let $U = \{y \in I \mid D(xy) - xy \in Z(R)\}$, $V = \{y \in I \mid D(xy) + xy \in Z(R)\}$. Then it can be easily that U and V are additive subgroups of I whose union is I. Thus by Brauer's trick, either $U = I$ or $V = I$. Further, using similar arguments we find that $I = \{x \in I \mid I = U\}$ or $I = \{x \in I \mid I = V\}$. Therefore, R is commutative by Theorem 2.3 and Theorem 2.4.   □

Theorem 2.6. *Let R be a prime ring and I be a nonzero ideal of R. If R admits a nonzero left derivation $D : R \to R$ such that*

(i) $[D(x), x] = 0$ *for all $x, y \in I$, or*
ii) $[D(x), y] = [x, y]$ *for all $x, y \in I$, or*
(iii) $D(x \circ y) = x \circ y$ *for all $x, y \in I$, or*
(iv) $D(x) \circ y = x \circ y$ *for all $x, y \in I$,*

then R is commutative

Proof. (i) We have

$$(2) \qquad\qquad [D(x), x] = 0 \text{ for all } x \in I.$$

Linearizing (2), we obtain $[D(x), y] + [D(y), x] = 0$ for all $x, y \in I$. Replacing x by yx, we find that $[yD(x) + xD(y), y] + [D(y), yx] = 0$, that is, $[x, y]D(y) = 0$. Again replace x by zx to get

$$(3) \qquad\qquad [z, y]xD(y) = 0 \text{ for all } x, y, z \in I;$$

and applying Lemma 2.2 and the fact that $(I, +)$ is not the union of two of its proper subgroups shows that either $[z, y] = 0$ or $D(y) = 0$. If $D(y) = 0$ for all $y \in I$, then using the similar arguments as used in the proof of Theorem 2.3, $D = 0$, a contradiction. On the other hand if $[x, y] = 0$, then I is commutative, and so R is commutative by Lemma 2.1.

(ii) If $D = 0$, then $[x, y] = 0$ and hence R is commutative by Lemma 2.1. Henceforth, we shall assume that $D \neq 0$. For any $x, y \in I$, we have $[D(x), y] = [x, y]$. Replacing x by yx in above relation, we find that $x[D(y), y] + [x, y]D(y) = 0$ for all $x, y \in I$. Now, Replacing x by zx in above, we obtain that $[z, y]xD(y) = 0$ for all $x, y, z \in I$. Notice that the arguments given in the last paragraph of the proof of (i) are still valid in the present situation, and hence repeating the same process, we get the required result.

(iii) If $D = 0$, then we have $x \circ y = 0$ for all $x, y \in I$. Replacing y by yz, we obtain $[x, y]z = 0$ for all $x, y, z \in I$ that is, $[x, y]I = \{0\}$, and so $[x, y] = 0$ for all $x, y \in I$ and hence by Lemma 2.1, R is commutative. Henceforth, we shall assume that $D \neq 0$. For any $x, y \in I$, we have $D(x \circ y) = x \circ y$ for all $x, y \in I$. Replacing y by xy, we find $D(x \circ xy) = x \circ xy$ for all $x, y \in I$. This implies that $D(x(x \circ y)) = x(x \circ y)$ for all $x, y \in I$, that is, $xD(x \circ y) + (x \circ y)D(x) = x(x \circ y)$ for all $x, y \in I$. Now, using our hypothesis,, we find that $(x \circ y)D(x) = 0$ for all $x, y \in I$. Replacing y by zy in above, we obtain $z(x \circ y)D(x) + [x, z]yD(x) = 0$ for all $x, y, z \in I$ which implies that $[x, z]ID(x) = \{0\}$ for all $x, y \in I$. The last expression is same as the equation (3) and hence the result follows.

(iv) If $D = 0$, then $x \circ y = 0$ for all $x, y \in I$. Using the same arguments as we have used in the proof of (iii), we get the required result. Therefore, we shall assume that $D \neq 0$. For any $x, y \in I$, we have $D(x) \circ y = x \circ y$. Replacing y by xy, we obtain $[D(x), x]y = 0$ for all $x, y \in I$ that is $[D(x), x]I = \{0\}$ and hence by Lemma 2.2 and (ii), we get the required result. $\qquad \square$

Proceeding on the same lines with necessary variations one can proved the following result:

Theorem 2.7. *Let R be a prime ring and I be a nonzero ideal of R. If R admits a nonzero left derivation $D : R \to R$ such that*

(I) $[D(x), y] + [x, y] = 0$ *for all $x, y \in I$, or*
(II) $D(x \circ y) + x \circ y = 0$ *for all $x, y \in I$, or*
(III) $D(x) \circ y + x \circ y = 0$ *for all $x, y \in I$,*

then R is commutative.

Acknowledgments. The author would like to thank the organizers of the sixth China-Japan-Korean ICRT held at Kyung Hee University, Suwon, Korean, for warm hospitality during the conference. The Support received from Aligarh Muslim University to attend this meeting is gratefully acknowledged.

References

[1] M. Ashraf and N. Rehman, *On derivations and commutativity in prime rings* , East-West J. Math. **3**(1) (2001), 87-91.
[2] H. E. Bell and M. N. Daif, *On commutativity and strong commutativity preserving maps*, Canad. Math. Bull. **37** (1994), 443-447.
[3] M. Bresar, *Centralizing mappings and derivations in prime rings*, J. Algebra **156** (1993), 385-394.
[4] M. Bresar and J. Vukman, *On left derivations and related mappings*, Proc. Amer. Math. Soc. **110** (1990), 7-16.

[5] M. N. Daif and H. E. Bell, *Remarks on derivations on semiprime rings*, Internal.
     J. Math. & Math. Sci. **15** (1992), 205-206.

[6] Q. Deng and M. Ashraf, *On strong commutativity preserving mappings*, Results
     in Math. **30** (1996), 259-263.

[7] M. Hongan, *A note on semiprime rings with derivation*, Internat. J. Math. &
     Math. Sci. **2** (1997), 413-415.

[8] Kill-Wong Jun and Byung-Do Kim , *A note on Jordan left derivations*, Bull.
     Korean Math. Soc. **33** (1996) No. 2, 221-228.

[9] J. H. Mayne, *Centralizing mappings of prime rings*, Canad. Math. Bull. **27**
     (1984), 122-126.

[10] E. C. Posner, *Derivations in prime rings*, Proc. Amer. -Math. Soc. **8** (1957),
     1093-1100.

[11] J. Vukman, *Jordan left derivations on semi-prime rings*, Math. J. Okayama **39**
     (1997), 1-6.

DEPARTMENT OF MATHEMATICS
ALIGARH MUSLIM UNIVERSITY
ALIGARH 202002, INDIA
E-mail address: rehman100@gmail.com

Proceedings of the Sixth China-Japan-Korea
International Conference on Ring Theory
June 27-July 2, 2011 Suwon, Korea

ON RINGS IN WHICH EVERY IDEAL IS PRIME

HISAYA TSUTSUI

ABSTRACT. This is a modified manuscript of my talk entitled
"Open Questions on Fully Prime Rings" presented at the 6th
CJK-ICRT. A brief survey of results on fully prime and related
rings is given and it is followed by a few results and problems that
may be helpful to investigate the main conjecture: *under the assumption that the ring is right and left Noetherian, a fully prime
ring is right primitive.*

1. Introduction

Definition *A ring R (possibly without identity) in which every ideal is
prime (i.e. every proper ideal is a prime ideal and (hence) R is a prime
ring) is called a fully prime ring.*

Our research on determining the structure of fully prime rings was initially motivated by the well-known fact that a commutative ring with
identity in which every ideal is prime is a field. We are currently interested in determining the structure of noncommutative right and left
Noetherian rings in which every ideal is prime. We first published a series of papers on the subject in 1994 and 1996 (Blair-Tsutsui[1], Tsutsui
[7]). Conditions similar to the fully prime condition have received attention in literature. Hirano [3] studied those rings in which every ideal is
completely prime. Courter [2] studied those rings in which every ideal
is semiprime, and Koh [6] studied those rings in which every right ideal
is prime. More recently, Hirano-Tsutsui [5] studied rings in which every
ideal is n-primary, and Hirano-Poon-Tsutsui [4] studied those rings in
which every ideal is weakly prime.

The main objective for the talk was to introduce the subject to a wider
audience and present a few open problems that we have worked on.

Throughout this article, we assume a ring to be associative but not
necessarily commutative. Due to a consideration that an ideal of a ring

---

2010 Mathematics Subject Classification: 16N60, 16P40.
Keywords: Fully prime, Noetherian, primitive.

being a ring of its own, we do not assume the existence of a multiplicative identity on a ring in Section 2, unless otherwise so stated.

2. Four basic theorems on fully prime rings (from Blair-Tsutsui [1])

Theorem 1 *A ring R is fully prime if and only if every (two sided) ideal is idempotent and the set of ideals is totally ordered under inclusion.*

Examples of a fully prime ring include the ring of endomorphism of a vector space V over a division ring D. This fact can easily be verified using Theorem 1. Denote the cardinality of a denumerable set by $\aleph_0$, and for any integer $n \geq 1$, let $\dim_D V = \aleph_{n-1}$. Then the ring of endomorphism $Hom_D(V, V)$ is a fully prime ring with exactly k non-zero proper ideals $I_{\aleph_0} = \{f \in Hom_D(V, V) | \dim f(V) < \aleph_n\}$, $n = 0, 2, \ldots, k - 1$. If $\dim_D V = \aleph_{\omega_0}$ where ω_0 is the first limit ordinal, then $Hom_D(V, V)$ is a fully prime ring that has countably many ideals.

Every right ideal, and hence every ideal of a regular ring is idempotent and for a regular self-injective rings R, an ideal P is prime if and only if R/P is totally ordered. Since there exists a regular self-injective ring T with a prime ideal P such that the set of ideals of T/P is not well-ordered, the set of ideals of a fully prime ring is not necessarily well-ordered.

It is well known that every ideal of a ring is semiprime if and only if every ideal is idempotent. Other analogous results to Theorem 1 for rings that assume a similar condition to the fully prime condition are listed below:

(1) Hirano [3]: Every ideal of a ring R is completely prime if and only if $< a >=< a^2 >$ for every $a \in R$ and the set of ideals is linearly ordered.

(2) Koh [6]: Every right ideal of a ring R is prime if and only if R is simple and $a \in aR$ for every $a \in R$.

(3) Blair-Tsutsui [1]: Every right ideal of a ring R with identity is idempotent and the set of right ideal is linearly ordered if and only if R is a division ring.

(4) Hirano-Tsutsui [5]: Every ideal of a ring R with identity is right n-primary for some positive integer $n \leq k$ (such a ring is called called fully right k primary ring) if and only if $I^k = I^{k+1}$ for all ideals I of R and $I = IJ$, $J = JI$, or $I^k = J^k$ for any ideals I and J of R and a fixed integer k.

(5) Hirano-Poon-Tsutsui [4]: Every ideal of a ring R is weakly prime if and only if for any ideals I and J of $R, IJ = I, IJ = J$, or $IJ = 0$.

Theorem 2 *The center of a fully prime ring is either a field or zero.*

As an analogous result, we note that the center of a ring in which every ideal is semiprime is a regular ring (Courter [2]). It was also proved that if R is a ring with identity whose center is not a field, then R is a fully right k-primary ring if and only if R has a unique maximal ideal M and $M^k = 0$ (Hirano-Tsutsui [5]).

Theorem 3 *Every ideal of a fully prime ring R is fully prime when it is considered as a ring without identity. Every ideal of an ideal of a fully prime ring R is an ideal of R. Further, a proper ideal of a fully prime ring R cannot be a ring with identity.*

If P is a proper ideal of a fully prime ring R with identity and F is a subfield of the center of R, then $S_P = F + P$ is a fully prime ring whose maximal ideal P is also a maximal right and left ideal. Further, proper ideals of S are precisely those ideals of R that are contained in P.

Theorem 4 $S_P = F + P$ *is right primitive if and only if R is right primitive. S_P is semiprimitive if and only if R is semiprimitive*

Our current objective is to determine the primitivity of a right and left Noetherian fully prime ring. Theorem 4 may play an important role for this investigation. As such, we have the following conjecture and a question.

Conjecture 1 $S_P = F + P$ *is right Noetherian if and only if R is right Noetherian.*
Problem 1 *Under what conditions, would S_P be equal to R?*

3. Right Noetherian fully prime rings

In this section, we assume that a ring has a multiplicative identity.

Theorem 5 (Blair-Tsutsui [1]) *A fully prime, right fully bounded right Noetherian ring is simple Artinian.*

More generally, a prime right fully bounded right Noetherian ring, all of whose ideals are idempotent, is a simple Artinian ring (Tsutsui [7]).

Let F be a field, and R be the set of all infinite matrices over F that have the form

$$\begin{bmatrix} A & & & & \\ & b & & 0 & \\ & & c & & \\ & & & b & \\ & 0 & & c & \\ & & & & \ddots \end{bmatrix}$$

where A is an arbitrary $2n$ by $2n$ matrix and b and c are any elements of F. Then R is a prime ring all of whose ideals are idempotent that contains non-prime ideals.

Problem 2 *Under what conditions, would a prime ring in which every ideal is idempotent be a fully prime ring?*

As shown in Blair-Tsutsui [1], a fully prime ring is, in general, not semiprimitive. By Nakayama's lemma, it is evident, however, that a right Noetherian fully prime ring is semiprimitive. More generally, an induction argument on the Krull dimension yields the following result.

Theorem 6 *A fully prime ring with right Krull dimension is semiprimitive.*

Let $A_1(k)$ be the first Weyl algebra over a field of characteristic 0. Then $S = k + xA_1(k)$ is a right and left Noetherian fully prime ring with exactly one nonzero proper ideal $xA_1(k)$. Thus, a Noetherian fully prime ring is not necessarily a simple ring. Further, since S is semiprimitive, there exists a maximal right ideal $M \neq xA_1(k)$.

Theorem 7 *Let $S = k + xA_1(k)$ where the characteristic of k is zero. Then the idealizer of any maximal right ideal M of S different from $xA_1(k)$ is not a fully prime ring.*

Proof. Clearly $M \cap xA_1(k)$ is an ideal of the idealizer of M. So we have $0 \subset M \cap xA_1(k) \subset M$. But since the idealizer of M is a hereditary Noetherian ring, $M \cap xA_1(k)$ is not a prime ideal.

Let $R = k + xA_1(k) \otimes_k (k + xA_1(k))$. Then R is a fully prime domain that has exactly two nonzero proper ideals $xA_1(k) \otimes_k xA_1(k)$ *and* $xA_1(k) \otimes_k (k + xA_1(k))$. We provide a detailed proof of this fact following the proof of more general case given in Hirano [3].

Proof. Let $R_2 = A_1(k) \otimes_k (k + xA_1(k))$. Then R_2 has exactly three proper ideals 0, $A_1(k) \otimes_k xA_1(k)$, and $A_1(k) \otimes_k (k + xA_1(k))$ and they are prime ideals. Now let P be an ideal of $R = k + xA_1(k) \otimes_k (k + xA_1(k))$. Since $P \subseteq R_2$ and $M = xA_1(k) \otimes_k (k + xA_1(k))$ is a right ideal of R_2, $R_2PM =$

$(A_1(k) \otimes_k (k + xA_1(k))) P (xA_1(k) \otimes_k (k + xA_1(k)))$ is an ideal of R_2. Hence we must have:

(1) $R_2PM = A_1(k) \otimes_k (k + xA_1(k))$,
(2) $R_2PM = A_1(k) \otimes_k xA_1(k)$, or
(3) $R_2PM = 0$.

Case (1): $R_2PM = A_1(k) \otimes_k (k + xA_1(k))$.
This means $R_2PM = R_2$. Since M is a right ideal of R_2, $MR_2 = M$. So $R_2PM = R_2$. This implies that $MR_2PM = MR_2$, and hence $M = MPM \subseteq P$. Since M is a maximal ideal and $P \neq R$, we have $M = P$.

Case (2): $R_2PM = A_1(k) \otimes_k xA_1(k)$.
$(A_1(k) \otimes_k (k + xA_1(k))) P (xA_1(k) \otimes_k (k + xA_1(k))) = A_1(k) \otimes_k xA_1(k)$ implies that $(x \otimes_k 1)P (x \otimes_k 1) \in xA_1(k) \otimes_k xA_1(k)$. But since $R/xA_1(k) \otimes_k xA_1(k) \approx S$ is a domain (hence no zero divisor), and $x \otimes_k 1 \notin xA_1(k) \otimes_k xA_1(k)$, $P \subseteq xA_1(k) \otimes_k xA_1(k)$. But $xA_1(k) \otimes_k xA_1(k) = xR_2PM = MPM \subseteq P$. Hence we conclude that $P = xA_1(k) \otimes_k xA_1(k)$.

Case (3): $R_2PM = 0$.
Since R_2P and M are right ideals of R_2, $M \neq 0$, and 0 is prime, we must have $R_2P = 0$. hence $P = 0$.

Problem 3 *Is $R = k + xA_1(k) \otimes_k (k + xA_1(k))$ a right Noetherian ring?*

Theorem 8 *A fully prime ring that has a right Krull dimension and a minimal nonzero ideal is a right primitive ring.*

Proof: Suppose that $\cap P_i$, the intersection of all nonzero (prime) ideals of R, is nonzero. Then 0 must be a primitive ideal since otherwise $0 \neq \cap_{P_i \text{ prime}} P_i \subseteq \cap_{P_i \text{ primitive}} P_i = J(R) = 0$, a contradiction.

Conjecture 2 *A right and left Noetherian fully prime ring R with identity is a right primitive ring.*

We end this article by listing some propositions which might be helpful for working on the conjectures and problems listed in this paper. Proofs for the statements are mostly evident.

(1) *Every right ideal of a fully prime ring is eventually idempotent.*
(2) *For any idempotent right ideals I and J of a fully prime ring, either $IJ = I$, or $JI = J$.*
(3) *Every non-idempotent right ideal of a fully prime ring is contained in the maximal ideal.*

(4) Let I be a right ideal of a ring R such that $RI = R$, and let S be a subring of R containing I. Then if S is fully prime, so is R.

(5) Let R be a fully prime right Noetherian ring with the maximal ideal M. Then R is simple if and only if one of R/M is flat (as R-module) or injective.

(6) Let R be a fully prime right Noetherian ring with the maximal ideal M. Then R is simple if for any right ideal I contained in M, $IM = M$.

References

[1] W.D. Blair and H. Tsutsui, *Fully Prime Rings*, Comm. Algebra (1994).

[2] R. Courter, *Rings all of whose factor rings are semiprime*, Canad. Math Bull (1969).

[3] Y. Hirano, *On rings all of whose factor rings are domain*, J. Austral. Math.Soc. (1993).

[4] Y. Hirano, E. Poon, and H.Tsutsui, *On rings in which every ideal is weakly prime*, Bulletin, KMS Vol 47, No. 5 September, 2010.

[5] Y. Hirano and H.Tsutsui, *Fully k-primary rings*, Comm. Algebra (2009).

[6] K. Koh, *On one-sided ideals of prime type*, Proc. Amer. Math Soc. (1971).

[7] H. Tsutsui, *Fully Prime Rings II*, Comm. Algebra 22 (1996).

DEPARTMENT OF MATHEMATICS
EMBRY-RIDDLE AERONAUTICAL UNIVERSITY
PRESCOTT, AZ, U.S.A.
E-mail address: Hisaya.Tsutsui@erau.edu

Proceedings of the Sixth China-Japan-Korea
International Conference on Ring Theory
June 27-July 2, 2011 Suwon, Korea

SOME COMMUTATIVITY THEOREMS CONCERNING ADDITIVE MAPPINGS AND DERIVATIONS ON SEMIPRIME RINGS

SHAKIR ALI, BASUDEB DHARA, AND AJDA FOŠNER*

ABSTRACT. Let R be a ring with its center $Z(R)$ and I a nonzero ideal of R. The purpose of this paper is to investigate identities satisfied by additive mappings on prime and semiprime rings. More precisely, we prove the following result. Let R be a semiprime ring, and let $F, d : R \to R$ be two additive mappings such that $F(xy) = F(x)y + xd(y)$ for all $x, y \in R$. If $F(xy) \pm xy \in Z(R)$ for all $x, y \in I$, then $[d(x), x] = 0$ for all $x \in I$. Further, if d is a derivation such that $d(I) \neq (0)$, then R contains a nonzero central ideal. Moreover, if R is prime and d is a derivation such that $d(I) \neq (0)$, then R is commutative.

1. Introduction

Let R be an associative ring with center $Z(R)$. For $x, y \in R$, the symbol $[x, y]$ will denote the commutator $xy - yx$ and the symbol $x \circ y$ will denote the anticommutator $xy + yx$. A ring R is called 2-torsion free, if $2x = 0$, $x \in R$, implies $x = 0$. Recall that a ring R is prime if for any $a, b \in R$, $aRb = (0)$ implies $a = 0$ or $b = 0$, and is semiprime if for any $a \in R$, $aRa = (0)$ implies $a = 0$. For any subset S of R, we will denote by $r_R(S)$ the right annihilator of S in R, that is, $r_R(S) = \{x \in R | Sx = 0\}$ and by $l_R(S)$ the left annihilator of S in R, that is, $l_R(S) = \{x \in R | xS = 0\}$. If $r_R(S) = l_R(S)$, then $r_R(S)$ is called an annihilator ideal of R and is written as $ann_R(S)$. We know that if R is a semiprime ring and I is an ideal of R, then $r_R(I) = l_R(I)$. An additive mapping $d : R \longrightarrow R$ is called a derivation if $d(xy) = d(x)y + xd(y)$ holds for all $x, y \in R$. Let S be a nonempty subset of R. A mapping $f : R \longrightarrow R$ is called centralizing on S if $[f(x), x] \in Z(R)$ for all $x \in S$ and is called commuting on S if $[f(x), x] = 0$ for all $x \in S$.

Let $F, g : R \to R$ be two mappings such that $F(xy) = F(x)y + xg(y)$ holds for all $x, y \in R$. If F is additive and g is a derivation of R, then

2010 Mathematics Subject Classification : 16N60, 16W25, 16R50.

Keywords : Prime ring, semiprime ring, ideal, derivation, generalized derivation.

*Corresponding author.

F is said to be a generalized derivation of R. The notion of generalized derivation was introduced by Brešar [8]. In [15], Hvala gave the algebraic study of generalized derivations of prime rings. Recall that if R has the property $Rx = (0)$ implies $x = 0$ and $F : R \to R$ is an additive mapping such that $F(xy) = F(x)y + xh(y)$ for all $x, y \in R$ and some function $h : R \to R$, then F is uniquely determined by h. Moreover, h must be a derivation by [8, Remark 1]. Obviously, every derivation is a generalized derivation of R. Thus, generalized derivations cover both the concept of derivations and the concept of left centralizers. An additive mapping $F : R \to R$ is a left centralizer if $F(xy) = F(x)y$ for all $x, y \in R$ (see [22] and [23] for details).

Over the last few decades many authors have investigated the relationship between the commutativity of the ring R and specific types of derivations on R. The first result in this direction is due to E. C. Posner [19] who proved that if a prime ring R admits a nonzero derivation d such that $[d(x), x] \in Z(R)$ for all $x \in R$, then R is commutative. This result was subsequently refined and extended by a number of algebraists (we refer to [7], [9], [16], [18], and the references therein). Recently, some authors have obtained commutativity of prime and semiprime rings with derivations and generalized derivations satisfying certain polynomial identities (see, for example, [1], [2], [4], [5], [6], [11], [12], [13], [17], [19], [20] and [21]). The aim of this paper is to investigate commutativity of prime and semiprime rings satisfying certain identities involving additive mappings.

2. Preliminaries

We shall do a great deal of calculation with commutators and routinely using the following basic identities: For all $x, y, z \in R$; we have

$$[xy, z] = x[y, z] + [x, z]y \text{ and } [x, yz] = y[x, z] + [x, y]z.$$

Moreover,

$$[[x, y], z] + [[y, z], x] + [[z, x], y] = 0.$$

The next statements are well-known results which we will use in the next section.

Lemma 2.1. [7, Theorem 3] *Let R be a semiprime ring and I a nonzero left ideal of R. If R admits a derivation d which is nonzero on I and centralizing on I, then R contains a nonzero central ideal.*

Lemma 2.2. [7, Theorem 4] *Let R be a prime ring and I a nonzero left ideal of R. If R admits a nonzero derivation d which is centralizing on I, then R is commutative.*

Lemma 2.3. [10, Lemma 2]

(a) *If R is a semiprime ring, then the center of a nonzero one-sided ideal is contained in the center of R. In particular, any commutative one-sided ideal is contained in the center of R.*

(b) *If R is prime with a nonzero central ideal, then R is commutative.*

Lemma 2.4. [14, Corollary 2] *If R is a semiprime ring and I is an ideal of R, then $I \bigcap ann_R(I) = 0$.*

3. The Results

The following result is a generalization of [3, Theorem 2.1].

Theorem 3.1. *Let R be a semiprime ring, I a nonzero ideal of R, and let $F, d : R \to R$ be two additive mappings such that $F(xy) = F(x)y + xd(y)$ for all $x, y \in R$. If $F(xy) - xy \in Z(R)$ for all $x, y \in I$, then $[d(x), x] = 0$ for all $x \in I$. Moreover, if d is a derivation such that $d(I) \neq (0)$, then R contains a nonzero central ideal.*

Proof. By the assumption, we have

$$(1) \qquad F(xy) - xy \in Z(R)$$

for all $x, y \in I$. Replacing y by yz in (1), we obtain

$$
\begin{aligned}
F(xyz) - xyz &= F(xy)z + xyd(z) - xyz \\
(2) &= (F(xy) - xy)z + xyd(z) \in Z(R),
\end{aligned}
$$

for all $x, y, z \in I$. Using (1), we get

$$(3) \qquad [xyd(z), z] = 0$$

for all $x, y, z \in I$. Substituting rx for x in (3), we obtain

$$(4) \quad 0 = [rxyd(z), z] = r[xyd(z), z] + [r, z]xyd(z) = [r, z]xyd(z)$$

for all $x, y, z \in I$ and $r \in R$. Replacing x by $d(z)x$ in (4) we get

$$(5) \qquad 0 = [r, z]d(z)xyd(z)$$

for all $x, y, z \in I$ and $r \in R$. This implies

$$(6) \qquad [r, z]d(z)Rxyd(z) = (0)$$

for all $x, y, z \in I$ and $r \in R$. Interchanging the role of x and y in the last relation, we find that

$$(7) \qquad [r, z]d(z)Ryxd(z) = (0)$$

for all $x, y, z \in I$ and $r \in R$. Subtracting (7) from (6), we obtain

$$(8) \qquad [r, z]d(z)R[x, y]d(z) = (0)$$

for all $x, y, z \in I$. In particular, $[x, z]d(z)R[x, z]d(z) = (0)$ for all $x, z \in I$. The semiprimeness of R yields that

(9) $$[x, z]d(z) = 0$$

for all $x, z \in I$. Right multiplication of (9) by z yields that

(10) $$[x, z]d(z)z = 0$$

for all $x, z \in I$. Replacing x by xz in (9) we get

(11) $$0 = [xz, z]d(z) = [x, z]zd(z) + x[z, z]d(z) = [x, z]zd(z)$$

for all $x, z \in I$. Equations (10) and (11) together imply that

(12) $$[x, z][d(z), z] = 0$$

for all $x, z \in I$. Replacing x by $d(z)x$ in the last expression, we obtain

$$[d(z), z]x[d(z), z] + d(z)[x, z][d(z), z] = 0$$

for all $x, z \in I$. By (12) we have $[d(z), z]x[d(z), z] = 0$ for all $x, z \in I$. That is, $I[d(z), z]RI[d(z), z] = (0)$ for all $z \in I$. Hence, the semiprimeness of R forces that $I[d(z), z] = (0)$ for all $z \in I$. Thus, $[d(z), z] \in Ann_R(I)$ for all $z \in I$. Since I is an ideal, we conclude that $[d(z), z] \in I$ for all $z \in I$. This implies that $[d(z), z] \in I \cap Ann_R(I)$ for all $z \in I$. In view of Lemma 2.4, $[d(z), z] = 0$ for all $z \in I$. Further, if d is derivation such that $d(I) \neq (0)$, then by Lemma 2.1, R contains a nonzero central ideal. Thereby, the proof is completed.   □

Theorem 3.2. *Let R be a semiprime ring, I a nonzero ideal of R, and let $F, d : R \to R$ be two additive mappings such that $F(xy) = F(x)y + xd(y)$ for all $x, y \in R$. If $F(xy) + xy \in Z(R)$ for all $x, y \in I$, then $[d(x), x] = 0$ for all $x \in I$. Moreover, if d is a derivation such that $d(I) \neq (0)$, then R contains a nonzero central ideal.*

Proof. If we replace F with $-F$ and d with $-d$ in Theorem 3.1, we conclude that $(-F)(xy) - xy \in Z(R)$ implies $[(-d)(x), x] = 0$ for all $x \in I$. Since $F(xy) + xy \in Z(R)$ for all $x, y \in I$, we have $[d(x), x] = 0$ for all $x \in I$. Moreover, if d is derivation such that $d(I) \neq (0)$, then by Lemma 2.1, R contains a nonzero central ideal.   □

Following corollaries are the immediate consequences of the above theorems.

Corollary 3.3 ([3, Theorem 2.2]). *Let R be a prime ring and I a nonzero ideal of R. If R admits a generalized derivation F associated* *with a nonzero derivation d such that $F(xy) + xy \in Z(R)$ for all $x, y \in I$, then R is commutative.*

Corollary 3.4. *Let R be a prime ring, I a nonzero ideal of R, and let $F, d : R \to R$ be two additive mappings such that $F(xy) = F(x)y + xd(y)$ for all $x, y \in R$. If $F(xy) \pm xy \in Z(R)$ for all $x, y \in I$, then $[d(x), x] = 0$ for all $x \in I$. Moreover, if d is a derivation, then one of the following holds:*

(1) *$d = 0$ and $F(x) = \mp x + \zeta(x)$ for all $x \in I$, where $\zeta : I \to Z(R)$ is an additive mapping.*
(2) *R is commutative.*

Proof. By Theorem 3.1 and Theorem 3.2, we conclude that $[d(x), x] = 0$ for all $x \in I$. In particular, if d is a derivation, then by Lemma 2.2, we conclude that either $d = 0$ or R is commutative. If R is commutative, we obtain our conclusion (2). So assume that R is noncommutative. Then $d = 0$. In this case, for all $r, s \in R$, we get $F(rs) = F(r)s + rd(s) = F(r)s$. In other words, F is a left centralizer on R. By the hypothesis, we obtain $(F(x) \pm x)y \in Z(R)$ for all $x, y \in I$. This implies that $[(F(x) \pm x)y, r] = 0$ for all $x, y \in I$ and $r \in R$. In particular, we obtain $[(F(x) \pm x)y, (F(x) \pm x)] = 0$ for all $x, y \in I$ and hence

(13)
$$(F(x) \pm x)[y, (F(x) \pm x)] = 0$$

for all $x, y \in I$. Replacing y with yr, where $r \in R$, we get

$$\begin{aligned} 0 &= (F(x) \pm x)[yr, (F(x) \pm x)] \\ (14) \quad &= (F(x) \pm x)[y, (F(x) \pm x)]r + (F(x) \pm x)y[r, (F(x) \pm x)]. \end{aligned}$$

Using (13) and (14) we obtain

(15)
$$0 = (F(x) \pm x)y[r, (F(x) \pm x)]$$

for all $x, y \in I$ and $r \in R$. Replacing y with ry, we get

(16)
$$0 = (F(x) \pm x)ry[r, (F(x) \pm x)].$$

Multiplying (15) by r from left, we get

(17)
$$0 = r(F(x) \pm x)y[r, (F(x) \pm x)].$$

Now, (16) and (17) together imply

(18)
$$0 = [r, (F(x) \pm x)]y[r, (F(x) \pm x)]$$

for all $x, y \in I$ and $r \in R$. Therefore, $y[r, (F(x) \pm x)]Ry[r, (F(x) \pm x)] = (0)$ for all $x, y \in I$ and $r \in R$. Since R is a prime ring, $y[r, (F(x) \pm x)] = 0$ for all $x, y \in I$ and $r \in R$. Since I is an ideal, it follows that $IR[r, (F(x) \pm x)] = (0)$ for all $x \in I$ and for all $r \in R$. Again, by primeness of R, we have $[R, (F(x) \pm x)] = (0)$, that is $F(x) \pm x \in Z(R)$ for all $x \in I$. Let us write $F(x) \pm x = \zeta(x) \in Z(R)$ for all $x \in I$. Then $F(x) = \mp x + \zeta(x)$ for all $x \in I$, where $\zeta : I \to Z(R)$. Since F is additive, ζ is also additive. Hence, we obtain our conclusion (1).   □

Theorem 3.5. *Let R be a semiprime ring, I a nonzero ideal of R, and let $F, d : R \to R$ be two additive mappings such that $F(xy) = F(x)y + xd(y)$ for all $x, y \in R$. If $F(xy) - yx \in Z(R)$ for all $x, y \in I$, then $[[d(x), x], x] = 0$ for all $x \in I$. Moreover, if d is a derivation such that $d(I) \neq (0)$, then R contains a nonzero central ideal.*

Proof. For all $x, y \in I$, we have

(19) $$F(xy) - yx \in Z(R).$$

Replacing x with xy in (19) we get

(20) $$F(xy^2) - yxy \in Z(R)$$

for all $x, y \in I$. Replacing y with y^2 in (19) we get

(21) $$F(xy^2) - y^2 x \in Z(R)$$

for all $x, y \in I$. Subtracting (21) from (20), we obtain

$$y[y, x] \in Z(R)$$

for all $x, y \in I$.

Now, replacing y with yx in (19), we get

$$F(xyx) - yx^2 = F(xy)x + xyd(x) - yx^2$$
(22) $$= (F(xy) - yx)x + xyd(x) \in Z(R).$$

This implies that $[(F(xy) - yx)x + xyd(x), r] = 0$ for all $x, y \in I$, $r \in R$ and hence in particular we obtain $[(F(xy) - yx)x + xyd(x), x] = 0$ for all $x, y \in I$. Application of relation (19) yields that $[xyd(x), x] = 0$ for all $x, y \in I$. Therefore, we find that,

(23) $$0 = xy[d(x), x] + x[y, x]d(x)$$

for all $x, y \in I$. By using the fact $y[y, x] \in Z(R)$ for all $x, y \in I$, we have from above that

(24) $$0 = xy[d(x), x] + d(x)x[y, x]$$

for all $x, y \in I$. Now, putting $y = yx$ in (24), we see that

(25) $$0 = xyx[d(x), x] + d(x)x[y, x]x$$

for all $x, y \in I$. Right multiplying (24) by x gives us

(26) $$0 = xy[d(x), x]x + d(x)x[y, x]x$$

for all $x, y \in I$. Subtracting (25) from (26), we get

$$(27) \qquad 0 = xy[[d(x), x], x]$$

for all $x, y \in I$. This implies $[[d(x), x], x]y[[d(x), x], x] = 0$ for all $x, y \in I$. Thus, $([[d(x), x], x]I)^2 = (0)$ for all $x \in I$. Since semiprime ring contains no nonzero nilpotent ideal, we conclude that $[[d(x), x], x]I = (0)$ for all $x \in I$. Therefore, for all $x \in I$ we have $[[d(x), x], x] \in I \bigcap ann_R(I)$. Since R is semiprime, by Lemma 2.4, we conclude that $[[d(x), x], x] = 0$ for all $x \in I$.

If d is a derivation, then by [16, Main Theorem], we can conclude that either $d(I) = (0)$ or R contains a nonzero central ideal. $\square$

With the same arguments which we have used in the proof of Theorem 3.2, we can show the following result.

Theorem 3.6. *Let R be a semiprime ring, I a nonzero ideal of R, and let $F, d : R \to R$ be two additive mappings such that $F(xy) = F(x)y + xd(y)$ for all $x, y \in R$. If $F(xy) + yx \in Z(R)$ for all $x, y \in I$, then $[[d(x), x], x] = 0$ for all $x \in I$. Moreover, if d is a derivation such that $d(I) \neq (0)$, then R contains a nonzero central ideal.*

Corollary 3.7 ([3, Theorem 2.3]). *Let R be a prime ring and I be a nonzero ideal of R. If R admits a generalized derivation F associated with a nonzero derivation d such that $F(xy) - yx \in Z(R)$ for all $x, y \in I$, then R is commutative.*

Corollary 3.8. *Let R be a prime ring, I a nonzero ideal of R and let $F, d : R \to R$ be two additive mappings such that $F(xy) = F(x)y + xd(y)$ for all $x, y \in R$. If $F(xy) \pm yx \in Z(R)$ for all $x, y \in I$, then $[[d(x), x], x] = 0$ for all $x \in I$. Moreover, if d is a derivation, then R is commutative.*

Proof. In a view of Theorems 3.5 and 3.6 we have $[[d(x), x], x] = 0$ for all $x \in I$. In particular, if d is a derivation, then by [16, Main Theorem], either $d = 0$ or R is commutative. If R is commutative, then we are done. So, let R be noncommutative. Then $d = 0$. In this case, for any $r, s \in R$, we have $F(rs) = F(r)s + rd(s) = F(r)s$. In other words, F is a left centralizer on R. By the hypothesis

$$(28) \qquad F(x)y - yx \in Z(R)$$

for all $x, y \in I$. Replacing x with x^2 and y with xy, respectively, and then subtracting so obtained relations one from another, we see that $(F(x^2)y - yx^2) - (F(x)xy - xyx) \in Z(R)$. Since F is a left centralizer, it follows that $[x, y]x \in Z(R)$ for all $x, y \in I$. Now, replacing y with

$[x, y]$ in (28) and then using the fact that $[x, I]x \subseteq Z(R)$ for all $x \in I$, we obtain $F(x)[x, y] \in Z(R)$ for all $x, y \in I$. Writing yx instead of y we get $F(x)[x, y]x \in Z(R)$ for all $x, y \in I$. Since $[x, y]x \in Z(R)$, for each $x \in I$, either $[x, I]x = (0)$ or $F(x) \in Z(R)$. For a prime ring R, it is easy to check that $[x, I]x = (0)$ implies $x \in Z(R)$. Thus, either $x \in Z(R)$ or $F(x) \in Z(R)$ for all $x \in I$. Since R is prime ring and I is an ideal of R, so I is prime as well. Thus, the standard argument gives us that either $I \subseteq Z(R)$ or $F(I) \subseteq Z(R)$. If $I \subseteq Z(R)$, then R is commutative by Lemma 2.3(b), a contradiction. But if $F(I) \subseteq Z(R)$, then $I^2 \subseteq Z(R)$, since $F(xy) \pm yx \in Z(R)$ for all $x, y \in I$. Since I^2 is a nonzero central ideal of R, by Lemma 2.3(b), R is commutative, again a contradiction. Hence R must be commutative. Thereby the proof is completed.   □

Acknowledgments. The first and third authors would like to thank the organizers of the sixth China-Japan-Korea ISORT held at Suwon, Korea, for providing warm hospitality during the symposium. The support received from UGC(India) and MHEST(Slovenia) to attend this meeting is gratefully acknowledged.

References

[1] S. Ali, S. Huang, *On derivations in semiprime rings*, Algebra and Representation Theory, Punlished online 24 February 2011(DOI 10.1007/s10468-011-9271-9).

[2] N. Argac, *On prime and semiprime rings with derivations*, Algebra Colloq. **13** (3) (2006), 371-380.

[3] M. Ashraf, A. Ali, S. Ali, *Some commutativity theorems for rings with generalized derivations*, Southeast Asian Bull. Math. *31* (2007), 415-421.

[4] M. Ashraf, N. Rehman, *On commutativity of rings with derivations*, Results Math. *42* (2002), 3-8.

[5] M. Ashraf, N. Rehman, *On derivations and commutativity in prime rings*, East-West J. Math. *3* (1) (2001), 87-91.

[6] H. E. Bell, M. N. Daif, *On commutativity and strong commutativity preserving maps*, Canad. Math. Bull. *37* (1994), 443-447.

[7] H. E. Bell, W. S. Martindale III, *Centralizing mappings of semiprime rings*, Canad. Math. Bull. *30* (1987), 92-101.

[8] M. Brešar, *On the distance of the composition of two derivations to the generalized derivations*, Glasgow Math. J. *33* (1991), 89-93.

[9] M. Brešar, *Centralizing mappings and derivations in prime rings*, J. Algebra *156* (1993), 385-394.

[10] M. N. Daif *Commutativity results for semiprime rings with derivations*, Internt. J. Math. & Math. Sci. *21* (3) (1998,) 471-474.

[11] M. N. Daif, H. E. Bell, *On remark on derivations on semiprime rings*, Internt. J. Math. & Math. Sci. *15* (1) (1992), 205-206.

[12] B. Dhara, A. Pattanayak, *Generalized derivations and left ideals in prime and semiprime rings*, ISRN Algebra **2011**, Article ID 750382, 5 pages.

[13] B. Dhara, *Remarks on generalized derivations in prime and semiprime rings*, Internt. J. Math. & Math. Sci. **2010**, Article ID 646587, 6 pages.

[14] I. N. Herstein, Rings with involution, Univ. of Chicago Press, 1976.

[15] B. Hvala, *Generalized derivations in rings*, Comm. Algebra *26* (4) (1998), 1147-1166.

[16] C. Lanski, *An Engel condition with derivation for left ideals*, Proc. Amer. Math. Soc. *125* (2) (1997), 339-345.

[17] J. H. Mayne, *Centralizing mappings of prime rings*, Canad. Math. Bull. *27* (1984), 122-126.

[18] M. Marubayashi, M. Ashraf, N. Rehman, S. Ali, *On generalized* (α, β)-*derivations in prime rings*, Algebra Colloq., *17* (2010), 865-874.

[19] E. C. Posner, *Derivations in prime rings*, Proc. Amer. Math. Soc. *8* (1957), 1093-1100.

[20] M. A. Quadri, M. S. Khan, N. Rehman, *Generalized derivations and commutativity of prime rings*, Indian J. Pure Appl. Math. **34** (9) (2003), 1393-1396.

[21] N. Rehman, *On commutativity of rings with generalized derivations*, Math. J. Okayama Univ. **44** (2002), 43-49.

[22] J. Vukman *Centralizers in prime and semiprime rings*, Comment. Math. Univ. Carol. **38** (1997), 231-240.

[23] B. Zalar, *On centralizers of semiprime rings*, Comment. Math. Univ. Carol. **32** (1991), 609-614.

DEPARTMENT OF MATHEMATICS
ALIGARH MUSLIM UNIVERSITY
ALIGARH 202002, INDIA
E-mail address: shakir.ali.mm@amu.ac.in

DEPARTMENT OF MATHEMATICS
BELDA COLLEGE
BELDA, PASCHIM MEDINIPUR-721424, INDIA
E-mail address: basu_dhara@yahoo.com

FACULTY OF MANAGEMENT
UNIVERSITY OF PRIMORSKA
CANKARJEVA 5, SI-6104 KOPER, SLOVENIA
E-mail address: ajda.fosner@fm-kp.si

Proceedings of the Sixth China-Japan-Korea
International Conference on Ring Theory
June 27-July 2, 2011 Suwon, Korea

STUDY ON THE ALGEBRAIC STRUCTURES IN TERMS OF GEOMETRY AND DEFORMATION THEORY

FUMIYA SUENOBU AND FUJIO KUBO*

ABSTRACT. In this paper, we study the geography of the algebraic set $\mathfrak{C}$. We have three subjects. The first is the point on $\mathfrak{C}$ closest to a given point. This point represents the closest associative algebra structure to any given algebra structure. The second is a path on $\mathfrak{C}$ from a point to another point, which represents an algebraic deformation of an associative algebra. The third is the orbits. We discuss how the 7 orbits of pair-wise non-isomorphic associative algebras over $\mathbb{R}$ of 2-dimension cover our algebraic set $\mathfrak{C}$. First two of them are based on our papers [5] and [6], and the last is our further work.

1. Introduction

Let $\mathbb{R}$ be the field of real numbers and V be the finite-dimensional vector space over $\mathbb{R}$ with the fixed basis $\{e_1, \ldots, e_n\}$. A point $(c_{ijk}) \in \mathbb{R}^{n^3}$ defines a multiplication on V by

$$e_i \circ e_j = \sum_{k=1}^{n} c_{ijk} e_k \quad (i, j = 1, \ldots, n),$$

and one has the algebra $A = (V, (c_{ijk}), \circ)$ with the underlying vector space V and the set of structure constants (c_{ijk}).

When the multiplication "$\circ$" satisfies some conditions, the sets of structure constants (c_{ijk}) must have some relations among c_{ijk}. For example, if the multiplication "$\circ$" is associative multiplication, the set of structure constants satisfies the equations (1) given in the following section. These draw an algebraic variety in $\mathbb{R}^{n^3}$ and we denote this by $\mathfrak{C}$.

It is natural to ask a question:

2010 Mathematics Subject Classification : 16B99, 16S80, 17B99.

Keywords : Associative algebras, Lie algebras, closest structures, algebraic deformation.

*Corresponding author.

what is the closest associative algebra structure to a
given algebra structure?

When we face a new multiplication, being caused by noise and so on, it
must be useful to compute with the closest associative multiplication to
such a perturbed one. Our question turns out be

how to find systematically the point on $\mathfrak{C}$ closest to a
given point?

Let a point (a_{ijk}) be the set of structure constants of a multiplication
"$*$" which does not satisfy the associative law, in other words, the point
(a_{ijk}) does not stay on $\mathfrak{C}$. We find the point (c_{ijk}) on $\mathfrak{C}$ closest to this
point.

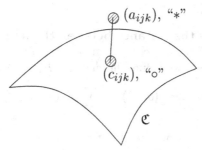

In §2 we will answer to this question in case of the two dimensional
associative algebras by giving an algorithm to find such a point.

In §3 we trace the points of the sets of structure constants of the
deformed algebras.

We are interested in the 7 associative algebras which appear in the
classification of 2-dimensional associative algebras over $\mathbb{R}$ ([1]). In §4 we
figure out the orbits of the sets of structure constants of these algebras,
and express the algebraic set $\mathfrak{C}$ in terms of these orbits.

The final section touches on the closest Lie algebra structures.

2. Closest associative algebra structures

2.1. The set of structure constants of associative algebras   If
A is an associative algebra, the set of structure constants (c_{ijk}) satisfies
the n^4 equations

$$(1) \qquad \sum_{p=1}^{n}(c_{ijp}c_{pkq} - c_{jkp}c_{ipq}) = 0 \quad (i,j,k,q = 1,\ldots,n).$$

We denote by $\mathfrak{C}$ the algebraic set defined by (1).

2.2. Expression of $\mathfrak{C}$ We parameterize $\mathfrak{C}$ in the case of $n = 2$ by
using the Groebner basis and elimination-extension method.

Theorem ([5]). *Let $\mathfrak{C}$ be the algebraic set defined by (1) in the case of $n = 2$ in $\mathbb{R}^8$. Then $\mathfrak{C}$ is expressed by*

$$\begin{aligned}
\mathfrak{C} &= \mathfrak{C}_1 \cup \mathfrak{C}_2 \cup \mathfrak{C}_3 \cup \mathfrak{C}_4 \cup \mathfrak{C}_5; \\
\mathfrak{C}_1 &= \{(\alpha, 0, \beta, 0, 0, \alpha, 0, \beta) \,|\, \alpha, \beta \in \mathbb{R}\}, \\
\mathfrak{C}_2 &= \{(\alpha, 0, 0, 0, 0, 0, 0, \beta) \,|\, \alpha, \beta \in \mathbb{R}\}, \\
\mathfrak{C}_3 &= \{(\alpha, 0, 0, \alpha, \beta, 0, 0, \beta) \,|\, \alpha, \beta \in \mathbb{R}\}, \\
\mathfrak{C}_4 &= \{(\alpha, \beta, \gamma, 0, \gamma, 0, 0, \gamma) \,|\, \alpha, \beta, \gamma \in \mathbb{R}\}, \\
\mathfrak{C}_5 &= \{(\alpha - p\gamma + p^2\beta, p\alpha, p\beta, \alpha, p\beta, \alpha, \beta, \gamma) \,| \\
&\quad\ \alpha, \beta, \gamma, p \in \mathbb{R}\}.
\end{aligned}$$

Note: $\mathfrak{C}_2 \subseteq \mathfrak{C}_4 \cup \mathfrak{C}_5$.

2.3. Definition of the distance between the multiplications For a pair of the sets of the structures constants (c_{ijk}) of the multiplication "$\circ$" and (a_{ijk}) of "$*$", we define the *distance* $D((c_{ijk}), (a_{ijk}))$ by

$$D((c_{ijk}), (a_{ijk}))^2 = \sum_{i,j,k} (c_{ijk} - a_{ijk})^2.$$

Note. We define a norm on V of $x = \sum_{k=1}^{n} \alpha_k e_k \in V$ by $\|x\| = \sqrt{\sum_{k=1}^{n} \alpha_k^2}$. Then we have

$$\frac{\|x \circ y - x * y\|^2}{\|x\|^2 \|y\|^2} \leq D((c_{ijk}), (a_{ijk}))^2.$$

Thus, when the distance $D((c_{ijk}), (a_{ijk}))$ is small, we can regard as the multiplication "$\circ$" *close* to multiplication "$*$".

2.4. The closest associative structure For a point $a = (a_{ijk})$, we first find the point $c_i = (c_{ijk})$ on $\mathfrak{C}_i$ minimize $D(c_i, a)$ for each $i = 1, \ldots, 5$. Then we choose the point c_0 among them with $D(c_i, a)$ minimum. The point c_0 corresponds to the closest associative multiplication to that represented by a.

Theorem ([5]). *Let $\mathfrak{C}_1, \ldots, \mathfrak{C}_5$ be the algebraic set given in Theorem of §2.2 and (a_{ijk}) be a given point in $\mathbb{R}^8$. Then the point c_i on $\mathfrak{C}_i$ closest to the point (a_{ijk}) is of the following form:*

(1) $c_1 = \left(\dfrac{a_{111} + a_{212}}{2}, 0, \dfrac{a_{121} + a_{222}}{2}, 0, 0, \dfrac{a_{111} + a_{212}}{2}, 0, \dfrac{a_{121} + a_{222}}{2}\right),$

(2) $c_2 = (a_{111}, 0, 0, 0, 0, 0, 0, a_{222}),$

(3) $c_3 = \left(\dfrac{a_{111} + a_{122}}{2}, 0, 0, \dfrac{a_{111} + a_{122}}{2}, \dfrac{a_{211} + a_{222}}{2}, 0, 0, \dfrac{a_{211} + a_{222}}{2}\right),$

(4) $c_4 = \left(a_{111}, a_{112}, \dfrac{a_{121} + a_{211} + a_{222}}{3}, 0, \dfrac{a_{121} + a_{211} + a_{222}}{3}, 0, 0, \dfrac{a_{121} + a_{211} + a_{222}}{3}\right),$

(5) $c_5 = (\alpha - p\gamma + p^2\beta, p\alpha, p\beta, \alpha, p\beta, \alpha, \beta, \gamma)$. Here the real numbers p, α, β, γ are chosen as follows : Let us consider

$$A(p) = \begin{pmatrix} 1+p^2 & -p^3 & -p \\ -p^3 & (1+p^2)^2 & p^2 \\ -p & p^2 & 3+p^2 \end{pmatrix}, \quad C = \textstyle\sum_{i,j,k} a_{ijk}^2,$$

$$b(p) = \begin{pmatrix} -a_{222} + pa_{111} \\ -a_{221} - pa_{211} - pa_{121} - p^2a_{111} \\ -a_{212} - a_{122} - pa_{112} - a_{111} \end{pmatrix},$$

and denote the transpose of the matrix or the vector X by $\,^t X$. Then
(5-1) $\,^t(\gamma, \beta, \alpha) = -A(p)^{-1}b(p)$.
(5-2) p minimizes the function

$$f(p) = -\,^t b(p) A(p)^{-1} b(p) + C.$$

2.5. Example of the closest associative algebra structure Our point $a \in \mathbb{R}^8$ is $a = (2.1, 1.2, 1.2, 2.0, 1.2, 2.0, 2.0, 1.0)$, and the corresponding multiplication table is

$*$	e_1	e_2	
e_1	$2.1e_1 + 1.2e_2$	$1.2e_1 + 2.0e_2$	(A)
e_2	$1.2e_1 + 2.0e_2$	$2.0e_1 + 1.0e_1$	.

The multiplication "$*$" represented by a does not satisfy the associative law, for example, $(e_2 * e_2) * e_1 = 5.4e_1 + 4.4e_2$ and $e_2 * (e_2 * e_1) = 5.44e_1 + 4.4e_2$. Following to the algorithm given in Theorem of §2.4, we find the point c_i listed below on $\mathfrak{C}_i$ closest to the point a.

$c_1 = (2.05, 0, 1, 1.1, 0, 0, 2.05, 0, 1.1)$; $D(c_1, a)^2 = 10.905$
$c_2 = (2.1, 0, 0, 0, 0, 0, 0, 1.0)$; $D(c_2, a)^2 = 16.32$
$c_3 = (2.05, 0, 0, 2.05, 1.1, 0, 0, 1.1)$; $D(c_3, a)^2 = 10.905$
$c_4 = (2.1, 1.2, 1.13333, 0,$
 $1.13333, 0, 0, 1.13333)$; $D(c_4, a)^2 = 12.0267$
$c_5 = (2.11066, 1.19643, 1.19808, 1.99574,$
 $1.19808, 1.99574, 1.99848, 1.00639)$; $D(c_5, a)^2 = 0.000213081$

For example, as for c_5, our computations are based on the following data:

$b(p) = \,^t(-1 + 2.1p, -2 - 4p - 2.1p^2, -4.5 - 1.2p)$
$f(p) = \{7.54667\,(p^2 - 1.9773p + 0.97819)\,(p^2 - 0.685296p + 1.39919)$
 $\cdot (p^2 + 0.64493p + 0.966216)\}/(p^2 + 1)^3$.

Since $D(c_5, a)$ is the smallest among the $D(c_i, a)$'s, the closest point c_0 on $\mathfrak{C}$ to a is c_5 and the corresponding multiplication table is

$\circ$	e_1	e_2
e_1	$2.11066e_1 + 1.19643e_2$	$1.19808e_1 + 1.99574e_2$
e_2	$1.19808e_1 + 1.99574e_2$	$1.99848e_1 + 1.00639e_2$

(B)

When one computes according to the multiplication (A), we shall suggest replacing it with (B).

3. Associative structures obtainable by deformations

3.1. Deformation of associative algebras Let A be an associative algebra over $\mathbb{R}$ with a bilinear multiplication $\alpha : A \times A \to A$. We consider the formal power series $\alpha_t : A[[t]] \times A[[t]] \to A[[t]]$ of the form

$$\alpha_t(x, y) := \alpha_0(x, y) + t\alpha_1(x, y) + t^2\alpha_2(x, y) + \cdots,$$

where each $\alpha_i : A \times A \to A$ is a $\mathbb{R}$-bilinear map and extended to that of the power series ring $\mathbb{R}[[t]]$ and $\alpha_0 = \alpha$. When α_t satisfies the associative law, we call α_t a *deformation* of α ([2], [3]).

These equations can be analyzed using by Hochschild cohomology. Let $Z^2(A, \circ)$ be the set of all Hochschild 2-cocycle of A with a multiplication "$\circ$". Then $C \in Z^2(A, \circ)$ satisfies the equation

$$\delta C(x, y, z) = x \circ C(y, z) - C(x \circ y, z) + C(x, y \circ z) - C(x, y) \circ z = 0$$

for $x, y, z \in A$ and Hochschild coboundary operator δ.

3.2. Step deformation Let A be an associative algebra with a multiplicaion $x \circ y$ and "$*_t^n$" be the multiplication defined by

$$x *_t^n y = x \circ y + t^n C_n(x, y).$$

Then the multiplication "$*_t^n$" satisfies the associative law if $C_n(-, -) \in Z^2(A, \circ)$ and $C_n(-, -)$ satisfies the associative law ([5]).

This allows us to define the new associative multiplication $x *_t^\bullet y$ on $A[t]$:

$x *_t^1 y = x \circ y + tC_1(x, y)$
"$*_t^1$": $A[t] \times A[t] \to A[t]$, $C_1 : A \times A \to A$.
$(C_1(-, -) \in Z^2(A, \circ)$ and $C_1(C_1(x, y), z) = C_1(x, C_1(y, z)))$,
$x *_t^2 y = x *_t^1 y + t^2 C_2(x, y)$
"$*_t^2$": $A[t] \times A[t] \to A[t]$, $C_2 : A[t] \times A[t] \to A[t]$.
$(C_2(-, -) \in Z^2(A[t], *_t^1)$ and $C_2(C_2(x, y), z) = C_2(x, C_2(y, z)))$,

and so on. Then we can define the associative algebra multiplication,

$$x *_t^n y = x \circ y + tC_1(x, y) + t^2 C_2(x, y) + \cdots + t^n C_n(x, y),$$

we call this a *step deformation* of $(A, \circ)$.

To find a 2-cocycle, we state the useful

Lemma ([5]). *Let V be a vector space over $\mathbb{R}$ with a basis and A be the associative algebra with the underlying vector space V. If the associative multiplication of A on V is given by the differentiable functions $c_{ijk} = c_{ijk}(t_1, \ldots, t_p)$ on $\mathbb{R}^p$ in $t_1, \ldots, t_p$ so that*

$$e_i e_j = \sum_{k=1}^{n} c_{ijk}(t_1, \ldots, t_p) e_k \quad (i, j, k = 1, \ldots, n),$$

then for a derivation D of the set of all differentiable functuions on $\mathbb{R}^p$, $\phi(e_i, e_j) = \sum_{k=1}^{n} (Dc_{ijk}) e_k$ is a 2-cocycle of the algebra A.

Now we can find a path from one multiplication to another.

Theorem ([5]). *Let "$\circ$" and "$*$" be the associative multiplication whose structure constants stay on the same set $\mathfrak{C}_i$. Then the multiplication "$*$" can be obtainable from the multiplication "$\circ$" by a step deformation.*

Here, we illustrate a construction of a step deformation from a multiplication "$\circ$" corresponding to a point $(c_{ijk})(\alpha, \beta, \gamma, p)$ to a multiplication "$*$" corresponding to a point $(c_{ijk})(\alpha', \beta', \gamma', p')$:

$$
\begin{aligned}
&\text{``}\circ\text{''} & \cdots \quad \cdots \quad & (\alpha - p\gamma + p^2\beta, p\alpha, p\beta, \alpha, p\beta, \alpha, \beta, \gamma) \\
&\quad \downarrow \quad C_1 & = (p' - p)(\tfrac{\partial}{\partial p} + \beta\tfrac{\partial}{\partial \gamma}) & \text{ at } (\alpha, \beta, \gamma, p) \\
&\text{``}*_t^1\text{''} & \cdots \quad \cdots \quad & (\alpha - \hat{p}\hat{\gamma} + \hat{p}^2\beta, \hat{p}\alpha, \hat{p}\beta, \alpha, \hat{p}\beta, \alpha, \beta, \hat{\gamma}) \\
&\quad \downarrow \quad C_2 & = (\alpha' - \alpha)\tfrac{\partial}{\partial \alpha} + (\beta' - \beta)\tfrac{\partial}{\partial \beta} & \\
& & \quad + (\gamma' - \hat{\gamma})\tfrac{\partial}{\partial \gamma} & \text{ at } (\alpha, \beta, \hat{\gamma}, \hat{p}) \\
&\text{``}*_t^2\text{''} & \cdots \quad \cdots \quad & (\alpha'' - p''\gamma'' + p''^2\beta'', p''\alpha'', p''\beta'', \alpha'', \\
& & & p''\beta'', \alpha'', \beta'', \gamma'').
\end{aligned}
$$

We have a step deformation $x *_t^2 y = x \circ y + tC_1(x, y) + t^2 C_2(x, y)$ with $x * y = x *_1^2 y$. Here, $\hat{p}, \hat{\gamma}, p'', \alpha'', \beta'', \gamma''$ are of the forms $\hat{p} = p + t(p' - p)$, $\hat{\gamma} = \gamma + t(p' - p)$ and $p'' = \hat{p}$, $\alpha'' = \alpha + t^2(\alpha' - \alpha)$, $\beta'' = \beta + t^2(\beta' - \beta)$, $\gamma'' = \hat{\gamma} + t^2(\gamma' - \hat{\gamma})$.

4. Orbits of pair-wise non-isomorphic associative algebras in our variety $\mathfrak{C}$

4.1. Classification of two dimensional real associative algebras

Bermúdez et al. classify 2-dimensional real associative algebras into the 7 non-isomorphic associative algebras ([1]). The sets of structure constants of them are given by

$$
\begin{array}{rcll}
\beta_1 &=& (1,0,0,1,0,1,-1,0) & (\in \mathfrak{C}_5), \\
\beta_2 &=& (1,0,0,1,0,1,1,0) & (\in \mathfrak{C}_5), \\
\beta_3 &=& (1,0,0,1,0,1,0,0) & (\in \mathfrak{C}_5), \\
\beta_4 &=& (0,0,0,0,0,0,0,1) & (\in \mathfrak{C}_5), \\
\beta_5 &=& (0,1,0,0,0,0,0,0) & (\in \mathfrak{C}_4), \\
\beta_6 &=& (1,0,0,1,0,0,0,0) & (\in \mathfrak{C}_3), \\
\beta_7 &=& (1,0,0,0,0,1,0,0) & (\in \mathfrak{C}_1).
\end{array}
$$

4.2. Orbit of β_i We express the set of structure constants (c_{ijk}) as a 4×2 matrix

$$
\begin{pmatrix}
c_{111} & c_{112} \\
c_{121} & c_{122} \\
c_{211} & c_{212} \\
c_{221} & c_{222}
\end{pmatrix}.
$$

The isomorphic multiplication β_i' to β_i is expressed by

$$
\beta_i' = (A \otimes A)\beta_i A^{-1}
$$

for some 2×2 real non-singular matrix A, here $A \otimes A$ is kronecker product of A and A. For example, the explicit form of $(A \otimes A)\beta_1 A^{-1}$ is given by

$$
\frac{1}{ut - vs}
\begin{pmatrix}
u^2 & uv & uv & v^2 \\
us & ut & vs & vt \\
us & vs & ut & vt \\
s^2 & ts & ts & t^2
\end{pmatrix}
\begin{pmatrix}
1 & 0 \\
0 & 1 \\
0 & 1 \\
-1 & 0
\end{pmatrix}
\begin{pmatrix}
t & -v \\
-s & u
\end{pmatrix}
$$

$$
(2) \quad =\quad \frac{1}{ut - vs}
\begin{pmatrix}
u^2 t - v^2 t - 2uvs & u^2 v + v^3 \\
-vt^2 - vs^2 & v^2 t + u^2 t \\
-vt^2 - vs^2 & v^2 t + u^2 t \\
-s^2 t - t^3 & -vs^2 + vt^2 + 2ust
\end{pmatrix}.
$$

For each β_i, we denote by

$$
\mathcal{O}(\beta_i) = \{(A \otimes A)\beta_i A^{-1} \mid A \in \mathrm{GL}(2, \mathbb{R})\},
$$

and call this the orbit of β_i as usual.

Theorem. *Let us denote by $\mathfrak{C}_i^* = \mathfrak{C}_i \backslash \{0\}$ for $i = 1, \ldots, 5$. Let $\beta_1, \ldots, \beta_7$ be the set of structure constants given in the beginning of this section, and $A = \begin{pmatrix} u & v \\ s & t \end{pmatrix}$ be a non sigular real matrix, and $\mathcal{O}(\beta_i)$ be the orbit of β_i $(i = 1, \ldots, 7)$ under the action $(A \otimes A)\beta_i A^{-1}$. Then $\mathfrak{C}_i^*$ is expressed* *in terms of the orbits as follows:*

$$\mathfrak{C}_1^* = \mathcal{O}(\beta_7),$$
$$\mathfrak{C}_3^* = \mathcal{O}(\beta_6),$$
$$\mathfrak{C}_4^* = \mathcal{O}(\beta_1)_{t=0} \cup \mathcal{O}(\beta_2)_{t=0} \cup \mathcal{O}(\beta_3)_{t=0} \cup \mathcal{O}(\beta_4)_{t=0} \cup \mathcal{O}(\beta_5)_{s=0},$$
$$\mathfrak{C}_5^* = \mathcal{O}(\beta_1)_{t\neq0} \cup \mathcal{O}(\beta_2)_{t\neq0} \cup \mathcal{O}(\beta_3)_{t\neq0} \cup \mathcal{O}(\beta_4)_{t\neq0} \cup \mathcal{O}(\beta_5)_{s\neq0},$$

and $\mathfrak{C}_2^* \subseteq \mathcal{O}(\beta_2) \cup \mathcal{O}(\beta_4)$. *Here,* $\mathcal{O}(\beta_i)_{t=0}$, $\mathcal{O}(\beta_i)_{s=0}$, $\mathcal{O}(\beta_i)_{t\neq0}$ *and* $\mathcal{O}(\beta_i)_{s\neq0}$ *are the sets of elements of* $\mathcal{O}(\beta_i)$ *given by* A *satisfying* $t = 0, s = 0, t \neq 0, s \neq 0$ *respectively.*

Proof. Let us show only that $\mathfrak{C}_4^* = \mathcal{O}(\beta_1)_{t=0} \cup \mathcal{O}(\beta_2)_{t=0} \cup \mathcal{O}(\beta_3)_{t=0} \cup \mathcal{O}(\beta_4)_{t=0} \cup \mathcal{O}(\beta_5)_{s=0}$. A point of $\mathcal{O}(\beta_1), \ldots, \mathcal{O}(\beta_4)$ of $t = 0$ is of the form

$$\begin{pmatrix} 2u & -\frac{u^2+v^2}{s} \\ s & 0 \\ s & 0 \\ 0 & s \end{pmatrix} \in \mathcal{O}(\beta_1)_{t=0}, \qquad \begin{pmatrix} 2u & -\frac{u^2-v^2}{s} \\ s & 0 \\ s & 0 \\ 0 & s \end{pmatrix} \in \mathcal{O}(\beta_2)_{t=0},$$

$$\begin{pmatrix} 2u & -\frac{u^2}{s} \\ s & 0 \\ s & 0 \\ 0 & s \end{pmatrix} \in \mathcal{O}(\beta_3)_{t=0}, \qquad \begin{pmatrix} v & -\frac{u}{s} \\ 0 & 0 \\ 0 & 0 \\ 0 & 0 \end{pmatrix} \in \mathcal{O}(\beta_4)_{t=0},$$

and a point of $\mathcal{O}(\beta_5)_{s=0}$ is of the form $\begin{pmatrix} 0 & \frac{u^2}{t} \\ 0 & 0 \\ 0 & 0 \\ 0 & 0 \end{pmatrix}$. For example, for

$\mathcal{O}(\beta_1)_{t=0}$, we have the above form by substituting $t = 0$ into the matrix

of (2). By the form $\begin{pmatrix} \alpha & \beta \\ \gamma & 0 \\ \gamma & 0 \\ 0 & \gamma \end{pmatrix}$ of an element of $\mathfrak{C}_4^*$, we easily see that

$\mathcal{O}(\beta_1)_{t=0} \cup \mathcal{O}(\beta_2)_{t=0} \cup \mathcal{O}(\beta_3)_{t=0} \cup \mathcal{O}(\beta_4)_{t=0} \cup \mathcal{O}(\beta_5)_{s=0} \subseteq \mathfrak{C}_4^*$.

On the other hand, we can find an A and β_i such that

$$(A \otimes A)\beta_i A^{-1} = \begin{pmatrix} \alpha & \beta \\ \gamma & 0 \\ \gamma & 0 \\ 0 & \gamma \end{pmatrix}.$$

For the case $\gamma \neq 0$, $\alpha^2 + 4\beta\gamma < 0$, A is of $u = \alpha/2$, $v^2 = -(\alpha^2 + 4\beta\gamma)/4$, $s = \gamma$, $t = 0$ and $\beta_i = \beta_1$.

For the case $\gamma \neq 0$, $\alpha^2 + 4\beta\gamma > 0$, A is of $u = \alpha/2$, $v^2 = (\alpha^2 + 4\beta\gamma)/4$, $s = \gamma$, $t = 0$ and $\beta_i = \beta_2$.

For the case $\gamma \neq 0$, $\alpha^2 + 4\beta\gamma = 0$, A is of $u = \alpha/2$, $s = \gamma$, $v \neq 0$, $t = 0$ and $\beta_i = \beta_3$.

For the case $\gamma = 0$, $\alpha \neq 0$, A is of $v = \alpha$, $u = -\beta s$, $s \neq 0$, $t = 0$ and $\beta_i = \beta_4$.

For the case $\gamma = 0$, $\alpha = 0$, $\beta \neq 0$, A is of $t = u^2/\beta$, $u \neq 0$, $v \in \mathbb{R}$, $s = 0$ and $\beta_i = \beta_5$. $\square$

5. Closest Lie algebras structure

5.1. The case of Lie algebras   If A is a Lie algebra, the set of structure constants (c_{ijk}) has the following relations among c_{ijk}.

$$(3) \qquad c_{ijk} = -c_{jik} \quad (i, j, k = 1, \ldots, n),$$

$$(4) \qquad \sum_{p=1}^{n} (c_{ijp} c_{pkq} + c_{jkp} c_{piq} + c_{kip} c_{pjq}) = 0 \quad (i, j, k, q = 1, \ldots, n).$$

We denote by $\mathfrak{L}$ the algebraic set defined by (3), (4). As in the case of the associative algebras, we want to parameterize $\mathfrak{L}$ in the case of $n = 3$. However, the Groebner basis does not work in this case. So, we use the Jacobson's method of determination of the low dimensional Lie algebras ([4]). Then we have the parameterization of $\mathfrak{L}$ as in the following theorem.

Let us illustrate this method. Let L be a Lie algebra over $\mathbb{R}$ with a basis $\{e_1, e_2, e_3\}$ and (c_{ijk}) be the set of structure constants of L. We only consider the simplest case, that is of $\dim[L, L] = 3$. The Jacobi identity gives

$$[[e_1, e_2], e_3] + [[e_2, e_3], e_1] + [[e_3, e_1], e_2]$$
$$= (c_{311} - c_{232})[e_1, e_2] + (c_{122} - c_{313})[e_2, e_3] + (c_{233} - c_{122})[e_3, e_1] = 0.$$

Since $[e_1, e_2], [e_2, e_3]$ and $[e_3, e_1]$ are linearly independent, we see $c_{311} = c_{232}$, $c_{122} = c_{313}$, $c_{233} = c_{121}$. These equations lead to the algebraic set $\mathfrak{L}_7$ given in the following theorem.

5.2. Expression of $\mathfrak{L}$ The set of structure constants (c_{ijk}) is of the form

$$(c_{121}, c_{122}, c_{123}, c_{231}, c_{232}, c_{233}, c_{311}, c_{312}, c_{313},$$
$$c_{111}, c_{112}, c_{113}, c_{221}, c_{222}, c_{223}, c_{331}, c_{332}, c_{333},$$
$$c_{211}, c_{212}, c_{213}, c_{321}, c_{322}, c_{323}, c_{131}, c_{132}, c_{133}).$$

The first nine entries determine all the entries of (c_{ijk}), we write

$$(c_{ijk}) = (c_{121}, c_{122}, \ldots, c_{313}, -, -),$$

for short, here the first "$-$" $= (0, \ldots, 0)$, the second "$-$" $= (-c_{121}, \ldots, -c_{313})$.

Theorem ([6]). *Let $\mathfrak{L}$ be the algebraic set defined by (3), (4) in $\mathbb{R}^{27}$. Then the elements (c_{ijk}) of $\mathfrak{L}$ are expressed by*

$$\mathfrak{L} = \mathfrak{L}_1 \cup \mathfrak{L}_2 \cup \mathfrak{L}_3 \cup \mathfrak{L}_4 \cup \mathfrak{L}_5 \cup \mathfrak{L}_6 \cup \mathfrak{L}_7;$$

$$\mathfrak{L}_1 = \{(\alpha, \beta, \gamma, p\alpha, p\beta, p\gamma, q\alpha, q\beta, q\gamma, -, -) \,|\, \alpha, \beta, \gamma, p, q \in \mathbb{R}\},$$

$$\mathfrak{L}_2 = \{(0, 0, 0, \alpha, \beta, \gamma, p\alpha, p\beta, p\gamma, -, -) \,|\, \alpha, \beta, \gamma, p \in \mathbb{R}\},$$

$$\mathfrak{L}_3 = \{(0, 0, 0, 0, 0, 0, \alpha, \beta, \gamma, -, -) \,|\, \alpha, \beta, \gamma \in \mathbb{R}\},$$

$$\mathfrak{L}_4 = \{(\alpha, q\alpha + p\beta, \beta, \gamma, q\gamma + p\delta, \delta, p\alpha + q\gamma,$$
$$p^2\beta + pq(\alpha + \delta) + q^2\gamma, p\beta + q\delta, -, -) \,|\, \alpha, \beta, \gamma, \delta, p, q \in \mathbb{R}\},$$

$$\mathfrak{L}_5 = \{(p\alpha, \beta, \alpha, p^2\alpha, p\beta, p\alpha, p\gamma, \delta, \gamma, -, -) \,|\, \alpha, \beta, \gamma, \delta, p \in \mathbb{R}\}$$

$$\mathfrak{L}_6 = \{(0, 0, 0, \alpha, \beta, 0, \gamma, \delta, 0, -, -) \,|\, \alpha, \beta, \gamma, \delta \in \mathbb{R}\},$$

$$\mathfrak{L}_7 = \{(\alpha, \beta, \gamma, \delta, \epsilon, \alpha, \epsilon, \zeta, \beta, -, -) \,|\, \alpha, \beta, \gamma, \delta, \epsilon, \zeta \in \mathbb{R}\}.$$

5.3. Example of the closest Lie algebra structure Let us consider a three dimensional algebra with the multiplication table

$\cdot$	e_1	e_2	e_3
e_1	0	$e_1 - e_3$	$-e_3$
e_2	$-e_1 + e_3$	0	$-e_1 + e_2$
e_3	e_3	$e_1 - e_2$	0

Although this multiplication satisfies the skew symmetry, it does not satisfy the Jacobi identity, in fact, $(e_1 \cdot e_2) \cdot e_3 + (e_2 \cdot e_3) \cdot e_1 + (e_3 \cdot e_1) \cdot e_2 = -e_2$. We find the Lie product "$[-, -]$" closest to this multiplication by the algorithm given in [6], that is,

$$[e_1, e_2] = 1.10653e_1 - 0.0525276e_2 - 1.02323e_3,$$

$$[e_2, e_3] = -1.02323e_1 + 0.741678e_2 - 0.152996e_3,$$

$$[e_3, e_1] = -0.0525276e_1 - 0.434549e_2 + 0.741678e_3.$$

We note that the closest Lie product to "$\cdot$" stays on $\mathfrak{L}_4$ and $D(c_4, a)^2 = 0.727296$.

References

[1] J. M. A. Bermúdez, J. Fresán, J. S. Hernández, *On the variety of two dimensional real associative algebras*, Int. J. Contemp. Math. Sciences, **26** (2007), 1293-1305.

[2] M. Gerstenhaber, *On the cohomology structure of an associative ring*, Ann. of Math., **78** (1963), 59-103.

[3] M. Gerstenhaber, S. D. Schack, *Algebraic cohomology and deformation theory*, In: M. Hazewinkel, M. Gerstenhaber, eds., Deformation Theory of Algebras and Structures and Applications, Kluwer Academic Publishers, Dordrecht, 1988, 11-264.

[4] N. Jacobson, *Lie Algebras*, Interscience, New York, 1962, 11-14.
[5] F. Kubo, F. Suenobu, *On the associative algebra structures closest to algebra structures*, Journal of Algebra and its Applications, **10** (2011), 365-376.
[6] F. Suenobu, F. Kubo, *On the Lie algebra structures closest to algebra structures*, JP Journal of Algebra, Number Theory and Applications, **17** (2010), 117-128.

DEPARTMENT OF APPLIED MATHEMATICS
GRADUATE SCHOOL OF ENGINEERING
HIROSHIMA UNIVERSITY
HIGASHI HIROSHIMA 739-8527, JAPAN
E-mail address: suenobu@amath.hiroshima-u.ac.jp

DEPARTMENT OF APPLIED MATHEMATICS
GRADUATE SCHOOL OF ENGINEERING
HIROSHIMA UNIVERSITY
HIGASHI HIROSHIMA 739-8527, JAPAN
E-mail address: remakubo@hiroshima-u.ac.jp

Proceedings of the Sixth China-Japan-Korea
International Conference on Ring Theory
June 27-July 2, 2011 Suwon, Korea

ON THE FAITH CONJECTURE

KIYOICHI OSHIRO

ABSTRACT. Is a semiprimary right self-injective ring a quasi-Frobenius
(QF) ring? Faith raised this problem and conjectured "no". In
this paper we study this problem and show "yes" for some important cases.

1. Background

In 1939, Nakayama intruduced a quasi-Frobenius ring (QF-ring)
as an artinian ring with the so called Nakayama permutation. Usually
we use "QF-ring" as a ring which satisfies the condition (3) below.

Theorem 1.1. *(Nakayama [12], Ikeda [8]). The following are equivalent.*

(1) R is QF.
(2) R is right or left artinian and right or left self-injective.
(3) R is right and left artinian and right and left self-injective.

The following characterizations are well known.

Theorem 1.2. *(Osofsky [14] (1966), Kato [9] (1967)). The following
are equivalent:*
(1) R is QF
(2) R is right perfect and right and left self-injective

This theorem implies the following question.

Problem A. *Is a perfect right self-injective ring a QF-ring?*

In 1976, in his book "Algebra II", Faith picked up this problem and
conjectured "no" even for a semiprimary ring, that is, he raised

Problem B. *Is a semiprimary right self-injective ring a QF-ring?*

2010 Mathematics Subject Classification : 16S99, 16U80.
Keywords : Semiprimary right self-injective ring, Faith conjecture, division algebra D, (D, D)-space, Erdos-Kaplansky's theorem, quasi-Frobenius ring, Nakayama
permutation.

and claimed "no" for this problem. Since then, almost half century spent (see [7] and [13]). Though, negrettably, there are no decisive results on this problem, following results give good imformations.

Theorem 1.3. *(Clark-Huynh [4]). Let R be a right perfect right self-injective ring. If the second socle $S_2(R_R)_R$ of R is finitely generated, then R is QF.*

Remark. (1) For a semiprimary ring, this theorem is also shown in Ara and Park [1].

(2) It is known as Osofsky's lemma that if R is a left perfect ring such that J/J_R^2 is finitely generated, then R is artinian, where J is the Jacobson radical of R. Therefore, if R is a semiprimary right self-injective ring such that J/J_R^2 is finitely generated, then R is QF. In addition, for later use, we note that if R is a semiprimary ring such that J/J_R^2 is α-generated, then $S_2(R_R)_R$ is α-generated.

Theorem 1.4. *(Baba-Oshiro [2]). For a semiprimary ring R, the following are equivalent:*
(1) R is right self-injective.
(2) R is right simple injective.

This theorem urges the study of the Faith conjecture for a "local" semiprimary ring R with Jacobson radical cubu $J^3 = 0$. The Faith conjecture for this special case is translated with the following problem (see, Chapter 10 of [3]):

Problem C. *Let D be a division ring and let $_DV_D$ be (D, D)-space with $\alpha = dim V_D \geq \aleph_0$ and let $pD = Dp$ be a left and right one dimensional (D, D)-space. Put $V^* = Hom_D(V_D, pD_D)$. Then V^* is (D, D)-space. Put $\beta = dim_D V^*$. Then by Ërdos-Kaplansky's theorem, $\beta = dim_D V^* =| D |^\alpha$, and hence we see that $\beta = dim_D V^* = | D |$. ($| D |$ is cardinal of D, and $dim_D V^*$ denotes the dimension of $_DV^*$.). Then does there exist a (D, D)-isomorphism*

$$_DV_D \cong _DV_D^* ?$$

This problem is not only a problem concerning the Faith conjecture but also an important one for fundamental mathematics.

Let us explain this situation. Let R be a local semiplimary ring right self-injective ring with $J^3 = 0$. Put $D = R/J$ and $V = J/J^2$. Then D is a division ring and the left socle and right socle of R coincide and they are simple modules, so it is a one dimensional left and right D-space in the sense that it is a form: $pD = Dp$. Put $V^* = Hom_D(V_D, pD_D)$. Then $_DV_D \cong _DV_D^*$ and it is shown in Chapter 10 of [2] that R is QF if and only if $dimV_D$ is finite.

Therefore, by this fact, Problem C is translated as a problem of (D, D)-spaces.

After we first study this problem, and show following theorem by dividing it into several cases.

Main Theorem. *Let R be a semiprimary right self injective ring. For each primitive idempotent e, $D(e)$ denotes the division ring eRe/eJe (J: Jacobson radical). By $dimD(e)$, we denote the dimension of the division algebra $D(e)$ over its center, and by $\mid D(e) \mid$ we denote the cardinal of $D(e)$. If $dimD(e) < \mid D(e) \mid$ for all primitive idempotent e, then R is QF.*

Now, throughout this paper we assume that R is a basic semiprimary right self injective ring. We use following symbols:

(1) $J = J(R)$: The Jacobson radical of R.

(2) $S(R_R)$: The socle of R_R.

(3) $S_2(R_2)_2$: The second socle of R_R.

(4) It is known that $S(R_R) = S(_RR)$ and $S_2(R_R) = S_2(_RR)$. So we denote these by $S(R)$ and $S_2(R)$, respectively.

(5) In the case when R is a local ring, we put $D = J/J^2$. Then D is a division ring and $S(R)$ is a one dimensional left and right D-space. We put it by $S(R) = pD = Dp$.

Following facts are known (see [2], [10], [13]):

Fact 1. $_RHom_R(J/J_R^2, S(R)_R)_R \cong _RS_2(R)/S(R)_R$,

Fact 2. When R is a local ring, $_DHom_D(J/J_D^2, pD_D)_D \cong _DS_2(R)/S(R)_D$,

Fact 3. $_RHom_R(S_2(R)/S(R)_R, S(R))_R \cong _RJ/l(S_2)_R$, where $l(S_2(R))$ is the left annihilator of $S_2(R)$.

Fact 4. When R is a local ring, $_DHom_R(S_2(R)/S(R)_D, pD_D)_D \cong _DJ/l(S_2)_D$.

2. Local Semiprimary Right Self-Injective Rings with $J^3 = 0$

We return to Problem C:
$$_DV_D \cong {}_DV_D^*?$$
First, we consider this problem for a special case that D is a commutative field and V is generated by one element as a (D,D)-space: $V = DxD$ such that
$$_DDxD_D \cong {}_D(DxD)_D^* , \alpha = dimDxD_D \geq \aleph_0$$
Put $(DxD)^* = D\varphi D$ and $\beta = dim_D(DxD)^*$. Then, by Erdos-Kaplansky's theorem, we see that
$$| D |= \beta > \alpha = dimV_D.$$
Put
$$(DxD)^* = D\varphi D = \sum_\Lambda \oplus D\varphi s_i \ (| \Lambda |= \beta)$$
$$DxD = \sum_\Gamma \oplus u_j xD \ (| \Gamma |= \alpha)$$
$$= xD \oplus u_2 xD \oplus u_3 xD \oplus \cdots \oplus u_\omega xD \oplus \cdots \ (u_1 = 1).$$
We claim that $\{s_i x\}_\Lambda$ is independent, which contradicts $\beta > \alpha = dimV_D \geq \aleph_0$. In fact, let $s_1, ..., s_n \in \{s_i\}_\Lambda$ and let $b_1, ..., b_n \in D$ and assume that $\sum_i^n s_i xb_i = 0$.

For any $u \in \{u_i\}$,
$$\begin{aligned} 0 &= \varphi(\sum_i^n us_i xb_i) \\ &= \sum_i^n \varphi(s_i ux)b_i \\ &= \sum_i^n pw_i b_i \qquad (\varphi(s_i ux) := pw_i) \\ &= \sum_i^n pb_i w_i \\ &= \sum_i^n b_i' pw_i \qquad (pb_i = b_i'p) \\ &= \sum_i^n b_i' \varphi(s_i ux). \end{aligned}$$

Consequently, $0 = \sum_i^n b_i' \varphi(s_i ux)$ for all $u \in \{u_i\}_\Gamma$. Therefore

$$\sum_i^n b_i' \varphi s_i = 0.$$

Since $\{D\varphi s_1, ..., D\varphi s_n\}$ is independent, this implies that each $b_i' = 0$ and hence each $b_i = 0$ as desired. Thus the following case does not occur.
$$_DDxD_D \cong {}_D(DxD)_D^*, \alpha = dimDxD_D \geq \aleph_0.$$
By this observation, we obtain

Theorem 2.1. *Let R be a local semiprimary right self-injective ring with $J^3 = 0$. If $D = R/J$ is a commutative field and the (D,D)-space J/J^2 is the form DxD, then $dimDxD_D < \aleph_0$ and hence R is QF.*

Next we observe Problem C without assuming $V = DxD$. In this case we show the following crucial lemma:

Lemma A. If $dimV_D \geq \aleph_0$, then there exists $0 \neq x \in V$ such that $dimDxD_D \geq \aleph_0$

Proof. Assume that $dimDxD_D < \aleph_0$ for any $x \in V$. For $x \in V$, put $_D(DxD)^\star{}_D = Hom_D(DxD_D, pD_D)$. Then $_D(DxD)^\star$ becomes a cofinite dimensional left D-subspace of $V^\star$. Here we use "cofinite dimensional" as $dim_D(DxD)^\star = {}_D(DxD)^\star + {}_DW$ for some finite dimensional left D-subspace W.

Put $V = x_1D \oplus x_2D \oplus \cdots \oplus x_nD \oplus \cdots$ where $\mid \{x_i\} \mid = \alpha$. For a subset Y of V, we put $l(Y) = \{\phi \in V^\star \mid \phi(Y) = 0\}$. Noting $_DV_D^\star \cong {}_DV_D$ and $\beta > \alpha$ we can take $\varphi \in V^\star$ such that $dim_D D\varphi D = \beta$.

Then $A_1 \cap D\varphi D \supseteq A_2 \cap D\varphi D \supseteq A_3 \cap D\varphi D \supseteq \cdots$. Further we see that $dim_D(A_n \cap D\varphi D)$ is infinite, which implies $A_n \cap D\varphi D \neq 0$ for all n. Since $dim(A_n \cap D\varphi D_D)$ is finite as a right D-space, there must exists k such that $dim(A_k \cap D\varphi D_D) = dim(A_{k+1} \cap D\varphi D_D) = \cdots$, and hence $A_k \cap D\varphi D = A_{k+1} \cap D\varphi D = \cdots$. This show that $(A_k \cap D\varphi D)(V) = 0$ and hence $A_k \cap D\varphi D = 0$, contradiction. Consequently we can take some $x \in \{x_i\}$ such that $dimDxD_D \geq \aleph_0$.

Using Lemma A, we extend Theorem 2.1 as follows:

Theorem 2.2. *Let R be a local semiprimary right self-injective ring with $J^3 = 0$. If $D = R/J$ is a commutative field, then R is QF.*

Proof. Put $V = J/J^2$. We want to show that $dimV$ is finite. Assume that $dimV_D = \alpha \geq \aleph_0$. Then $_DV_D \cong {}_DV_D^\star$ and $\beta = \mid D \mid > \alpha$. By Lemma A, there exists $x \in V$ such that $dimDxD_D \geq \aleph_0$. Put $V = DxD \oplus W$ for some subspace W_D. Put

$$(DxD)^\star = Hom_D(DxD_D, pD_D).$$

Then, any $\phi \in (DxD)^\star$ can be extended to $\phi^\star$ in $V^\star$ by $\phi^\star(W) = 0$, and hence $(DxD)^\star$ can be embedded in $V^\star$. Therefore noting that $dim(DxD)_D^\star \leq \alpha$ together with $\beta = dim_D(DxD)^\star$, we can take $\varphi \in (DxD)^\star$ and $\{s_i\}_\Lambda \subseteq D$ such that $\beta = \mid \Lambda \mid$ and $\{D\varphi s_i\}_\Lambda$ is independent. Then by same argument as in the proof of Theorem 2.1, we can show that $\{s_ixD\}_\Lambda$ is independent, which contradicts $\beta > dimV_D$. Thus $dimV_D$ is finite, as claimed. $\qquad\square$

Next we observe Problem C in the case when D is a two dimensional division algebra over its center F, say $D = d_1F \oplus d_2F$. Then $\mid F \mid = \mid D \mid = \beta > \alpha$.

By the argument above, we can take $x \in V$ and $\varphi \in (DxD)^*$ such that $dim_D D\varphi D = \beta$. Hence $dim_D D\varphi d_1 F = \beta$ or $_D D\varphi d_2 F = \beta$. We may assume that $dim_D D\varphi d_1 F = \beta$. Then we can take $\{s_i\}_\Lambda \subseteq D$ such that $\{D\varphi d_1 s_i\}_\Lambda$ is independent.

Put $\phi = \varphi d_1$. Then $\{F\phi s_i\}_\Lambda$ is also independent. Noting this fact, by the same argument above, we can show that $\{s_i x F\}_\Lambda$ is independent (as F-spaces). Then this implies that $\{s_i x d_1 F\}_\Lambda$ and $\{s_i x d_2 F\}_\Lambda$ are also independent. Since $s_i x D = s_i x d_1 F \oplus s_i x d_2 F$, $s_i x D$ is a two dimensional F-space for any $i \in \Lambda$. On the other hand, $\sum_\Lambda \oplus s_i x d_1 F$ and $\sum_\Lambda \oplus s_i x d_2 F$ are β dimensional. Noting these facts, we can take a subset Ω of Λ such that $|\Omega| > \alpha$ and $\{s_i x D\}$ is independent, which is a contradiction since $dim V_D = \alpha$. Thus $dim V_D$ must be finite.

Further our argument works for $dim D_F < \beta$, so if $dim D_F < \beta$ and

$$_D V_D \cong {}_D V_D^*$$

then $dim V_D$ must be finite.

Thus we obtain the following theorem.

Theorem 2.3. *Let R be a local semiprimary right self-injective ring with $J^3 = 0$. Put $D = R/J$ and let F be its center. If $dim D_F <| D |$, then R is QF.*

3. Local Semiprimary Right Self-Injective Rings with $J^{n-1} \neq 0$ and $J^n = 0$

Again let D be a division ring and let $_D V_D$ be a (D, D)-space with $\alpha = dim V_D \geq \aleph_0$ and let $pD = Dp$ be a left and right one dimensional (D, D)-space. Put $V^* = Hom_D(V_D, pD_D)$ and $\beta = dim_D V^*$. Then $\beta = dim_D V^* =| D |^\alpha$ and $\beta = dim_D V^* =| D |$. Then in Problem C we raised the question: Does there exist a (D, D)-isomorphism

$$_D V_D \cong {}_D V_D^*?$$

If this holds, then $dim V_D^* = \alpha < \beta$, and we often used this fact.

So, from this view point, we need to study a new problem:

$$dim V_D^* < \beta?$$

In view of the proof of Lemma A, we can improve the lemma as follows.

Lemma B. *If $dim DxD_D < \aleph_0$ any $x \in V$, then there exists $\varphi \in V^*$ such that $dim D\varphi D_D \geq \aleph_0$.*

Now let R is a local semiprimary right self-injective ring with $J^{n-1} \neq 0$ and $J^n = 0$. Put $V = J/J^2$ and $W = S_2(R)/S(R)$. We note that $dim V_D \geq dim W_D$. Put $S(R) = pD = Dp$, $V^\star = Hom_D(V_D, pD_D)$ and $W^\star = Hom_D(W_D, pD_D)$. Further put $\beta = dim V_D$ and $\gamma = dim W_D$.

In order to show that $dim V_D$ or $dim W_D$ is finite, we assume that

$$dim V_D = \beta \geq \aleph_0 \text{ and } dim W_D = \gamma \geq \aleph_0.$$

Then, as mentioned above, following facts hold.
(1) $_D V_D^\star \cong {}_D W_D$,
(2) $_D W_D^\star \cong {}_D J/l(S_2(R))_D$,
(3) So, there is a canonical epimorphism: $_D V_D \twoheadrightarrow {}_D W_D^\star$.

Here using Lemma B, we can implies following cases:
(a) There exists $x \in V$ such that $dim DxD_D \geq \aleph_0$
or
(b) There exists $y \in W$ such that $dim {}_D DyD_D \geq \aleph_0$.

First we observe the case (a). Put $(DxD)^\star = Hom_D(DxD_D, pD_D)$. Then $dim_D(DxD)^\star = \beta$ and $dim(DxD)_D^\star < \beta$. Noting these we can take $\{s_i\}_\Lambda \subseteq D$ such that $|\Lambda| = \beta$ and $\{_D Dxs_i\}_\Lambda$ is independent. Here if we assume that $dim_D F <| D |$, then we can show a contradiction by same arguments as in Section 2.

Next in the case (b), we put $(DyD)^\star = Hom_D(DyD_D, pD_D)$. Then $dim_D(DyD)^\star = \beta$. Since there exists an epimorphism $_D V_D \twoheadrightarrow {}_D W_D^\star$, we also see that $dim(DyD)_D^\star < \beta$. Noting this, by the same arguments, we can show a contradiction. Accordingly dim V_D or dim W_D must be finite.

Thus we obtain the following theorem

Theorem 3.1. *Let R be a local semiprimary right self-injective ring. If $dim_D F < | D | = dim_D(J/J^2)$, then R is QF.*

4. Semiprimary Right Self-Injective Rings with $J^{n-1} \neq 0$ and $J^n = 0$

In this section, R is a basic semiprimary right self-injective ring, and J denotes the Jacobson radical $J(R)$.

Let $Pi(R)$ be a complete set of orthogonal primitive idempotents of R. $Pi(R)$ has a Nakayama permutation, that is, for any $e_i \in Pi(R)$, there exists a unique $f_i \in Pi(R)$ such that $S(e_i R_R)_R \cong f_i R/f_i J$, and $_R S(_R R f_i) \cong Re_i/Je_i$. Then we say that $(e_i R, R f_i)$ is an i-pair. $\{f_1, \cdots, f_n\}$ is a permutation of $\{e_1, \cdots, e_n\}$ and it is called a Nakayama permutation. We note that (eR, Rf) is an i-pair if and only if $eSf \neq 0$. For any $e \in Pi(R)$, we put $D(e) = eRe/eJe$ (division ring).

We use following results as known ones

Lemma C.

(1) For an i-pair (eR, Rf),
$$S(eR_R) = S(_R Rf) = {}_{eRe}S(eRf) = S(eRf)_{fRf}.$$
Put it by $S(eRf)$. Then
$$dim_{D(e)}S(eRf) = 1 = dimS(eRf)_{D(f)}.$$

(2) $S(R_R) = S(_R R)$. Put it by $S(R)$. Then
$$S(R) = S(e_1 R_R) \oplus \cdots \oplus e_n R_R = S(_R Re_1) \oplus \cdots \oplus S(_R Re_n)$$
and each $S(e_i R_R)_R$ and each $_R S(_R Re_i)$ are simple.

In order to observe $Hom_R(J/J_R^2, S(R)_R)$, we express J/J_R^2 as
$$J/J^2 = X_1 \oplus \cdots \oplus X_n$$
where
$$X_i = \sum_{\lambda \in \Lambda_i} \oplus T_\lambda, \quad T_{\lambda D(f_i)} \cong S(e_i R f_i)_{D(f_i)}$$
for $i = 1, ..., n$. Here we put $\alpha = Max\{dim X_{i D(f_i)} \mid i = 1, \cdots, n\}$. Then $\alpha = dim X_k$ for some X_k. Now, assume that
$$\alpha \geq \aleph_0.$$
Put $e = e_k$, $f = f_k$ and $X = X_k$. We can take $h \in Pi(R)$ such that dim $hX_{D(f)} = \alpha$.

Put
$$V = hX, \quad V^\star = Hom_{D(f)}(_{D(h)}V_{D(f)}, gS(R)f_{D(f)})$$
where (gR, Rf) is an i-pair. Then V is a $(D(h), D(f))$-space and $V^\star$ is a $(D(g), D(h))$-space, and note that
$$_{D(g)}V^\star_{D(h)} \cong \overline{gS_2(R)h} = gS_2(R)h/gS(R)h.$$

Put $\beta = dim_{D(g)}V^\star$. Then $\beta > \alpha$ and dim $\overline{gS_2(R)h}_{D(h)} \leq \alpha$. Under this situation, we can take $x \in V$ or $y \in \overline{gS_2(R)h}$ such that
$$dim\ D(h)xD(f)_{D(f)} \geq \aleph_0 \text{ or } dim\ D(g)yD(h)_{D(h)} \geq \aleph_0.$$
We denote by $F(f)$ the center of $D(f)$. Here if $dimD(f)_{F(f)} < \mid D(f) \mid$, by using same arguments as in Sections 2 and 3, we can show a contradiction. Namely, if $dimD(f)_{F(f)} < \mid D(f) \mid$, then it must hold that $\alpha < \aleph_0$.

Thus we can obtain our main theorem.

Theorem 4.1. *Let R be a semiprimary right self injective ring. If $dimD(e)_{F(e)} < \mid D(e) \mid$ for each $e \in Pi(R)$, then R is QF.*

Remark. (1) This theorem shows that if each $D(e)$ is a finite dimensional division algebra or a countable dimensional division algebra, then R is QF.

(2) P.M.Cohn's example: For any infinite cardinals $\beta > \alpha \geq \aleph_0$, there exist division rings $K \subset L$ such that $dim_K L = \beta > dim L_K = \alpha$, and the dimension of K over its center is $\beta =| L |$ (the center of K is a countable field). In this example, when we consider V and D as L and K, respectively, can we take $x \in L$ such that $dim D x D_D \geq \aleph_0$? If we can not take such x, Cohn's example seems not work for making a counter example for Faith's problem.

(3) We again return to Problem C:

$$_D V_D \cong {}_D V_D^* \ ?$$

If the center F of D is $\{1.0\}$, then $\dim D_F =| D |$. Therefore to observe Problem C for this case seems to be meaningful.

References

[1] P.Ara and J.K Park, *On continuos semiprimar rings*, Comm. Algebra **19**(1991), 1945-1957.

[2] Y. Baba and K. Oshiro, *On a Theorem of Fuller*, J. Algebra **154** (1993), 86-94.

[3] Y. Baba and K. Oshiro, "Classical Artinian Rings and Related Topics", World Scientific Publishing Co. Pte. Ltd., 2009.

[4] J. Clark and D. V. Huynh, *A note on perfect self-injective rings*, Quart. J. Math. (Oxford) **45** (1994), 13-17.

[5] P. M. Cohn, "Skew Field Constructions", London Mathematical Society, Lecture note Series **27**, 1977.

[6] C. Faith, "Algebra II. Ring Theory", Grundlehren Math. Wiss. **192**, Springer-Verlag, Berlin/Heidelberg/New York, 1976.

[7] C. Faith and D. V. Huynh, *When self-injective rings are QF: A report on a problem*, J. Algebra Appl.1, (2002), 72-105.

[8] M. Ikeda, *A characterization of quasi-Frobenius rings*, Osaka J. Math.4(1952), 203-209.

[9] T. Kato, *Self-injective rings*, Tohoku Math. J. 19 (1967), 485-494.

[10] K. Koike, *On self-injective semiprimary rings*, Comm. Algebra **28** (2000), 4303-4319.

[11] J. Lawrence, *A countable self-injective ring is quasi-Frobenius*, Proc. Amer.Math. Soc. 65(1977),217-220.

[12] T. Nakayama, *On Frobenius algebras II*, Ann. Math. **42** (1941), 1-21.

[13] W. K. Nicholson and M. F. Yousif, "Quasi-Frobenius Rings", Cambridge Tracts in Mathematics **158**, Cambridge University Press, Cambridge, 2003.

[14] B. Osofsky, *A generalization of quasi-Frobenius rings*, J. Algebra 4 (1966), 373-387.

DEPARTMENT OF MATHEMATICS
YAMAGUCHI UNIVERSITY
YAMAGUCHI, JAPAN
E-mail address: oshiro@yamaguchi-u.ac.jp

Proceedings of the Sixth China-Japan-Korea
International Conference on Ring Theory
June 27-July 2, 2011 Suwon, Korea

A SHORT PROOF THAT CONTINUOUS MODULES
ARE CLEAN

V.P. Camillo, D. Khurana, T.Y. Lam, W.K. Nicholson*, and Y. Zhou

ABSTRACT. A ring is called clean if every element is the sum of
an idempotent and a unit, and a module is called clean if its endomorphism ring is a clean ring. In this note we give a short proof
of the fact that every continuous or discrete module is clean. The
key idea is to show first that every nonsingular injective module
is clean by extending the argument for vector spaces.

Let R be an associative ring with unity. An element $a \in R$ is called
clean if it is the sum of an idempotent and a unit, and R is called a *clean
ring* if every element of R is clean. A right module M_R is called a *clean
module* if its endomorphism ring, $\text{End}(M)$, is a clean ring. It is an open
question which modules are clean.

In 2006, we characterized when an endomorphism $\alpha \in \text{End}(M)$ is
clean, and used it to show that every continuous module is clean [3]. The
goal of the present paper is to provide a different, shorter proof of this
result. This is accomplished by showing directly that every nonsingular
injective module is clean, using an analog of the proof for vector spaces
given by M.Ó Searcóid in [8]. With this we quickly derive the main
corollaries from [3]: Every continuous or discrete module is clean.

We write $J(R)$ for the Jacobson radical of the ring R. Modules are
right modules, and homomorphisms are written on the left. For a module
M, the injective hull of M is denoted by $\widehat{M}$. We write $N \leq M$ when
N is a submodule of M, and we write $N \leq_e M$ if N is essential in M.
Throughout, a ring R will be called regular if it is regular in the sense
of von Neumann.

The proof that every nonsingular injective module is clean requires
three preliminary lemmas.

Lemma 1. *Let M_R be a module. If $\alpha \in \text{End}(M)$ is an automorphism
then there exists an automorphism $\hat{\alpha} \in \text{End}(\widehat{M})$ such that $\hat{\alpha}|_M = \alpha$. If*

2010 Mathematics Subject Classification : 16S50, 16D50, 16U60.

Keywords : Clean endomorphism rings, injective modules, continuous and discrete
modules.

*Corresponding author.

in addition M is nonsingular, then such an $\hat{\alpha}$ is unique in the sense that if $\lambda \in \text{End}(\widehat{M})$ is such that $\lambda|_M = \alpha$ then $\lambda = \hat{\alpha}$.

Proof. Because $\widehat{M}$ is injective, there exists $\hat{\alpha} \in \text{End}(\widehat{M})$ such that $\hat{\alpha}|_M = \alpha$. This means that $\text{Ker}(\hat{\alpha}) \cap M = \text{Ker}(\alpha) = 0$, so $\text{Ker}(\hat{\alpha}) = 0$ (since $M \leq_e \widehat{M}$). Thus, $\hat{\alpha}\widehat{M} \cong \widehat{M}$ is injective, showing that $\hat{\alpha}\widehat{M}$ is a direct summand of $\widehat{M}$. But $\hat{\alpha}\widehat{M} \leq_e \widehat{M}$ because $M = \alpha M \subseteq \hat{\alpha}\widehat{M} \subseteq \widehat{M}$. It follows that $\widehat{M} = \hat{\alpha}\widehat{M}$ and hence that $\hat{\alpha} : \widehat{M} \to \widehat{M}$ is an automorphism. If M is nonsingular, let $\lambda \in \text{End}(\widehat{M})$ satisfy $\lambda|_M = \alpha$. Then, using [5, 7.6], $\widehat{M}/\text{Ker}(\hat{\alpha} - \lambda)$ is nonsingular [it embeds in $\widehat{M}$] and also singular [$M \subseteq \text{Ker}(\hat{\alpha} - \lambda)$ and $\text{Ker}(\hat{\alpha} - \lambda) \leq_e \widehat{M}$]. Hence $\text{Ker}(\hat{\alpha} - \lambda) = \widehat{M}$, so $\lambda = \hat{\alpha}$. $\square$

Lemma 2. *Let M_R be a nonsingular module. If $\alpha^2 = \alpha \in \text{End}(M)$ there exists $\hat{\alpha}^2 = \hat{\alpha} \in \text{End}(\widehat{M})$ such that $\hat{\alpha}|_M = \alpha$.*

Proof. There exists $\hat{\alpha} \in \text{End}(\widehat{M})$ such that $\hat{\alpha}|_M = \alpha$. This means that $(\hat{\alpha}^2 - \hat{\alpha})|_M = \alpha^2 - \alpha = 0$, so $M \subseteq \text{Ker}(\hat{\alpha}^2 - \hat{\alpha}) \subseteq \widehat{M}$. This shows that $\widehat{M}/\text{Ker}(\hat{\alpha}^2 - \hat{\alpha})$ is singular. But it is also nonsingular because $\widehat{M}/\text{Ker}(\hat{\alpha}^2 - \hat{\alpha}) \cong \text{Im}(\hat{\alpha}^2 - \hat{\alpha}) \subseteq \widehat{M}$. This implies that $\widehat{M} = \text{Ker}(\hat{\alpha}^2 - \hat{\alpha})$, so $\hat{\alpha}^2 = \hat{\alpha}$. $\square$

Lemma 3. *Let M_R be a module, and assume that $\widehat{M} \subseteq E$ where E_R nonsingular. If $\alpha \in \text{End}(E)$ and $\alpha M \subseteq M$, then $\alpha \widehat{M} \subseteq \widehat{M}$.*

Proof. As $\widehat{M}$ is injective, there exists $\hat{\alpha} \in \text{End}(\widehat{M})$ such that $\hat{\alpha}|_M = \alpha$. Then $\alpha - \hat{\alpha} : \widehat{M} \to E$ satisfies $M \subseteq \text{Ker}(\alpha - \hat{\alpha})$, so $\widehat{M}/\text{Ker}(\alpha - \hat{\alpha})$ is singular. It is also nonsingular (as $\widehat{M}/\text{Ker}(\alpha - \hat{\alpha}) \cong \text{Im}(\alpha - \hat{\alpha}) \subseteq E$), so it follows that $\widehat{M} = \text{Ker}(\alpha - \hat{\alpha})$. Hence $\alpha = \hat{\alpha}$, and so $\alpha \widehat{M} = \hat{\alpha}\widehat{M} \subseteq \widehat{M}$. $\square$

To prove that every injective module is clean, the following special case is crucial.

Lemma 4. *Every nonsingular injective module M_R is clean.*

Proof. To show that any $\alpha \in \text{End}(M)$ is clean, define

$$\mathcal{S} = \{(K, \beta) \mid K \leq M, \, \alpha K \subseteq K, \, \beta^2 = \beta \in \text{End}(K), \text{ and } \alpha|_K - \beta \text{ is a unit of } \text{End}(K)\}.$$

Define a partial ordering on $\mathcal{S}$ by

$$(K', \beta') \leq (K, \beta) \qquad \text{if and only if} \qquad K' \subseteq K \text{ and } \beta' = \beta|_{K'}.$$

If $\{(K_i, \beta_i) : i \in I\}$ is a chain in $\mathcal{S}$, let $K = \bigcup\{K_i : i \in I\}$ and define $\beta \in \text{End}(K)$ by setting $\beta x = \beta_i x$ for any $x \in K_i$. One verifies that

$(K, \beta) \in \mathcal{S}$ is a upper bound of the chain. Thus $\mathcal{S}$ is an inductive set so, by Zorn's Lemma, let (W, η) be a maximal element of $\mathcal{S}$. We prove Lemma 4 by showing that $W = M$.

Claim 1. W is injective.

Proof. Since M is injective, W has an injective hull $\widehat{W}$ contained in M. By Lemma 2, there exists $\hat{\eta}^2 = \hat{\eta} \in \text{End}(\widehat{W})$ such that $\hat{\eta}|_W = \eta$. By Lemma 3, $\alpha\widehat{W} \subseteq \widehat{W}$. Moreover, since $(\alpha - \eta)|_W = \alpha|_W - \eta$ is a unit of $\text{End}(W)$, $\alpha|_{\widehat{W}} - \hat{\eta} \in \text{End}(\widehat{W})$ is a unit by Lemma 1. Thus $(\widehat{W}, \hat{\eta}) \in \mathcal{S}$, so $W = \widehat{W}$ by the maximality of (W, η). Hence W is injective, proving Claim 1.

It follows from Claim 1 that $M = W \oplus N$ for some submodule N of M; we prove Lemma 4 by showing that $N = 0$. Let $\pi_W : M \to W$ and $\pi_N : M \to N$ be the projections of M onto W and N respectively, and let $\phi = \pi_N \alpha|_N \in \text{End}(N)$.

Claim 2. Suppose $X \leq N$ satisfies $\phi X \subseteq X$ and $\phi - \varepsilon$ is a unit of $\text{End}(X)$ for some $\varepsilon^2 = \varepsilon \in \text{End}(X)$. Then $X = 0$.

Proof. Given X, define $\theta : W \oplus X \to W \oplus X$ by $\theta(w + x) = \eta w + \varepsilon x$ for $w \in W$, $x \in X$. We assert that the following hold [this proves Claim 2 because then $(W \oplus X, \theta) \in \mathcal{S}$].

(i) $\theta^2 = \theta \in \text{End}(W \oplus X)$, and $\theta|_W = \eta$.

(ii) $W \oplus X$ is α-invariant.

(iii) $\alpha|_{W \oplus X} - \theta$ is a unit in $\text{End}(W \oplus X)$.

The proof of condition (i) is routine; and (ii) holds because, given $x \in X$,

$$\alpha x = \pi_W \alpha x + \pi_N \alpha x = \pi_W \alpha x + \phi x \in W \oplus X. \tag{*}$$

As to (iii), let $w \in W$ and $x \in X$. Then (*) and the fact that $\theta x = \varepsilon x$ give

$$\begin{aligned}
(\alpha - \theta)(w + x) &= (\alpha - \theta)w + (\pi_W \alpha x + \phi x) - \varepsilon x \\
&= [(\alpha - \theta)w + \pi_W \alpha x] + (\phi - \varepsilon)x \in W \oplus N.
\end{aligned} \tag{**}$$

Thus if $(\alpha - \theta)(w + x) = 0$, then $(\alpha - \theta)w + \pi_W \alpha x = 0$ and $(\phi - \varepsilon)x = 0$. Since $\phi - \varepsilon$ is a unit of $\text{End}(X)$, it follows that $x = 0$ and hence that $0 = (\alpha - \theta)w = (\alpha - \eta)w$. But $\alpha - \eta$ is a unit of $\text{End}(W)$, so $w = 0$. This proves that $\alpha - \theta : W \oplus X \to W \oplus X$ is monic.

To see that it is epic, observe first that

$$W = (\alpha - \eta)W = (\alpha - \theta)W \subseteq (\alpha - \theta)(W \oplus X).$$

For any $x \in X$, $(\alpha - \theta)x = \pi_W \alpha x + (\phi - \varepsilon)x$ using (**). It follows that

$$(\phi - \varepsilon)x \in W + (\alpha - \theta)X \subseteq (\alpha - \theta)(W \oplus X).$$

Thus, $X = (\phi - \varepsilon)X \subseteq (\alpha - \theta)(W \oplus X)$, so $\alpha - \theta : W \oplus X \to W \oplus X$ is epic. This proves (iii), and so verifies Claim 2.

Now taking $X = \mathrm{Ker}(\phi)$ and $\varepsilon = 1_X$ in Claim 2 shows that $X = 0$, so $\phi : N \to N$ is monic. Hence ϕN is a direct summand of N because N is injective, say $N = Y \oplus \phi N$. Applying ϕ to N, we have $\phi N = \phi Y \oplus \phi^2 N$, so $N = Y \oplus \phi Y \oplus \phi^2 N$. A simple induction shows that $\{Y, \phi Y, \phi^2 Y, \ldots\}$ is an independent family of submodules of N. Write $V = \oplus_{n=0}^{\infty} \phi^n Y$, so that $\phi V \subseteq V$. We next follow an argument of Ó Searcóid [8] to construct a desirable idempotent in $\mathrm{End}(V)$. For $n \geq 0$, define

$$\lambda_{2n} : \phi^{2n} Y \to V \text{ by } \phi^{2n} y \mapsto \phi^{2n} y \text{ for } y \in Y.$$
$$\lambda_{2n+1} : \phi^{2n+1} Y \to V \text{ by } \phi^{2n+1} y \mapsto \phi^{2n+2} y - \phi^{2n} y \text{ for } y \in Y.$$

Each λ_{2n+1} is well-defined because ϕ is monic, so all λ_n are well-defined. Thus we can define $\lambda = \oplus_{n \geq 0} \lambda_n : V \to V$. It can be easily verified (from the definition of λ) that $\lambda^2 = \lambda \in \mathrm{End}(V)$. Now write $\sigma = \phi|_V - \lambda$. Then, for each $y \in Y$, we have

$$\sigma(\phi^{2n} y) = \phi^{2n+1} y - \phi^{2n} y \quad \text{and} \quad \sigma(\phi^{2n+1} y) = \phi^{2n} y.$$

It follows that σ is a unit of $\mathrm{End}(V)$ whose inverse is $1_V + \sigma$. So taking $X = V$ and $\varepsilon = \lambda$ in Claim 2 shows that $V = 0$, and hence that $N = \phi N$. Therefore, $\phi \in \mathrm{End}(N)$ is a unit. Now a final appeal to Claim 2 (with $X = N$ and $\varepsilon = 0$) shows that $N = 0$, proving Lemma 4. $\qquad\square$

With this we can show that *every* injective module is clean. In fact, with a little more effort we obtain much more. Following [7], a module M_R is called *continuous* if:

(C1) Every submodule is essential in a direct summand; and

(C2) Every submodule that is isomorphic to a summand is a summand.

Dually, M is called *discrete* if:

(D1) Given $X \subseteq M$ then $M = A \oplus B$ where $A \subseteq X$ and $X \cap B$ is small in M; and

(D2) If M/K is isomorphic to a summand of M then K is a summand.

Clearly every (quasi-) injective module is continuous. A module is called *semiperfect* [6] if every image has a projective cover. The "duals" of the injective modules are the projective semiperfect modules, and they are all discrete by [7, 4.43]. The proof that all continuous or discrete modules are clean requires the following basic lemma [4].

Lemma 5. *Let R be a ring and I be an ideal of R with $I \subseteq J(R)$. Then R is clean if and only if R/I is clean and idempotents lift modulo I.*

Theorem 6. *Every continuous or discrete module is clean.*

Proof. Let M_R be a module and write $S = \mathrm{End}(M)$. First assume that M is continuous. Then $S/J(S) = T_1 \times T_2$ by [7, Theorem 3.11, Corollary 3.13], where T_1 is a regular, left self-injective ring and T_2 is a reduced regular ring. The ring T_1 is clean by Lemma 4; the reduced regular ring T_2 is unit regular, and so is clean by [2]. Hence $S/J(S)$ is clean. Moreover, idempotents of $S/J(S)$ can be lifted to idempotents of S by [7, Proposition 3.5 and Lemma 3.7]. So S is clean by Lemma 5.

If M is discrete then, by [1, Theorem 2.3], the ring $S/J(S)$ is left continuous, and so $S/J(S)$ is clean by the above discussion. By [7, Lemma 5.3 and Theorem 5.4], idempotents of $S/J(S)$ can be lifted to idempotents of S. Therefore S is clean, again by Lemma 5.   □

References

[1] M.H. Ali and J.M. Zelmanowitz, *Discrete implies continuous*, J. Algebra **183** (1996), 186-192.

[2] V.P. Camillo and D. Khurana, *A characterization of unit regular rings*, Comm. Algebra **29** (2001), 2293-2295.

[3] V.P. Camillo, D. Khurana, T.Y. Lam, W.K. Nicholson and Y. Zhou, *Continuous modules are clean*, J. Algebra **304** (2006), 94-111.

[4] J. Han and W.K. Nicholson, *Extensions of clean rings*, Comm. Algebra **29** (2001), 2589-2596.

[5] T.Y. Lam, "Lectures on Modules and Rings". Graduate Texts in Math. 189, Springer-Verlag, New York, 1999.

[6] E.A. Mares, *Semi-perfect modules*, Math. Zeitschr. **82** (1963), 347-360.

[7] S. Mohamed and B.J. Müller, "Continuous and Discrete Modules". Cambridge Univ. Press, Cambridge, UK, 1990.

[8] M.Ó Searcóid, *Perturbation of linear operators by idempotents*, Irish Math. Soc. Bull. **39** (1997), 10-13.

DEPARTMENT OF MATHEMATICS
THE UNIVERSITY OF IOWA
IOWA CITY, IA 52246, U.S.A.
E-mail address: camillo@math.uiowa.edu

DEPARTMENT OF MATHEMATICS
PANJAB UNIVERSITY
CHANDIGARH 160014, INDIA
E-mail address: dkhurana@pu.ac.in

DEPARTMENT OF MATHEMATICS
UNIVERSITY OF CALIFORNIA
BERKELEY CA 94720, U.S.A.
E-mail address: lam@berkeley.edu

DEPARTMENT OF MATHEMATICS
UNIVERSITY OF CALGARY
CALGARY T2N 1N4, CANADA
E-mail address: wknichol@ucalgary.ca

DEPARTMENT OF MATHEMATICS
MEMORIAL UNIVERSITY OF NEWFOUNDLAND
ST. JOHN'S A1C 5S7, CANADA
E-mail address: zhou@mun.ca

Proceedings of the Sixth China-Japan-Korea
International Conference on Ring Theory
June 27-July 2, 2011 Suwon, Korea

STRUCTURES ON S.G. NEAR-RINGS AND
⟨R, S⟩-GROUPS

Yong Uk Cho

ABSTRACT. In this paper, we denote that R is a near-ring and G
an R-group. We initiate a study of the substructures of R and
G. Next, we define the concepts of s.g. near-rings, ⟨R, S⟩-groups,
S-subgroups and S-homomorphisms, and then we will investigate
some properties of them.

1. Introduction

A near-ring R is an algebraic system $(R, +, \cdot)$ with two binary operations $+$ and $\cdot$ such that $(R, +)$ is a group (not necessarily abelian) with
neutral element 0, $(R, \cdot)$ is a semigroup and $a(b + c) = ab + ac$ for all
a, b, c in R. We note that obviously, $a0 = 0$ and $a(-b) = -ab$ for all a, b
in R, but in general, $0a \neq 0$ and $(-a)b \neq -ab$.

If R has a unity 1, then R is called *unitary*. An element d in R is
called *distributive* if $(a + b)d = ad + bd$ for all a and b in R.

For example, if R is a near-ring with unity 1, then 0 and 1 are clearly
distributive elements.

An *ideal* of R is a subset I of R such that (i) $(I, +)$ is a normal
subgroup of $(R, +)$, (ii) $aI \subset I$ for all $a \in R$, (iii) $(I + a)b - ab \subset I$ for
all $a, b \in R$. If I satisfies (i) and (ii) then it is called a *left ideal* of R. If
I satisfies (i) and (iii) then it is called a *right ideal* of R.

On the other hand, an *R-subgroup* of R is a subset H of R such that
(i) $(H, +)$ is a subgroup of $(R, +)$, (ii) $RH \subset H$ and (iii) $HR \subset H$.
If H satisfies (i) and (ii) then it is called a *left R-subgroup* of R. If H
satisfies (i) and (iii) then it is called a *right R-subgroup* of R. In case,
$(H, +)$ is normal in above, we say that *normal R-subgroup*, *normal left
R-subgroup* and *normal right R-subgroup* instead of R-subgroup, left R-
subgroup and right R-subgroup, respectively. Note that the normal left
R-subgroups of R are equivalent to the left ideals of R.

2010 Mathematics Subject Classification: 16Y30.

Keywords : Representations, zero symmetric, constant part, normal R-subgroups,
R-homomorphisms, monogenic R-groups, d.g. near-rings, s.g. near-rings, S-subgroups
and S-homomorphisms, ⟨R, S⟩-groups.

We consider the following substructures of near-rings: Given a near-ring R, $R_0 = \{a \in R \mid 0a = 0\}$ which is called the *zero symmetric part* of R,

$$R_c = \{a \in R \mid 0a = a\} = \{a \in R \mid ra = a, \text{ for all } r \in R\}$$
$$= \{0a \in R \mid a \in R\}$$

which is called the *constant part* of R, and

$$R_d = \{a \in R \mid a \text{ is distributive}\}$$

which is called the *distributive part* of R.

A non-empty subset S of a near-ring R is said to be a *subnear-ring* of R, if S is a near-ring under the operations of R, equivalently, for all a, b in S, $a - b \in S$ and $ab \in S$. This is called a *criterion of subnear-rings*. Sometimes, we denote it by $S < R$.

We note that R_0 and R_c are subnear-rings of R, on the other hand, R_d is a multiplicative semigroup of R (see following Lemma 2.2), but not a subnear-ring of R. A near-ring R with the extra axiom $0a = 0$ for all $a \in R$, that is, $R = R_0$ is said to be *zero symmetric*, also, in case $R = R_c$, R is called a *constant* near-ring, and in case $R = R_d$, R is called a *distributive* near-ring.

Note that R_0 is a right ideal of R, but not generally ideal of R, also R_c is a right R-subgroup of R, but in general neither a right nor a left ideal of R.

Let $(G, +)$ be a group (not necessarily abelian). We may obtain some useful examples of near-rings as following, there are three kinds of trivial near-rings:

Examples 1.1 Let G be an additive group.

(1) If we define a multiplication on G by $xy = 0$ for all x, y in G, then $(G, +, \cdot)$ becomes a near-ring, which is called the trivial near-ring on G.

(2) If we define a multiplication on G by $xy = y$ for all x, y in G, then $(G, +, \cdot)$ becomes a near-ring, because $(xy)z = z = x(yz)$ and $x(y + z) = y + z = xy + xz$, for all x, y, z in G, but in general, $0x = 0$ and $(x + y)z = xz + yz$ are not true. This kind of near-ring is constant near-ring, which is called the *trivial constant near-ring on G*.

(3) If we define a multiplication on G by $xy = 0$ if $x = 0$, $= y$ otherwise in G, then $(G, +, \cdot)$ becomes a zero symmetric near-ring, which is called the *trivial zero symmetric near-ring on G*.

Next is in [6], in the set

$$M(G) = \{f \mid f : G \longrightarrow G\}$$

of all the self maps of G, if we define the sum $f + g$ of any two mappings f, g in $M(G)$ by the rule $x(f + g) = xf + xg$ for all $x \in G$ and the product $f \cdot g$ by the rule $x(f \cdot g) = (xf)g$ for all $x \in G$, then $(M(G), +, \cdot)$ becomes a near-ring. It is called the *self map near-ring* on G.

Also, we can define the substructures of $(M(G), +, \cdot)$ as following:

$$M_0(G) = \{f \in M(G) \mid 0f = 0\} \text{ and}$$

$$M_c(G) = \{f \in M(G) \mid f \text{ is constant}\}.$$

Then $(M_0(G), +, \cdot)$ is a zero symmetric near-ring. Moreover, $M_0(G) = M(G)_0$ and $M_c(G) = M(G)_c$.

The ideas in Examples 1.1 inspire the following notation which we will use.

Examples 1.2 Let G be an additive group.

(1) If $x \in G$, then we define $\alpha_x \in M(G)$ by $g\alpha_x = x$ for all $g \in G$.

(2) If $x \in G$, then we define $\beta_x \in M_0(G)$ by $o\beta_x = o$ and $g\beta_x = x$ for all $g \in G - \{o\}$.

Note that $\alpha_o \in M_0(G)$, but α_x is not in $M_0(G)$ for $x \neq o$. The following result ties this in with Examples 1.1.

Let R and S be two near-rings. Then a mapping θ from R to S is called a *near-ring homomorphism* if (i) $(a + b)\theta = a\theta + b\theta$, (ii) $(ab)\theta = a\theta b\theta$. Obviously, $R\theta < S$ and $T\theta^{-1} < R$ for any $T < S$.

We can replace homomorphism by monomorphism, epimorphism, isomorphism, endomorphism and automorphism, if these terms have their usual meanings as in ring theory [1]. If θ from R to S is a near-ring isomorphism, then we say that R *is isomorphic to* S, and denoted it by $R \cong S$.

Proposition 1.3 *Let G be an additive group, and let $x \in G$. Then*

$$\gamma\alpha_x = \alpha_x, \forall \gamma \in M(G).$$

Also, $\{\alpha_x \mid x \in G\} < M(G)$ and isomorphic to the trivial constant near-ring on G, and $\{\beta_x \mid x \in G\} < M_0(G)$ and isomorphic to the trivial zero symmetric near-ring on G.

Proof. Because of the definition of α_x, for any $t \in G$,

$$t\gamma\alpha_x = x = t\alpha_x, \forall \gamma \in M(G),$$

we see that $\gamma\alpha_x = \alpha_x$. So the first part is easily proved.

Next, since for any α_x and α_y in the given set, $\alpha_x - \alpha_y = \alpha_{x-y}$ and $\alpha_x\alpha_y = \alpha_y$, from the 1st part. Hence by the criterion of subnear-ring, $\{\alpha_x \mid x \in G\} < M(G)$. Similarly, we can prove that $\{\beta_x \mid x \in G\} < M_0(G)$.

Finally, the two maps which are given by $\alpha_x \to x$, $\beta_x, \to x$, are clearly isomorphisms.   □

We can check that $\{\alpha_x \mid x \in G\}$ is the unique maximal constant subnear-ring of $M(G)$, so we see that $\{\alpha_x \mid x \in G\} = M_c(G)$.

Let R be any near-ring and G an additive group. Then G is called an R-*group* if there exists a near-ring homomorphism

$$\theta : (R, +, \cdot) \longrightarrow (M(G), +, \cdot).$$

Such a homomorphism θ is called a *representation* of R on G. We denote it by G_R.

We may write that xr (as a scalar product in G by R) for $x(r\theta)$ for all $x \in G$ and $r \in R$. If R is unitary, then R-group G is called *unitary*. Thus an R-group is an additive group G satisfying (i) $x(a+b) = xa + xb$, (ii) $x(ab) = (xa)b$ and (iii) $x1 = x$ (if R has a unity 1), for all $x \in G$ and $a, b \in R$.

Naturally, we can define a new concept of R-group: An R-group G is called *distributive*, in case $(x + y)a = xa + ya$, for all $x, y \in G$ and for each $a \in R$. For example, every distributive near-ring R is a distributive R-group.

Evidently, every near-ring R can be given the structure of an R-group (unitary, if R is unitary) by right multiplication in R. Moreover, every group G has a natural $M(G)$-group structure, from the representation of $M(G)$ on G by applying the $f \in M(G)$ to the $x \in G$ as a scalar multiplication xf.

Let R be a near-ring, and let G and H be two R-groups. Then a mapping θ from G to H is called an R-*group homomorphism, or R-homomorphism* if
(i) $(x + y)\theta = x\theta + y\theta$, (ii) $(xa)\theta = (x\theta)a$.

We can replace R-homomorphism by R-monomorphism, R-epimorphism, R-isomorphism, R-endomorphism and R-automorphism, if these terms have their usual meanings as in module theory [1]. If θ from G to H is an isomorphism, then we say that G *is R-isomorphic to H*, and denoted it by $G \cong_R H$.

A representation θ of R on G is called *faithful* if $Ker\theta = \{0\}$, that is, θ is a monomorphism, equivalently, $xr = 0$ implies $r = 0$, for all $x \in G$ and $r \in R$. In this case, we say that G is called a *faithful R-group*.

For an R-group G, a subgroup T of G such that $TR \subset T$ is called an R-*subgroup* of G, a normal subgroup N of G such that $NR \subset N$ is called a *normal R-subgroup* of G, and an R-*ideal* of G is a normal subgroup N of G such that $(N + x)a - xa \subset N$ for all $x \in G$, $a \in R$. Also, note that the R-ideals of R_R are equivalent to the right ideals of R.

Let R be a near-ring and let G be an R-group. If there exists x in G such that $G = xR$, that is, $G = \{xr \mid r \in R\}$, then G is called a *monogenic R-group* and the element x is called a *generator* of G, more specially, if G is monogenic and for each $x \in G$, $xR = 0$ or $xR = G$, then G is called a *strongly monogenic R-group*.

For the remainder concepts and results on near-rings, we refer to G. Pilz [6].

2. Some properties of S-subgroups and S-homomorphisms

Distributive near-rings, which are near-rings satisfying both distributive laws, are very close to rings and will be considered a bit more closely later. In the mean time, we consider a larger class of near-rings which has a lot of distributivity built in.

In the period 1958-1962, A. Frohlich published some papers on distributively generated near-rings [3], [4], [5]. These mark the real beginning of these subjects.

A near-ring R is called a *distributively generated near-ring*, denoted by *d.g. near-ring*, if $(R, +)$ is generated as a group by a semigroup $(S, \cdot)$ of distributive elements.

A d.g. near-ring R which is generated by a semigroup S is denoted by (R, S).

Rings are special cases of d.g. near-rings, because of all elements of a ring are distributive.

We can define more generalization of d.g. near-rings which are s.g. near-rings, and give a double definitions which will be useful in the sequence work.

Let R be an near-ring and let G and H be two R-groups. Let S be a multiplicative subsemigroup of R.

(1) The near-ring R is called a *s.g. near-ring*, if $(R, +)$ is generated as a group by a semigroup $(S, \cdot)$ of R, which is denoted by $\langle R, S \rangle$.

(2) A subgroup K of G such that $KS \subseteq K$ is called a *S-subgroup* of G.

(3) A homomorphism of groups θ from G to H is called a *S-homomorphism* if $(xs)\theta = (x\theta)s$, $\forall x \in G$, $\forall ts \in S$.

Example 2.1 Let G be an additive group.

(1) The trivial near-ring on a group G has all its elements distributive.

(2) Consider the set $EndG = \{\alpha | \alpha \text{ is an endomorphism of } G\}$. Then $EndG \subseteq M_0(G)$, and $EndG$ consists of distributive elements of $M_0(G)$.

Lemma 2.2 *Let R be an near-ring. The set R_d of all distributive elements of R forms a multiplicative semigroup.*

Proof. Let $x, y \in R$ and $a, b \in R_d$. Then $(x + y)ab = (xa + ya)b = (xa)b + (ya)b = x(ab) + y(ab)$. $\square$

The proof of the following result is very similar to that of ring theoretical result.

Lemma 2.3 *Let R be a near-ring (not necessarily zero symmetric), and let $x \in R_d$. Then*

$$0x = 0, \ (-y)x = -yx \ \forall y \in R.$$

Proposition 2.4 *Let G be an additive group. Then*
$$M_0(G)_d = EndG.$$

Proof. We see that $EndG \subseteq M_0(G)_d$ is immediate from the definition of near-ring homomorphism. So consider that $f \in M_0(G)_d$. Since f is distributive, using Example 1.2, for any x, $y \in G$,

$$(\beta_x + \beta_y)f = \beta_x f + \beta_y f.$$

On the other hand, applying this equation to any non-zero element t in G yields the equation

$$(x + y)f = xf + yf, \ \forall x, \ y \in G,$$

that is $f \in EndG$. Hence $M_0(G)_d \subseteq EndG$. $\square$

Proposition 2.5 *Let R be a near-ring and S a semigroup of distributive elements of R. Then the d.g. near-ring generated by S is precisely the d.g. near-ring generated, as a near-ring, by S.*

Proof. Let (T, S) be the d.g. near-ring generated by S. It suffices to show that T is closed under products. Thus let $\epsilon_1 s_1 + \cdots + \epsilon_n s_n$, and $\eta_1 t_1 + \cdots + \eta_m t_m$ in T, where $\epsilon_i = \pm 1$, $\eta_j = \pm 1$, $s_i, t_j \in S$ $\forall i, j$, $1 \leq i \leq n, 1 \leq j \leq m$. Then

$$(\epsilon_1 s_1 + \cdots + \epsilon_n s_n)(\eta_1 t_1 + \cdots + \eta_m t_m)$$
$$= (\epsilon_1 s_1 + \cdots + \epsilon_n s_n)\eta_1 t_1 + \cdots + (\epsilon_1 s_1 + \cdots + \epsilon_n s_n)\eta_m t_m$$
$$= \eta_1(\epsilon_1 s_1 t_1 + \cdots + \epsilon_n s_n t_1) + \cdots + \eta_m(\epsilon_1 s_1 t_m + \cdots + \epsilon_n s_n t_m).$$

Since $s_i t_j \in S$ $\forall i, j$, $1 \leq i \leq n, 1 \leq j \leq m$, this proves our result. $\square$

Examples 2.6 Let G be an additive group and let $S \subseteq EndG$ a semigroup of endomorphisms of G. Then S is a semigroup of distributive elements of the near-ring $M_0(G)$. In this case, near-ring R generated by S is a d.g. near-ring (R, S). There are three kinds of d.g. near-rings which arise in this way.

(1) $S = EndG$, we denote the d.g. near-ring by $(E(G), EndG)$.

(2) $S = AutG$, the group of all automorphisms of G, we denote the d.g. near-ring by $(A(G), AutG)$.

(3) $S = InnG$, the group of all inner automorphisms of G, we denote the d.g. near-ring by $(I(G), InnG)$.

Note that $I(G) \subseteq A(G) \subseteq E(G)$, and $E(G) = EndG$, in case G is abelian.

Now, we consider that the substructures of R and G, also quotients of substructure relations between them.

Let G be an R-group and K, K_1 and K_2 be nonempty subsets of G. Define

$$(K_1 : K_2) := \{a \in R | K_1 a \subset K_2\}.$$

We abbreviate that for $x \in G$

$$(\{x\} : K) =: (x : K).$$

Similarly for $(K : x)$. $(K : o)$ is called the *annihilator* of K in R, denoted it by $Ann(K)$. We say that G is a *faithful R-group* or that R *acts faithfully* on G if $Ann(G) = \{0\}$, that is, $(G : o) = \{0\}$.

Also, we see that from the previous concepts to elementwise, a subgroup H of G such that $xa \in H$ for all $x \in H, a \in R$, is an R-*subgroup* of G, and an R-*ideal* of G is a normal subgroup N of G such that

$$(x + g)a - ga \in N$$

for all $g \in G, x \in N$ and $a \in R$.

Let (R, S) be the d.g. near-ring generated by S and let G be an R-group. Let θ be the representation which defines G as an R-group. We call G an (R, S)-*group* or *d.g. group* if

$$S\theta \subseteq EndG.$$

Also, let (R, S) be the d.g. near-ring. A representation θ of R is called a d.g. representation if $\theta : R \to M(G)$ satisfies $S\theta \subseteq EndG$, where G is an R-group associated with the representation θ.

Note that a d.g. representation is a representation associated with a d.g. group. In other words, G is an (R, S)-group if the elements of S induces endomorphisms on G. Examples 2.6 gives us some d.g. groups.

Some more examples of d.g. groups arise as following.

Lemma 2.7 *Let (R, S) be a d.g. near-ring generated by S. Then all R-subgroups and R-homomorphic images of an (R, S)-group are (R, S)-groups.*

Proof. These can be verified very easily from the definition of (R, S)-group. $\qquad\square$

Proposition 2.8 *Let R be a near-ring and G an R-group. Then we have the following statements:*

(1) $Ann(G)$ is a two-sided ideal of R. Moreover G is a faithful $R/Ann(G)$-group.

(2) For any $x \in G$, we get $xR \cong R/(x : o)$ as R-groups.

Proof. (1) Obviously, $Ann(G)$ is a two-sided ideal of R. We now make G an $R/Ann(G)$-group by defining, for $x \in G, r + Ann(G) \in R/Ann(G)$, the action $x(r + Ann(G)) = xr$. If $r + Ann(G) = r' + Ann(G)$, then $-r' + r \in Ann(G)$ hence $x(-r' + r) = 0$ for all x in G, that is to say, $xr = xr'$. This implies that

$$x(r + Ann(G)) = xr = xr' = x(r' + Ann(G));$$

thus the action of $R/Ann(G)$ on G is well defined. The verification of the structure of an $R/Ann(G)$-group is a routine fact.

Finally, to see that G is a faithful $R/Ann(G)$-group, we note that if $x(r + Ann(G)) = 0$ for all $x \in G$, then by the definition of $R/Ann(G)$-group structure, we have $xr = 0$. Hence $r \in Ann(G)$. This says that only the zero element of $R/Ann(G)$ annihilates all of G. Thus G is a faithful $R/Ann(G)$-group.

(2) For any $x \in G$, clearly xR is an R-subgroup of G. The map $\phi_x : R \to xR$ defined by $r\phi_x = xr$ is an R-epimorphism, so that from the isomorphism theorem in near-ring theory and the kernel of ϕ_x is $(x : o)$, we can induce that

$$xR \cong R/(x : o)$$

as R-groups. $\qquad \square$

Proposition 2.9 *If R is a near-ring and G an R-group, then $R/Ann(G)$ is isomorphic to a subnear-ring of $M(G)$.*

Proof. Let $a \in R$. We define $\tau_a : G \to G$ by $x\tau_a = xa$ for each $x \in G$. Then τ_a is in $M(G)$. Consider the mapping $\phi : R \to M(G)$ defined by $a\phi = \tau_a$. Then obviously, we see that $(a + b)\phi = a\phi + b\phi$ and $(ab)\phi = a\phi b\phi$, that is, ϕ is a near-ring homomorphism from R to $M(G)$.

Next, we must show that $Ker\phi = Ann(G)$. Indeed, if $a \in Ker\phi$, then $\tau_a = 0$, which implies that $Ga = G\tau_a = 0$, that is, $a \in Ann(G)$. On the other hand, if $a \in Ann(G)$, then by the definition of $Ann(G)$, $Ga = 0$ hence $0 = \tau_a = a\phi$, this implies that $a \in Ker\phi$. Therefore from the isomorphism theorem on R-groups, the image of R is a near-ring isomorphic to $R/Ann(G)$. Consequently, $R/Ann(G)$ is isomorphic to a subnear-ring of $M(G)$. $\qquad \square$

Thus we can obtain the following important statement as in ring theory.

Corollary 2.10 *If G is a faithful R-group, then R is embedded in $M(G)$.*

Proposition 2.11 *If (R, S) is a d.g. near-ring, then every monogenic R-group is an (R, S)-group.*

Proof. Let G be a monogenic R-group with x as a generator. Then $G = xR$ and the map $\phi_x : R_R \to G_R$ defined by $a\phi_x = xa$ is an R-epimorphism from R to G as R-groups. We see that by the Proposition 2.8,

$$G \cong R/Ann(x),$$

where $Ann(x) = (x : 0) = Ker\phi_x$. From the Lemma 2.7, we see that G is an (R, S)-group.   □

Corollary 2.12 *Let G be a monogenic R-group with x as a generator. Then we have the following isomorphic relation.*

$$G \cong R/(x : 0).$$

Proposition 2.13 *Let (R, S) be a d.g. near-ring generated by S. Then every left identity for S is a left identity for R.*

Proof. Let e be a left identity for S. Then $es = s$, $\forall s \in S$. We must show that $er = r$, $\forall r \in R$. Since R is a d.g. near-ring generated by S, r can be expressed as $r = \epsilon_1 s_1 + \cdots + \epsilon_n s_n$, where $\epsilon_i = \pm 1$, $s_i \in S$ $\forall i$, $1 \leq i \leq n$. Thus we have that

$$er = e(\epsilon_1 s_1 + \cdots + \epsilon_n s_n) = e(\epsilon_1 s_1) + \cdots + e(\epsilon_n s_n)$$

$$= \epsilon_1 e s_1 + \cdots + \epsilon_n e s_n = \epsilon_1 s_1 + \cdots + \epsilon_n s_n = r.$$

Hence e is a left identity for R.   □

Proposition 2.14 *Let R be a not necessarily zero symmetric near-ring and let S a semigroup of distributive elements of R. Then the d.g. near-ring generated by S is zero symmetric.*

Proof. Let (T, S) be the d.g. near-ring generated by S, and let $\epsilon_1 s_1 + \cdots + \epsilon_n s_n \in T$, where $\epsilon_i = \pm 1$, $s_i \in S$ $\forall i$, $1 \leq i \leq n$. Then

$$0(\epsilon_1 s_1 + \cdots + \epsilon_n s_n) = 0\epsilon_1 s_1 + \cdots + 0\epsilon_n s_n$$

$$= \epsilon_1 0 s_1 + \cdots + \epsilon_n 0 s_n = \epsilon_1 0 + \cdots + \epsilon_n 0 = 0,$$

using Lemma 2.3. Consequently, (T, S) is zero symmetric.   □

Proposition 2.15 *Let R be an near-ring and let G be an R-group. Let S be a multiplicative subsemigroup of R which generate $(R, +)$ as a group, that is, $\langle R, S \rangle$ is an s.g. near-ring generated by S.*

(1) A subgroup H of G is an R-subgroup if and only if it is an S-subgroup of G.

(2) A homomorphism θ from G to an R-group T is an R-homomorphism if and only if it is an S-homomorphism.

Proof. In both cases the implication one way, that which replaces R by S, is immediate from the definitions.

Let $r \in R$. Then $r = \epsilon_1 s_1 + \cdots + \epsilon_n s_n$, where $\epsilon_i = \pm 1$, $s_i \in S$ $\forall i$, $1 \leq i \leq n$. Let $h \in H$. It follows that

$$hr = h(\epsilon_1 s_1 + \cdots + \epsilon_n s_n) = \epsilon_1 h s_1 + \cdots + \epsilon_n h s_n = \epsilon_1 h_1 + \cdots + \epsilon_n h_n,$$

where $h s_i = h_i \in H$, $\forall i$, $1 \leq i \leq n$. Hence $hr \in H$, and H is an R-subgroup.

Next, suppose that $\theta : G \longrightarrow T$ is a S-homomorphism. Let $x \in G$, $r \in R$ as above. Then

$$(xr)\theta = [x(\epsilon_1 s_1 + \cdots + \epsilon_n s_n)\theta = (\epsilon_1 x s_1 + \cdots + \epsilon_n x s_n)\theta$$

$$= (\epsilon_1 x s_1)\theta + \cdots + (\epsilon_n x s_n)\theta = (\epsilon_1 x \theta s_1) + \cdots + (\epsilon_n x \theta s_n)$$

$$= (x\theta)\epsilon_1 s_1 + \cdots + (x\theta)\epsilon_n s_n = (x\theta)(\epsilon_1 s_1 + \cdots + \epsilon_n s_n) = (x\theta)r.$$

$\square$

Corollary 2.16 *Let (R, S) be a d.g. near-ring and G an R-group. Then*

(1) A subgroup H of G is an R-subgroup if and only if H is an S-subgroup of G.

(2) A homomorphism θ from G to an R-group T is an R-homomorphism if and only if it is an S-homomorphism.

The special role played by the semigroup of R which generates a d.g. near-ring extends to the representation of s.g. near-rings.

Let $\langle R, S \rangle$ be the s.g. near-ring generated by a subsemigroup S of R and let G be an R-group. Let θ be the representation which defines G as an R-group. We call G an $\langle R, S \rangle$-*group* or *s.g. group* if

$$S\theta \subseteq EndG.$$

Also, let $\langle R, S \rangle$ be the s.g. near-ring. A representation θ of R is called a s.g. representation if $\theta : R \to M(G)$ satisfies $S\theta \subseteq EndG$, where G is an R-group associated with the representation θ.

Note that a s.g. representation is a representation associated with a s.g. group. In other words, G is an $\langle R, S \rangle$-group if the elements of S induces endomorphisms on G. Examples 2.6 gives us some s.g. groups.

Some more examples of s.g. groups arise as following.

Lemma 2.17 *Let $\langle R, S \rangle$ be a s.g. near-ring generated by S. Then all R-subgroups and R-homomorphic images of an $\langle R, S \rangle$-group are $\langle R, S \rangle$-groups.*

Proof. These can be verified very easily from the definition of $\langle R, S \rangle$-group.   □

Proposition 2.18 *If $\langle R, S \rangle$ is a s.g. near-ring, then every monogenic R-group is an $\langle R, S \rangle$-group.*

Proof. Let G be a monogenic R-group with x as a generator. Then $G = xR$ and the map $\phi_x : R_R \to G_R$ defined by $a\phi_x = xa$ is an R-epimorphism from R to G as R-groups. We see that by the Proposition 2.8,

$$G \cong R/Ann(x),$$

where $Ann(x) = (x : 0) = Ker\phi_x$. From the Lemma 2.17, we see that G is an $\langle R, S \rangle$-group.   □

The next theorem shows that $\langle R, S \rangle$-groups do have their nice side.

Theorem 2.19 *Let $\langle R, S \rangle$ be a s.g. near-ring and let G be an $\langle R, S \rangle$-group. Then a subgroup H of G is an R-ideal if and only if it is a normal S-subgroup.*

Proof. The necessity of the condition is immediate from the definition, because the condition (i) of the R-ideal H implies that $(H, +)$ is normal subgroup of $(G, +)$, and the condition (ii) of the R-ideal H implies that $HS \subseteq H$.

We prove the sufficiency. Assume that H is a normal S-subgroup of G. Then H satisfies the condition (i) of the R-ideal. To prove the condition (ii) of the R-ideal of H, let $x \in G$, $h \in H$, $r \in R$, where $r = \epsilon_1 s_1 + \cdots + \epsilon_n s_n$, here $\epsilon_i = \pm 1$, $s_i \in S$, $\forall i$, $1 \leq i \leq n$. .

We use induction on n to show that

$$(x + h)r - xr \in H.$$

If $n = 1$, then

$$(x + h)r - xr = (x + h)\epsilon_1 s_1 - x\epsilon_1 s_1 = \epsilon_1(xs_1 + hs_1) - \epsilon_1 xs_1.$$

This is either $xs_1 + hs_1 - xs_1$, if $\epsilon_1 = 1$, or $-hs_1 - xs_1 + xs_1 = -hs_1$, if $\epsilon_1 = -1$. Since $hs_1 \in H$ and H is a normal subgroup, we conclude that in both cases,

$$(x + h)r - xr \in H.$$

Assume that the result is true for n. Suppose that $r = \epsilon_1 s_1 + \cdots + \epsilon_n s_n + \epsilon_{n+1} s_{n+1}$. Putting $r' = \epsilon_1 s_1 + \cdots + \epsilon_n s_n$. Then

$$(x + h)r - xr = (x + h)(r' + \epsilon_{n+1} s_{n+1}) - x(r' + \epsilon_{n+1} s_{n+1})$$

$$= (x+h)r' + \epsilon_{n+1}(xs_{n+1} + hs_{n+1}) - \epsilon_{n+1} xs_{n+1} - xr' = (x + h)r' + h' - xr',$$

where $h' = \epsilon_{n+1}(xs_{n+1} + hs_{n+1}) - \epsilon_{n+1}xs_{n+1}$ in H. Since H is a normal subgroup, we can take the next step:

$$(x + h)r - xr = h'' + (x + h)r' - xr',$$

where $h'' \in H$. Consequently, the induction hypothesis enables us to conclude that

$$(x + h)r - xr \in H.$$

Hence H is an R-ideal of G. □

Some special cases of interest are stated explicitly.

Corollary 2.20 *Let $\langle R, S \rangle$ be an s.g. near-ring. Then*

(1) A subset I of R is a right ideal if and only if I is a normal right S-subgroup of R.

(2) A subset I of R_R is an R-ideal if and only if I is a normal S-subgroup of R_R.

(3) A subset I of R is an ideal if and only if I is a normal S-subgroup of R.

References

[1] F. W. Anderson and K.R. Fuller, *Rings and categories of modules*, Springer-Verlag, New York, Heidelberg, Berlin (1974).

[2] Y.U. Cho, *Isomorphic structures in R-groups*, FJMS **48**(2) (2011), 175-180.

[3] C.G. Lyons and J.D.P. Meldrum, *Characterizing series for faithful D.G. near-rings*, Proc. Amer. Math. Soc. **72** (1978), 221-227.

[4] S.J. Mahmood and J.D.P. Meldrum, *D.G. near-rings on the infinite dihedral groups*, Near-rings and Near-fields, Elsevier Science Publishers B.V. (North-Holland) (1987), 151-166.

[5] J.D.P. Meldrum, *Upper faithful D.G. near-rings*, Proc. Edinburgh Math. Soc. **26** (1983), 361-370.

[6] G. Pilz, *Near-rings*, North Holland Publishing Company, Amsterdam, New York, Oxford (1983).

DEPARTMENT OF MATHEMATICS, COLLEGE OF EDUCATION
SILLA UNIVERSITY
PUSAN 617-736, KOREA
E-mail address: yuchosilla.ac.kr

Proceedings of the Sixth China-Japan-Korea
International Conference on Ring Theory
June 27-July 2, 2011 Suwon, Korea

IMPRIMITIVE REGULAR ACTION IN THE RING OF INTEGERS MODULO n

JUNCHEOL HAN, YANG LEE, AND SANGWON PARK*

ABSTRACT. Let n be any positive integer and $\mathbb{Z}_n = \{0, 1, \ldots, n-1\}$ be the ring of integers modulo n. Let X_n be the set of all nonzero, nonunits of $\mathbb{Z}_n$, and G_n the group of all units of $\mathbb{Z}_n$. In this paper, by considering the regular representation $\pi : G_n \longrightarrow \mathrm{Sym}(X_n)$, the following are investigated as follows: (1) G_n is not fixed-point free; (2) If $Fix(g) = \{x \in X_n : gx = x\} \neq \emptyset$ for some $g \in G_n$, then $Fix(g)$ is a union of orbits under the regular action of G_n on X_n; (3) $B(\subseteq o(x_1) \cup \cdots \cup o(x_\ell))$ is a set of imprimitivity under π for some orbits $o(x_1), \ldots, o(x_\ell)$ if and only if $B = g_1 H_1 x_1 \cup \cdots \cup g_\ell H_\ell x_\ell$ for some subgroups $H_1, \ldots, H_\ell$ of G_n and some elements $g_1, \ldots, g_\ell \in G_n$ satisfying that if $(gB) \cap B \neq \emptyset$ for some $g \in G$, then $g \in stab(x_i) H_i$ for each $i = 1, \ldots, \ell$.

1. Introduction and basic definitions

Let n be any positive integer which is not a prime and $\mathbb{Z}_n$ be the ring of integers modulo n. Let X_n be the set of all nonzero, nonunits of $\mathbb{Z}_n$ and G_n the group of all units of $\mathbb{Z}_n$. In this paper, we will consider a group action on X_n by G_n given by $((g, x) \longrightarrow gx)$ from $G_n \times X_n$ to X_n, called the regular action on X_n by G_n. Under the regular action on X_n by G_n, we define the *orbit* of x by $o(x) = \{gx : \forall g \in G_n\}$ and the *stabilizer* of x by $stab(x) = \{g \in G_n : gx = x\}$ (refer [1]).

Let $\mathrm{Sym}(X_n)$ denote the symmetric group on X_n. Recall that if $\pi : G_n \longrightarrow \mathrm{Sym}(X_n)$ is a representation, then π is *faithful* if $\ker(\pi) = \{1\}$. In this paper, we will consider a representation of G_n into $\mathrm{Sym}(X_n)$ defined by $(\pi(g))(x) = gx$ for all $g \in G_n$ and $x \in X_n$ (refer [2]). Suppose that π is a representation of G_n into $\mathrm{Sym}(X_n)$. G_n is *fixed-point free* (or G_n has no fixed points) provided that if $g \in G_n$ and there exists an $x \in X_n$ with $(\pi(g))(x) = x$, then $g \in \mathrm{Ker}(\pi)$. Equivalently, G_n is fixed-point free if $\mathrm{Stab}(x) = \mathrm{Ker}(\pi)$ for all $x \in X_n$ (refer [2]).

In section 2, we will show that (1) every orbit under the regular action on X_n by G_n consists of $o(x) = \{y \in X_n : (x, n) = (y, n)\}$ for every

---

2000 Mathematics Subject Classification: 16W22.
Keywords : Regular group action, orbit, faithful, fixed-point free.
*Corresponding author.

divisor $x(\neq 1, n)$ of n where (s, t) denotes the greatest common divisor of any two positive integers s, t; (2) if $Fix(g) = \{x \in X_n : gx = x\} \neq \emptyset$ for some $g \in G_n$, then $Fix(g)$ is a union of orbits under the regular action of G_n on X_n and is equal to the set of all multiples of x in X_n; (3) in particular, if n is odd, then $Fix(g) \neq \emptyset$ for all $g \in G_n$.

Suppose that S is a set with more than one element, H is a group and $\pi : H \longrightarrow Sym(S)$ is a representation of H into the symmetric group, $Sym(S)$, on S. A nonempty subset B of S is called a subset of *imprimitivity* for the action of H on S if for any $g \in H$, either $(\pi(g))(B) = B$ or $(\pi(g))(B) \cap B = \emptyset$. B is said to be *trivial* if $|B| = 1$ or $B = S$. H is imprimitive on S if there exists a nontrivial subset B of imprimitivity for the action on H on S. Otherwise H is said to be *primitive* on S (refer [1]). Suppose that Q is a partition of S. Q is $H - invariant$ if H permutes the member of Q. Q is *nontrivial* if $Q \neq \{S\}$ and $Q \neq \{\{s\} : s \in S\}$. By [1], H is imprimitive on S if and only if there exists a nontrivial $H-$invariant partition Q of S.

In section 3, by investigating the imprimitive subset in X_n for any positive integer n, we will show that (1) if B is a subset of $o(x)$ for some $x \in X_n$ such that $|B| > \frac{|o(x)|}{2}$, then B is not a set of imprimitivity under the regular representation $\pi : G_n \longrightarrow Sym(X_n)$; (2) if $B \subseteq o(x_1) \cup \cdots \cup o(x_\ell)$ for some orbits $o(x_1), \cdots, o(x_\ell)$ in X_n and B is a set of imprimitivity under the regular representation $\pi : G_n \longrightarrow Sym(X_n)$, then there exists subgroups H_i of G_n and an element $g_i \in G_n$ such that $B = g_1 H_1 \cup \cdots \cup g_\ell H_\ell$ for $i = 1, \cdots, \ell$.

Throughout this paper, for $x, y \in X_n$ $x|y$ means that x is a divisor of y.

2. Fixed point set in $\mathbb{Z}_n$ under the regular action

In [2], the following proposition was given:

Proposition 2.1 Let A be a ring, G be the group of units of A and X is the nonempty set of all nonzero, nonunits in A. The regular representation of G into $Sym(X)$ is faithful if and only if A is not a local ring or A is a local ring and the left annihilator of the Jacobson radical J of A is the zero ideal.

Proof. Refer [2]. $\square$

Note that (1) $\mathbb{Z}_n$ is a local ring if and only if $n = p^\alpha$ for some prime p and some positive integer α; (2) the Jacobson radical of $\mathbb{Z}_{p^\alpha}$, $J = \{kp \in \mathbb{Z}_{p^\alpha}\}$, is a maximal ideal of $\mathbb{Z}_{p^\alpha}$ and so annihilator of J is nonempty since $p^{\alpha-1}J = (0)$. Therefore, by Proposition 2.1 we note that the regular representation $\pi : G_n \longrightarrow \text{Sym}(X_n)$ is faithful if and only if $\mathbb{Z}_n$ is not a local ring.

Theorem 2.2 Let n be any positive integer such that $X_n \neq \emptyset$. If $Fix(g) = \{x \in X_n : gx = x\} \neq \emptyset$ for some $g \in G_n$, then $Fix(g)$ is a union of orbits under the regular action of G_n on X_n.

Proof. Since $Fix(g) \neq \emptyset$, there exists $x \in Fix(g)$, i.e., $gx = x$. Since for all $y \in o(x)$, $y = hx$ for some $h \in G$, $gy = g(hx) = h(gx) = hx = y$, which implies that $y \in Fix(g)$ and so $Fix(g) \supseteq o(x)$. If $Fix(g) \supset o(x)$, then there exists $x_1 \in Fix(g)$ but $x_1 \notin o(x)$. Since $gx_1 = x_1$, $Fix(g) \supseteq o(x_1)$ by the similar argument, and so $Fix(g) \supseteq o(x) \cup o(x_1)$. Continuing in this process, we have the result. $\qquad \square$

Example 1 Consider $\mathbb{Z}_{36}$. Then we obtained all the $Fix(g)$'s in X_{36} as follow:

$$Fix(1) = X_{36},$$
$$Fix(5) = Fix(17) = Fix(29) = \{9, 18, 27\} = o(9) \cup o(18),$$
$$Fix(7) = Fix(31) = \{6, 12, 18, 24, 30\} = o(6) \cup o(12) \cup o(18),$$
$$Fix(11) = Fix(23) = Fix(35) = \{18\} = o(18),$$
$$Fix(13) = Fix(25) = \{3, 6, \ldots, 33\} = o(3) \cup o(6) \cup o(12) \cup o(18),$$
$$Fix(19) = \{2, 4, \ldots, 34\} = o(2) \cup o(4) \cup o(6) \cup o(12) \cup o(18).$$

Lemma 2.3 Let n be any positive integer such that $X_n \neq \emptyset$ and $x, y \in X_n$ be divisors of n such that $(x, y) = 1$. If $x \in Fix(g)$ for some $g \in G_n$, then $y \in Fix(g)$ for all $y \in X_n$ such that $x|y$.

Proof. Since $x \in Fix(g)$, $gx = x$. Then for all $y \in X_n$ such that $x|y$ $gy = gx(\frac{y}{x}) = x(\frac{y}{x}) = y$, and so $y \in Fix(g)$. $\qquad \square$

Lemma 2.4 Let n be any positive integer such that $X_n \neq \emptyset$ and $y \in X_n$ be arbitrary. Then there exists $x \in X$ such that $x|n$ and $(x, n) = (y, n)$.

Proof. Let $x = (y, n)$. Then clearly, $x|n$ and $(x, n) = ((y, n), n) = (y, n)$. $\qquad \square$

Lemma 2.5 Let k and n be any positive integers such that $k|n$. If $\bar{g} \in G_k$, then there exists $g \in G_n$ such that $g \equiv \bar{g} \pmod{k}$.

Proof. Note that since $k|n$, $\mathbb{Z}_n/ <k>$ is isomorphic to $\mathbb{Z}_k$ where $<k>$ is an ideal of $\mathbb{Z}_n$ generated by k. Let $n = p_1^{\alpha_1} p_2^{\alpha_2} \cdots p_t^{\alpha_t}$ be the prime factorization of n where $p_1, p_2, \cdots, p_t$ are distinct primes for some positive integer t. Then $k = p_1^{\beta_1} p_2^{\beta_2} \cdots p_t^{\beta_t}$ with $\alpha_i \geq \beta_i \geq 0$ for all $i = 1, \cdots, t$. Without loss of generality, we can assume that $\mathbb{Z}_n = \mathbb{Z}_{p_1^{\alpha_1}} \times \mathbb{Z}_{p_2^{\alpha_2}} \cdots \times \mathbb{Z}_{p_t^{\alpha_t}}$ (resp. $\mathbb{Z}_k = \mathbb{Z}_{p_1^{\beta_1}} \times \mathbb{Z}_{p_2^{\beta_2}} \cdots \times \mathbb{Z}_{p_t^{\beta_t}}$). Then we can consider a ring epimorphism $\pi : \mathbb{Z}_{p_1^{\alpha_1}} \times \mathbb{Z}_{p_2^{\alpha_2}} \cdots \times \mathbb{Z}_{p_t^{\alpha_t}} \to \mathbb{Z}_{p_1^{\beta_1}} \times \mathbb{Z}_{p_2^{\beta_2}} \cdots \times \mathbb{Z}_{p_t^{\beta_t}}$ given by $\pi(a_1, \cdots, a_t) = (\bar{a}_1, \cdots, \bar{a}_t)$ for all $(a_1, \cdots, a_t) \in \mathbb{Z}_{p_1^{\alpha_1}} \times \mathbb{Z}_{p_2^{\alpha_2}} \cdots \times \mathbb{Z}_{p_t^{\alpha_t}}$ where $\bar{a}_i$ is the remainder obtained from dividing a_i by $p_i^{\beta_i}$ for all i.

Case 1. Suppose that $\beta_i \geq 1$ for all $i = 1, \cdots, t$.

Let $\bar{g} = (\bar{g}_1, \cdots, \bar{g}_t) \in \mathbb{Z}_{p_1^{\beta_1}} \times \mathbb{Z}_{p_2^{\beta_2}} \cdots \times \mathbb{Z}_{p_t^{\beta_t}}$ be an arbitrary unit. Then there exists an element $g = (g_1, \cdots, g_t) \in \mathbb{Z}_{p_1^{\alpha_1}} \times \cdots \times \mathbb{Z}_{p_t^{\alpha_t}}$ such that $\pi(g) = \bar{g}$ i.e., $g_i \equiv \bar{g}_i \pmod{p_i^{\beta_i}}$ for all i. Since $\bar{g}$ is a unit in $\mathbb{Z}_{p_1^{\beta_1}} \times \mathbb{Z}_{p_2^{\beta_2}} \cdots \times \mathbb{Z}_{p_t^{\beta_t}}$, we have $(\bar{g}_i, p_i^{\beta_i}) = 1$ and so $(g_i, p_i^{\alpha_i}) = 1$ for all $i = 1, \cdots, t$, which implies that $g \in \mathbb{Z}_n$ is a unit.

Case 2. Suppose that $\beta_i = 0$ for some i.

Let $I_1 = \{i \in \{1, \cdots, t\} : \beta_i \geq 1\}$ and $I_2 = \{i \in \{1, \cdots, t\} : \beta_i = 0\}$. Consider $R = R_1 \times R_2$ where $R_1 = \prod_{i \in I_1} \mathbb{Z}_{p_i^{\beta_i}}$ and $R_2 = \prod_{i \in I_2} \{1_i\}$ where 1_i is the unity of $\mathbb{Z}_{p_i^{\beta_i}}$. By changing the order of the $\mathbb{Z}_{p_i^{\beta_i}}$ if necessary we can assume that $R = \mathbb{Z}_k = \mathbb{Z}_{p_1^{\beta_1}} \times \mathbb{Z}_{p_2^{\beta_2}} \cdots \times \mathbb{Z}_{p_t^{\beta_t}}$. Let $G(R)$ be the group of all units in R. Let $\bar{g} = (\bar{g}_1, \cdots, \bar{g}_{|I_1|}, 1_1, \cdots, 1_{|I_2|}) \in G(R)$ be arbitrary. Then by the similar argument given in Case 1, there exists a unit $g_i \in \mathbb{Z}_{p_i^{\alpha_1}}$ such that $g_i \equiv \bar{g}_i \pmod{p_i^{\beta_i}}$ for all $i = 1, \cdots, |I_1|$. Let $g = (g_1, \cdots, g_{|I_1|}, 1_1, \cdots, 1_{|I_2|}) \in \mathbb{Z}_{p_1^{\alpha_1}} \times \cdots \times \mathbb{Z}_{p_t^{\alpha_t}}$. Then g is a unit in $\mathbb{Z}_{p_1^{\alpha_1}} \times \cdots \times \mathbb{Z}_{p_t^{\alpha_t}}$ such that $\pi(g) = \bar{g}$. □

Theorem 2.6 Let n be any positive integer such that $X_n \neq \emptyset$. Then for all $x, y \in X_n$, $o(x) = o(y)$ if and only if $(x, n) = (y, n)$.

Proof. ($\Rightarrow$) Suppose that for all $x, y \in X_n$, $o(x) = o(y)$. Then $y = gx$ for some $g \in G_n$. Since $(g, n) = 1$, we have $(y, n) = (gx, n) = (x, n)$.

($\Leftarrow$) Suppose that for all $x, y \in X_n$, $(x, n) = (y, n)$. It is enough to consider $x|n$, i.e., $x = (x, n)$ by Lemma 2.4. Since $x|y$, $y = ax$ for some integer a. Since $x = (y, n)$, $x = by + cn$ for some integers b and c. Hence $x \equiv by \equiv bax \pmod{n}$, and then $1 \equiv ba \pmod{\frac{n}{x}}$. Let $\bar{a}$ be an element of $\mathbb{Z}_{\frac{n}{x}}$ so that $a \equiv \bar{a} \pmod{\frac{n}{x}}$. Then $1 \equiv b\bar{a} \pmod{\frac{n}{x}}$, which implies that $\bar{a} \in G_{\frac{n}{x}}$. By Lemma 2.5, there exists $a_0 \in G_n$ such that $a_0 \equiv \bar{a} \pmod{\frac{n}{x}}$. Since $a_0 = \bar{a} + k(\frac{n}{x})$ for some integer k, we have $a_0 x \equiv (\bar{a} + k(\frac{n}{x}))x \equiv \bar{a}x \equiv ax \equiv y \pmod{n}$, which implies that $o(x) = o(y)$. □

Remark 1. We note that the regular action on X_n by G_n is transitive, i.e., $X_n = o(x)$ for some $x \in X_n$ if and only if $n = p^2$ for some prime p.

Corollary 2.7 Let n be a positive integer and $x(\neq 1, n)$ be a divisor of n. Then $o(x) = \{gx : \forall g \in G_{\frac{n}{x}}\}$, and so $|o(x)| = |G_{\frac{n}{x}}|$.

Proof. Let $y \in o(x)$ be arbitrary. By Theorem 2.6, $(x, n) = (y, n)$. Since x is a divisor of n, $x = (x, n) = (y, n)$, and so $1 = (\frac{y}{x}, \frac{n}{x})$. Thus $\frac{y}{x} \in G_{\frac{n}{x}}$, and then $y = gx$ for some $g \in G_{\frac{n}{x}}$. Assume that there exist $g_1, g_2 \in G_{\frac{n}{x}}$ $(g_1 \neq g_2)$ such that $g_1 x = g_2 x$. Then $(g_1 - g_2)x \equiv 0 \pmod{n}$, which implies that $g_1 - g_2 \equiv 0 \pmod{\frac{n}{x}}$. Since $g_1 - g_2 \in \mathbb{Z}_{\frac{n}{x}}$, $g_1 - g_2 = 0$, a contradiction. Hence $o(x) = \{gx : g \in G_{\frac{n}{x}}\}$, and so $|o(x)| = |G_{\frac{n}{x}}|$. $\square$

Lemma 2.8 Let n be any positive integer such that $X_n \neq \emptyset$. If $Fix(g) \neq \emptyset$, then there exists $x \in Fix(g)$ such that $x|n$.

Proof. Since $Fix(g) \neq \emptyset$, there exists $y \in X_n$ such that $y = gy$. Let $x = (y, n)$. Clearly, $x|y$ and $o(x) = o(y)$ by Theorem 2.6, and so $y = hx$ for some $h \in G_n$. Therefore $hx = y = gy = g(hx) = h(gx)$, and so $x = gx$. Thus $x \in Fix(g)$ and $x|n$. $\square$

Theorem 2.9 Let n be any positive integer such that $X_n \neq \emptyset$. If $Fix(g) \neq \emptyset$ for some $g(\neq 1) \in G_n$, then there exists a divisor $x \in Fix(g)$ of n so that $< x > = \{kx : k = 1, \cdots\} = Fix(g)$.

Proof. By Lemma 2.8, we can choose the smallest divisor $x \in Fix(g)$ of n. Clearly, $Fix(g) \supseteq < x >$ by Theorem 2.2. Let $y \in Fix(g)$ be arbitrary. Then $y = gy$ and also $x = gx$. If $(x, y) = 1$, then there exist integers α and β such that $\alpha x + \beta y = 1$. Then $g = g1 = g(\alpha x + \beta y) = \alpha(gx) + \beta(gy) = \alpha x + \beta y = 1$, a contradiction. Hence $(x, y) \neq 1$. Let $(x, y) = d(\neq 1)$. Then there exist α_1 and β_1 such that $\alpha_1 x + \beta_1 y = d$. Thus we have $gd = g(\alpha x + \beta y) = \alpha(gx) + \beta(gy) = \alpha x + \beta y = d$, which implies that $d \in Fix(g)$. By the choice of x, we have $d = x$ and so $x|y$. Therefore $y \in < x >$, i.e., $Fix(g) \subseteq < x >$. $\square$

Theorem 2.10 Let $d(\neq 1, n)$ be a divisor of any odd positive integer n. Then $1 + \frac{n}{d} \in G_n$ or $1 - \frac{n}{d} \in G_n$. In this case, $d \in Fix(1 + \frac{n}{d})$ or $d \in Fix(1 - \frac{n}{d})$.

Proof. Let $n = p_1^{\alpha_1} \cdot p_2^{\alpha_2} \cdots p_s^{\alpha_s}$ be the prime factorization of n where p_i are all distinct primes and $\alpha_i \geq 1$ for all $i = 1, \cdots, s$. Let $d = p_1^{\beta_1} \cdot p_2^{\beta_2} \cdots p_s^{\beta_s}$.

Case 1. Suppose that $\alpha_i > \beta_i \geq 1$ for all $i = 1, \cdots, s$.

Then $\frac{n}{d}$ is nilpotent, and so $1 \pm \frac{n}{d} \in G_n$. Since $(1 \pm \frac{n}{d})d = d$, $d \in Fix(1 \pm \frac{n}{d})$.

Case 2. Suppose that $\alpha_i = \beta_i$ for some i.

If $\alpha_i = \beta_i$ for some i, then we have $(1 + \frac{n}{d}, p_i) = 1$ or $(1 - \frac{n}{d}, p_i) = 1$ for some i. Indeed, if $(1 \pm \frac{n}{d}, p_i) = p_i$, then $p_i | (1 + \frac{n}{d}) - (1 - \frac{n}{d}) = 2(\frac{n}{d})$. Since p_i is an odd prime, $p_i | \frac{n}{d}$, a contradiction. If $\alpha_j > \beta_j \geq 0$ for all $j \neq i$, then $(1 \pm \frac{n}{d}, p_j) = 1$. Indeed, if $(1 \pm \frac{n}{d}, p_j) = p_j$, then $p_j | (1 \pm \frac{n}{d})$ since $p_j | \frac{n}{d}$, and so $p_j | 1$, a contradiction. Therefore $(1 + \frac{n}{d}, p_i) = 1$ or $(1 - \frac{n}{d}, p_i) = 1$ for all $i = 1, \cdots, s$, which implies that $1 + \frac{n}{d} \in G_n$ or $1 - \frac{n}{d} \in G_n$, and so $d \in Fix(1 + \frac{n}{d})$ or $d \in Fix(1 - \frac{n}{d})$. $\qquad \square$

Remark 2. Note that Theorem 2.10 does not hold for some even positive integer n. For example, in $\mathbb{Z}_{180}$ there exists a divisor $d = 12$ of 180 such that $1 \pm \frac{180}{12} = 1 \pm 15 \notin G_{180}$. On the other hand, we can observe that for any even positive integer n, $\frac{n}{2} \in Fix(g)$ for all $g \in G_n$. Indeed, since n is even integer, every $g \in G_n$ is odd integer and so, $g = 1 + 2k$ for some positive integer k. Then $g(\frac{n}{2}) = (1 + 2k)(\frac{n}{2}) = \frac{n}{2}$, and so $\frac{n}{2} \in Fix(g)$. Also note that there exists $g \in G_n$ for some odd positive integer n such that $Fix(g) = \emptyset$. For example, $Fix(2) = \emptyset$ for some $2 \in G_{15}$.

Corollary 2.11 Let n be any odd positive integer such that $X_n \neq \emptyset$. Then the number of all distinct $Fix(g)'$ $(g \in G_n)$ is equal to the number of all the distinct proper divisors d of n.

Proof. It follows from Theorem 2.8 and Theorem 2.10. $\qquad \square$

Example 2 Consider $\mathbb{Z}_{225}$. Then $\{1, 3, 9, 5, 15, 25, 45, 75\}$ is the set of all proper divisors of 225. Thus we obtained all the $Fix(g)$'s in X_{225} as follow:

$Fix(1) = X_{225}$,

$Fix(76) = <3> = \{3, 6, \ldots, 222\}$,

$Fix(46) = <5> = \{5, 25, \ldots, 220\}$,

$Fix(26) = <9> = \{9, 18, \ldots, 216\}$,

$Fix(16) = <15> = \{15, 30, \ldots, 210\}$,

$Fix(10) = <25> = \{25, 50, \ldots, 200\}$,

$Fix(6) = <45> = \{45, 90, 135, 180\}$,

$Fix(4) = <75> = \{75, 150\}$.

Theorem 2.12 Let n be any positive integer such that $X_n \neq \emptyset$. Then (1) if $n = p^2$ for some prime p, then G_{p^2} is fixed-point free under the regular representation $\pi : G_n \longrightarrow \text{Sym}(X_n)$; (2) otherwise, G_n is not fixed-point free under the regular representation $\pi : G_n \longrightarrow \text{Sym}(X_n)$.

Proof. (1) $n = p^2$ for some prime p, then $X_{p^2} = \{p, 2p, \ldots, (p-1)p\}$. Since for all $1 + ip \in G_{p^2}$ $(p - 1 \geq i \geq 1)$ $(1 + ip)x = x$ for all $x \in X_{p^2}$, and so $Fix(1 + ip) = X_{p^2}$. On the other hand, we will show that $Fix(j + ip) = \emptyset$ for all $j + ip \in G_{p^2}$ $(p - 1 \geq i \geq 1, p - 1 \geq j \geq 2)$. Assume that $kp \in Fix(j + ip)$ for some $kp \in X_{p^2}$ and some $j + ip \in G_{p^2}$. Then $(j + ip)kp = kp$, and then $(j - 1)kp = 0$. Thus $p|(j - 1)k$, which implies that $p|(j - 1)$ or $p|k$, a contradiction. Hence G_{p^2} is fixed-point free under the regular representation $\pi : G_n \longrightarrow Sym(X_n)$.

(2) Suppose that $n \neq p^2$ for any prime p. Consider the following cases:

Case 1. n is even.

Note that $\frac{n}{2} \in Fix(-1)$ for some $\frac{n}{2} \in X_n$ and then $-1 \in stab(\frac{n}{2})$. Assume that G_n is fixed-point free under the regular representation $\pi : G_n \longrightarrow Sym(X_n)$. By assumption, $(-1)x = x$ for all $x \in X_n$. Choose k be an element of X_n so that $k < \frac{n}{2}$. Since $k(-1) = k$, $2k = 0$, a contradiction. Hence G_n is not fixed-point free under the regular representation $\pi : G_n \longrightarrow Sym(X_n)$.

Case 2. n is odd.

Let $d \in X_n$ be a divisor of n. Then by Theorem 2.10, there exists $g(\neq 1) \in G_n$ such that $d \in Fix(g)$. By Theorem 2.8, $Fix(g) = < d_0 >$ for some divisor $d_0(\neq 1)$ of n. Since $< d_0 > \neq X_n$, G_n is not fixed-point free under the regular representation $\pi : G_n \longrightarrow Sym(X_n)$. $\square$

3. Imprimitive set under the the regular action

Lemma 3.1 Let n be any positive integer such that $X_n \neq \emptyset$. If the regular action of G_n on X_n is not transitive, then (1) for all $x \in X_n$, the orbit of x $o(x)$ is the set of imprimitivity under the regular representation $\pi : G_n \longrightarrow Sym(X_n)$; (2) any union of orbits is a set of imprimitivity under the regular representation $\pi : G_n \longrightarrow Sym(X_n)$.

Proof. Clear. $\square$

Remark 3. By Theorem 2.2 and Lemma 3.1, for all $x \in X_n$ $Fix(x)$ is a set of imprimitivity under the regular represention $\pi : G_n \longrightarrow Sym(X_n)$.

Lemma 3.2 Let n be any positive integer such that $X_n \neq \emptyset$. Then $B \subseteq X_n$ is a set of imprimitivity if and only if $-B = \{-b : b \in B\}$ is a set of imprimitivity.

Proof. Suppose that B is a set of imprimitivity under the regular representation $\pi : G_n \longrightarrow Sym(X_n)$. If $(\pi(g))(B) = B$, then clearly

$(\pi(g))(-B) = -B$. If $(\pi(g))(B) \cap B = \emptyset$, then $(\pi(g))(-B) \cap (-B) = \emptyset$. By the similar argument, the converse also holds. $\qquad\square$

Theorem 3.3 Let n be any positive integer such that $X_n \neq \emptyset$. Then the following are equivalent:

(1) G_n is primitive on X_n under the regular representation $\pi : G_n \longrightarrow$ Sym(X_n);
(2) $n = p^2$ for some prime p;
(3) The regular action of G_n on X_n is transitive.

Proof. (1) $\Rightarrow$ (2). Assume that $n \neq p^2$ for any prime p. Since $n \neq 4$, the regular action of G_n on X_n is not trivial. Let p be any odd prime. Since $n \neq p^2$ for any prime p, there exists a proper divisors $x(\neq 1, n)$ of n such that $o(x) \neq \{x\}$ and $o(x) \neq X_n$. Since $o(x)$ is a set of imprimitivity under the regular representation $\pi : G_n \longrightarrow$ Sym(X_n) by Lemma 3.1, G_n is imprimitive on X_n under the regular representation $\pi : G_n \longrightarrow$ Sym(X_n), a contradiction. Hence $n = p^2$ for some prime p.

(2)$\Leftrightarrow$ (3). It follows from Remark 1.

(3)$\Rightarrow$ (1). Suppose that the regular action of G_n on X_n is transitive. Then $n = p^2$ for some prime p. Since for all $x \in X_n$, $|stab(x)| = \frac{|G_n|}{|o(x)|} = p$, $stab(x)$ is a maximal subgroup of G_n. Hence G_n is primitive on X_n under the regular representation $\pi : H \longrightarrow$ Sym(S) by [[1], Lemma 3.1]. $\qquad\square$

Lemma 3.4 Let n be any positive integer such that $X_n \neq \emptyset$. If $B \subseteq X_n$ is a set of imprimitivity under the regular representation $\pi : G_n \longrightarrow$ Sym(X_n), then $-B$ is a set of imprimitivity.

Proof. Clear. $\qquad\square$

Lemma 3.5 Let n be a positive integer such that $X_n \neq \emptyset$ and $x, y \in X_n$ such that $x|y$. If $|G_{\frac{n}{x}}| = |G_{\frac{n}{y}}|$ (equivalently $|o(x)| = |o(y)|$), then $stab(x) = stab(y)$.

Proof. Since $x|y$, $y = dx$ for some $d \in X_n$. Let $g \in stab(x)$ be arbitrary. Thus $gx = x$, and then $gy = g(dx) = (gx)d = xd = y$ and so $g \in stab(y)$. Hence $stab(x) \subseteq stab(y)$. Since $|stab(x)| = \frac{|G_n|}{|o(x)|}$, $|stab(y)| = \frac{|G_n|}{|o(y)|}$ and $|o(x)| = |o(y)|$, $|stab(x)| = |stab(y)|$, and so $stab(x) = stab(y)$. $\qquad\square$

Corollary 3.6 Let n be a positive integer such that $X_n \neq \emptyset$. If $y \in o(x)$ under the regular action on X_n by G_n, then $stab(x) = stab(y)$.

Proof. It follows from Lemma 3.5. $\qquad\square$

Lemma 3.7 Let n be a positive integer such that $X_n \neq \emptyset$, $x \in X_n$ such that $x|n$ and let $A_x = \{y \in X_n : x|y, y|n, |o(x)| = |o(y)|\}$. Then $A_x = \{x\}$ or $A_x = \{x, 2x\}$.

Proof. If $|A_x| = 1$, then clearly $A_x = \{x\}$. Suppose that $|A_x| \geq 2$. Let $y \in A_x$ such that $x \neq y$. Then $|G_{\frac{n}{x}}| = |G_{\frac{n}{y}}|$. Let $n = p_1^{\alpha_1} \cdots p_t^{\alpha_t}$ be the prime factorization of n. Then $x = p_1^{\beta_1} \cdots p_t^{\beta_t}$ and $y = p_1^{\gamma_1} \cdots p_t^{\gamma_t}$. Since $x|y(x < y)$, $\alpha_i \geq \gamma_i > \beta_i$ for some i. Note $|G_{\frac{n}{x}}| = |G_{p_1^{\alpha_1-\beta_1}}| \cdots |G_{p_t^{\alpha_t-\beta_t}}|$ (resp. $|G_{\frac{n}{y}}| = |G_{p_1^{\alpha_1-\gamma_1}}| \cdots |G_{p_t^{\alpha_t-\gamma_t}}|$). Since $|G_{\frac{n}{x}}| = |G_{\frac{n}{y}}|$, we have $p_i = 2$, $\beta_i + 1 = \gamma_i$ and $\beta_j = \gamma_j$ for all $j(j \neq i)$, and so $y = 2x$. Hence $A_x = \{x, 2x\}$. $\square$

Corollary 3.8 Let n be an even positive integer such that $X_n \neq \emptyset$. If $x \in X_n$ is odd such that $x|n$, then $A_x = \{y \in X_n : x|y, y|n, |o(x)| = |o(y)|\} = \{x\}$.

Proof. By Lemma 3.7, $A_x = \{x\}$ or $A_x = \{x, 2x\}$. Assume that $A_x = \{x, 2x\}$. Then $|G_{\frac{n}{x}}| = |G_{\frac{n}{2x}}|$. Let $n = 2^{\alpha_0}p_1^{\alpha_1} \cdots p_t^{\alpha_t}$ be the prime factorization of n where all p_i are odd primes. Then $x = p_1^{\beta_1} \cdots p_t^{\beta_t}$ and $2x = 2p_1^{\gamma_1} \cdots p_t^{\gamma_t}$. Since $|G_{\frac{n}{x}}| = 2^{\alpha_0-1}|G_{p_1^{\alpha_1-\beta_1}}| \cdots |G_{p_t^{\alpha_t-\beta_t}}|$ (resp. $|G_{\frac{n}{2x}}| = 2^{\alpha_0-2}|G_{p_1^{\alpha_1-\beta_1}}| \cdots |G_{p_t^{\alpha_t-\beta_t}}|$), $|G_{\frac{n}{x}}| > |G_{\frac{n}{2x}}|$, which is a contradiction. Hence $A_x = \{x\}$. $\square$

Corollary 3.9 Let n be an odd positive integer. If $x \in X_n$ is odd such that $x|n$, then $A_x = \{x\}$.

Proof. It follows by Lemma 3.7. $\square$

Lemma 3.9 Let n be a positive integer, $A_x = \{y \in X_n : x|y, y|n, |o(x)| = |o(y)|\}$ and $B_x = \{y \in X_n : x|y, y|n, stab(x) = stab(y)\}$ for some $x \in X_n$. Then $A_x = B_x$.

Proof. By Lemma 3.5, we have $A_x \subseteq B_x$. Let $y \in B_x$ be arbitrary. Then $|stab(x)| = |stab(x)|$, and so $|o(x)| = |o(y)|$. Thus $B_x \subseteq A_x$. $\square$

Lemma 3.10 Let n be a positive integer. Suppose that $stab(x) = stab(y)$ for some $x, y \in X_n$. Then $stab((x, y)) = stab(x)$.

Proof. Let $d = (x, y)$. Then $d = \alpha x + \beta y$ for some $\alpha, \beta \in \mathbb{Z}$. For all $g \in stab(x) = stab(y)$, since $stab(x) = stab(y)$, $gd = g(\alpha x + \beta y) = \alpha(gx) + \beta(gy) = \alpha x + \beta y = d$, and so $g \in stab(d)$. Hence $stab(x) \subseteq stab(d)$. Clearly, $stab(d) \subseteq stab(x)$. Hence $stab((x, y)) = stab(x)$. $\square$

Theorem 3.11 Let n be a positive integer. Then $B_x = \{y \in X_n : stab(x) = stab(y)\} = o(x)$ or $o(x) \cup o(2x)$.

Proof. Without loss of generality, we can assume that $x|y$ for all $y \in B_x$ by Lemma 3.10. Let $P = \{x_1, \cdots, x_r\}$ be the set of all proper divisors of n. Then $B_x = \{y \in \cup_{i=1}^r o(x_i) : stab(x) = stab(y)\} = \cup_{i=1}^r \{y \in o(x_i) : stab(x) = stab(y)\}$. By Lemma 3.5 and Lemma 3.7, $\{x_i \in P : stab(x_i) = stab(x) = stab(y)\} = \{x\}$ or $\{x, 2x\}$. Therefore, $B_x = \{y \in X_n : stab(x) = stab(y)\} = o(x)$ or $o(x) \cup o(2x)$. $\qquad\square$

Example 2 In $\mathbb{Z}_{40}$, $B_5 = \{y \in X_{40} : stab(5) = stab(y)\} = o(5) = \{5, 15, 25\}$ and $B_4 = \{y \in X_{40} : stab(4) = stab(y)\} = o(4) \cup o(8)$ where $o(4) = \{4, 12, 28, 36\}$ and $o(8) = \{8, 16, 24, 32\}$.

Theorem 3.12 Let n be a positive integer and $B \subseteq o(x)$ be a set of imprimitivity under the regular representation $\pi : G_n \longrightarrow \mathrm{Sym}(X_n)$ for some $x \in X_n$ such that $x < \frac{n}{2}$ and $x|n$. If $|B| = 2$, then $B = \{x_1, -x_1\}$ for some $x_1 \in X_n$.

Proof. Let $B = \{x_1, y_1\}$ ($x_1 \neq y_1$). Then $y_1 = gx_1$ for some $g \in G_n$. Since B is a set of imprimitivity, $x_1 = gy_1$, and so $x_1 = g^2 x_1$ and $y_1 = g^2 y_1$. Assume that $(-g)x_1 = y_1$. Then $-x_1 = (-g)y_1 = x_1$, and so $2x_1 = 0$, and then $2x = 0$, which is a contradiction. Therefore $(-g)x_1 = x_1$, and so $y_1 = -x_1$. Hence $B = \{x_1, -x_1\}$. $\qquad\square$

Theorem 3.13 Let n be a positive integer. If B is a set of imprimitivity in X_n under the regular representation $\pi : G_n \longrightarrow \mathrm{Sym}(X_n)$ such that $|B| = |B^c|$ where $B^c = X_n \setminus B$, then B^c is a set of imprimitivity under the regular representation π.

Proof. Note that $X_n = B \cup B^c = g(B \cup B^c) = gB \cup gB^c$ for all $g \in G_n$.

Case 1. $gB = B$.

Then $X_n = B \cup B^c = gB \cup gB^c = B \cup gB^c$. Thus $B^c \subseteq gB^c$. Since $|B^c| = |gB^c|$, $B^c = gB^c$.

Case 2. $gB \cap B = \emptyset$.

Then $gB \subseteq B^c$. Since $X_n = B \cup B^c = gB \cup gB^c \subseteq B^c \cup gB^c = X_n$, $B \subseteq gB^c$. Since $|B| = |B^c| = |gB^c|$, $B = gB^c$, and so $B^c \cap gB^c = \emptyset$. $\qquad\square$

Corollary 3.14 Let n be a positive integer and $B \subseteq o(x)$ for some $x \in X_n$. If B is a set of imprimitivity in X_n under the regular representation $\pi : G_n \longrightarrow \mathrm{Sym}(X_n)$ such that $|B| = |B^c|$ where $B^c = o(x) \setminus B$, then B^c is a set of imprimitivity under the regular representation π.

Proof. It follows from the similar argument given in the proof of Theorem 3.13. $\qquad\square$

Example 3 In $\mathbb{Z}_{40}$, $B = \{2, 18, 22, 38\} \subseteq o(2)$ is a set of imprimitivity in X_{40} under the regular representation $\pi : G_{40} \longrightarrow \text{Sym}(X_{40})$. Also $B^c = o(2) \setminus B = \{6, 14, 26, 34\}$ is a set of imprimitivity in X_{40} under the regular representation π.

Theorem 3.15 Let n be a positive integer and B be a subset of $o(x)$ for some $x \in X_n$ such that $|B| > \frac{|o(x)|}{2}$. Then B is not a set of imprimitivity under the regular representation $\pi : G_n \longrightarrow \text{Sym}(X_n)$.

Proof. Note that $X_n = B \cup B^c = g(B \cup B^c) = gB \cup gB^c$ for all $g \in G_n$. Assume that B is a set of imprimitivity under the regular representation π.

Case 1. $gB = B$.

Then $B \subseteq B^c$. Since $|o(x)| - |B| = |B^c| = |gB^c| \geq |B|$, $\frac{|o(x)|}{2} \geq |B|$, a contradiction.

Case 2. $gB \cap B = \emptyset$.

Then $gB \subseteq B^c$ and so $|B^c| = |gB| \geq |B|$. We have a contradiction by the similar argument given in the case 1. $\square$

Theorem 3.16 Let n be a positive integer and let $x \in X_n$ such that $x|n$, $|o(x)| \neq 1$ and $x < \frac{n}{2}$. Then there exists no set $B \subseteq o(x)$ of imprimitivity under the regular representation $\pi : G_n \longrightarrow \text{Sym}(X_n)$ such that $|B|$ is odd and $|B|$ is not a divisor of $|o(x)|$.

Proof. Assume that there exists a set $B \subseteq o(x)$ of imprimitivity under the regular representation π such that $|B|$ is odd and $|B|$ is a divisor of $|o(x)|$. Let $B = \{x_1, \cdots, x_k\}$.

Case 1. $gB = B$.

Since $|o(x)| \neq 1$, there exists $g \in G_n$ such that $gx \neq x$. Assume that $gx_i = x_i$ for some $x_i \in B$, then $g \in stab(x_i)$. Since $stab(x_i) = stab(x)$, $gx = x$, a contradiction. Hence $gx_i \neq x_i$ for all $x_i \in B$. Note that there exists a permutation $\sigma_g \in S_k$ where S_k is the symmetric group of degree k such that $gx_i = x_{\sigma_g(i)}$. Assume that σ_g is a cycle of length k for all $g \in G_n$. Then $g^k x_i = x_i$ and by assumption, $x_i = (-g)^k x_i = -x_i$. Thus $2x_i = 0$, and so $2x = 0$, which is a contradiction. Hence there exists $g \in G_n$ such that $gx_i = x_{\sigma_g(i)}$ for all x_i where σ_g is not a cycle of length k. Let $\sigma_g = \mu_1 \mu_2 \cdots \mu_l$ be a product of disjoint cycles $\mu_1, \mu_2, \cdots, \mu_l \in S_k$. Since k is odd, we can let $|\mu_1| = r, |\mu_2| = s$ with $1 < r < s$

where $|\mu_i|$ is the length of μ_i. Note that $g^r x_i = \mu_1^r x_i = x_i$ for some $x_i \in B$ and $g^s x_j = \mu_2^s x_j = x_j$ for some $x_j \in B$. Then $g^r \in stab(x_i)$. Since $stab(x_i) = stab(x) = stab(x_j)$, $g^r x_j = \mu_2^r x_j = x_j$, which is a contradiction.

Case 2. $gB \cap B = \emptyset$.

We have that $G_n = \{g \in G_n : gB \cap B = \emptyset\}$ by assumption and Case 1. Consider a set $S = \{gB : \forall g \in G_n\}$. Observe that $g_1 B \neq g_2 B$ for all $g_1, g_2 \in G_n$ ($g_1 \neq g_2$). Let $y \in o(x)$ be arbitrary. Then $y = gx$ for some $g \in G_n$. Since there exists $b \in B$ such that $b = hx$ for some $h \in G_n$, $y = gx = (gh)b$, and so $y \in (gh)B$, which implies that $o(x)$ is equal to the disjoint union of all $gB \in S$. Since $|B| = |gB|$ for all $g \in G_n$, $|o(x)| = |S||B|$, which is a contradiction. $\qquad\square$

Theorem 3.17 Let n be a positive integer and let H a subgroup of G_n. If $B = \{hx : \forall h \in H\} \subseteq o(x)$ for some orbit $o(x)$ in X_n, then B is a set of imprimitivity under the regular representation $\pi : G_n \longrightarrow \operatorname{Sym}(X_n)$.

Proof. Note that $B = hB$ for all $h \in H$. Suppose that $gB \cap B \neq \emptyset$ for some $g \in G_n \setminus H$. Then there exist $h_i, h_j \in H$ such that $gh_i x = h_j x$. Let $b \in B$ be arbitrary. Then $b = hx$ for some $h \in H$, and so $b = hx = hh_j^{-1}(h_j x) = hh_j^{-1}(gh_i x) = g(hh_j^{-1}h_i x) \in gB$. Thus $B \subseteq gB$. Similarly, we have $B \subseteq g^{-1}B$. Since $gB \subseteq g(g^{-1}B) = B$, $B = gB$. Hence we have the result. $\qquad\square$

Theorem 3.18 Let n be a positive integer and B be a subset of $o(x)$ in X_n containing x. If B is a set of imprimitivity under the regular representation $\pi : G_n \longrightarrow \operatorname{Sym}(X_n)$, then there exists a subgroup H of G_n such that $B = Hx$.

Proof. Let $B = \{x, g_1 x, \cdots, g_k x\}$ for some $g_1, \cdots, g_k \in G_n$. Consider $H = \langle g_1, \cdots, g_k \rangle$, a subgroup of G_n generated by $\{g_1, \cdots, g_k\}$. Clearly, $B \subseteq Hx$. Next, we will show that $Hx \subseteq B$. for all $h \in H$. Since B is a set imprimitivity under the regular representation π and $g_i x \in B \cap g_i B$ (resp. $g_i^{-1}x \in B \cap g_i B$), $g_i B = B$ (resp. $g_i^{-1}B = B$ for all $i = 1, \cdots, k$, and so $g_i^{\alpha_i} B = B$ for any integer α_i. Then we have $gx \in B$ for all $g \in H$, and so $Hx \subseteq B$. $\qquad\square$

Corollary 3.19 Let n be a positive integer and B be a subset of $o(x)$ in X_n. If B is a set of imprimitivity under the regular representation $\pi : G_n \longrightarrow \operatorname{Sym}(X_n)$, then there exists a subgroup H of G_n and an element $g \in G_n$ such that $B = gHx$.

Proof. Observe that if B is a set of imprimitivity under the regular representation $\pi : G_n \longrightarrow \mathrm{Sym}(X_n)$, then for all $g \in G_n$ gB is also a set of imprimitivity under the regular representation π. Indeed, if $g_i(gB) \cap gB \neq \emptyset$ for some $g_i \in G_n$, then $g_i(gb) = gb_1$ for some $b, b_1 \in B$, and so $g_i b = b_1 \in g_i B \cap B$. Since B is a set of imprimitivity under the regular representation π, $g_i B = B$, and so $g_i(gB) = gB$. Thus gB is also a set of imprimitivity under the regular representation π for all $g \in G_n$.

If $x \in B$, then $B = Hx$ for some subgroup H of G_n by Theorem 3.18. Suppose that $x \notin B$. Then we can take $g \in G_n$ so that $x \in gB$. By the observation, gB is also a set of imprimitivity under the regular representation π containing x. Hence $gB = Hx$ for some subgroup H of G_n by Theorem 3.18, and so $B = g^{-1}Hx$. $\qquad\square$

Corollary 3.20 Let n be a positive integer and $B \subseteq o(x_1) \cup \cdots \cup o(x_\ell)$ for some orbits $o(x_1), \cdots, o(x_\ell)$ in X_n. If B is a set of imprimitivity under the regular representation $\pi : G_n \longrightarrow \mathrm{Sym}(X_n)$, then there exist subgroups H_i of G_n and elements $g_i \in G_n$ such that $B = g_1 H_1 \cup \cdots \cup g_\ell H_\ell$ for $i = 1, \cdots, \ell$.

Proof. Let $B_{x_i} = B \cap o(x_i)$ for each $i = 1, \cdots, \ell$. We can observe that for each $i = 1, \cdots, \ell$, B_{x_i} is a set of imprimitivity under the regular representation π. Indeed, suppose that $B_{x_i} \cap gB_{x_i} \neq \emptyset$ for some $g \in G$. Then $b_1 = gb$ for some $b_1, b \in B_{x_i}$, and so $b_1 = gb \in B \cap gB (\neq \emptyset)$. Since B is a set of imprimitivity under the regular representation π, $B = gB$, $B_{x_i} = B \cap o(x_i) = gB \cap o(x_i) = g(B \cap o(x_i)) = gB_{x_i}$, which implies that B_{x_i} is a set of imprimitivity under the regular representation π for each $i = 1, \cdots, \ell$. By Corollary 3.19, we have the result. $\qquad\square$

References

[1] J.A. Cohen, K. Koh *Half-transitive group actions in a compact ring*, J. Pure Appl. Algebra **60** (1989), 139-153.
[2] J.A. Cohen, K. Koh and J.W. Lorimer, *On the primitive representations of the group of units of a ring to the symmetric group of the nonunits*, J. Pure Appl. Algebra **94** (1994), 285-306.

DEPARTMENT OF MATHEMATICS EDUCATION
PUSAN NATIONAL UNIVERSITY
PUSAN, 609-735, SOUTH KOREA
E-mail address: jchan@pusan.ac.kr

DEPARTMENT OF MATHEMATICS EDUCATION
PUSAN NATIONAL UNIVERSITY
PUSAN, 609-735, SOUTH KOREA
E-mail address: ylee@pusan.ac.kr

DEPARTMENT OF MATHEMATICS
DONG-A UNIVERSITY
PUSAN, 604-714, SOUTH KOREA
E-mail address: swpark@donga.ac.kr

Proceedings of the Sixth China-Japan-Korea
International Conference on Ring Theory
June 27-July 2, 2011 Suwon, Korea

ON SYMMETRIC BIDERIVATIONS OF
SEMIPRIME RINGS

ASMA ALI* AND FAIZA SHUJAT

ABSTRACT. Let R be a ring with centre $Z(R)$. A biadditive
symmetric mapping $D(.,.) : R \times R \longrightarrow R$ is called symmetric
biderivation if for any fixed $y \in R$, the mapping $x \mapsto D(x,y)$ is
a derivation. A mapping $f : R \longrightarrow R$ defined by $f(x) = D(x,x)$
is called the trace of D. In this paper we prove that a nonzero
Lie ideal L of a semiprime ring R of characteristic different from
two is central if it satisfies any one of the following properties:
(i) $f(xy) \mp [x,y] \in Z(R)$, (ii) $f(xy) \mp [y,x] \in Z(R)$, (iii) $f(xy) \mp
xy \in Z(R)$, (iv)$f(xy) \mp yx \in Z(R)$, (v) $f([x,y]) \mp [x,y] \in Z(R)$,
(vi) $f([x,y]) \mp [y,x] \in Z(R)$, (vii) $f([x,y]) \mp xy \in Z(R)$, (viii)
$f([x,y]) \mp yx \in Z(R)$, (ix) $f(xy) \mp f(x) \mp [x,y] \in Z(R)$, (x)
$f(xy) \mp f(y) \mp [x,y] \in Z(R)$, (xi) $f([x,y]) \mp f(x) \mp [y,x] \in Z(R)$,
(xii) $f([x,y]) \mp f(x) \mp [y,x] \in Z(R)$, (xiii) $f([x,y]) \mp f(y) \mp
[x,y] \in Z(R)$, (xiv) $f([x,y]) \mp f(y) \mp [y,x] \in Z(R)$, (xv) $f([x,y]) \mp
f(xy) \mp [x,y] \in Z(R)$, (xvi) $f([x,y]) \mp f(xy) \mp [y,x] \in Z(R)$, (xvii)
$f(x)f(y) \mp [x,y] \in Z(R)$, (xviii) $f(x)f(y) \mp [y,x] \in Z(R)$, (xix)
$f(x)f(y) \mp xy \in Z(R)$, (xx) $f(x)f(y) \mp yx \in Z(R)$, where f stands
for the trace of a biadditive symmetric mapping $D(.,.) : R \times R \longrightarrow
R$. Moreover, motivated by a well known theorem of Posner [11,
Theorem 2] and a result of Deng and Bell [6, Theorem 2], we
prove that if R admits a symmetric biderivation D such that the
trace f of D is n-centralizing on L, then f is n-commuting on L.

1. Introduction

Throughout the paper R will denote an associative ring with centre $Z(R)$. A ring R is said to be prime (resp. semiprime) if $aRb = \{0\}$
implies that either $a = 0$ or $b = 0$ (resp. $aRa = \{0\}$ implies that $a = 0$).
For each pair of elements $x, y \in R$ we shall write $[x,y]$ the commutator $xy - yx$. We make extensive use of basic commutator identities
$[xy, z] = [x, z]y + x[y, z]$ and $[x, yz] = [x, y]z + y[x, z]$. An additive subgroup L of R is said to be a Lie ideal of R if $[x, r] \in L$ for all $x \in L$
and $r \in R$. An additive mapping $d : R \longrightarrow R$ is called a derivation if

2010 Mathematics Subject Classification : 16N60, 16W25, 16U80.

Keywords : Semiprime rings, Lie ideals, symmetric biadditive mappings, symmetric biderivations.

*Corresponding author.

$d(xy) = d(x)y + xd(y)$, for all $x, y \in R$. A derivation d is inner if there exists an element $a \in R$ such that $d(x) = [a, x]$ for all $x \in R$. A mapping $D(.,.) : R \times R \longrightarrow R$ is said to be symmetric if $D(x, y) = D(y, x)$, for all $x, y \in R$. A mapping $f : R \longrightarrow R$ defined by $f(x) = D(x, x)$, where $D(.,.) : R \times R \longrightarrow R$ is a symmetric mapping, is called the trace of D. It is obvious that in the case $D(.,.) : R \times R \longrightarrow R$ is a symmetric mapping which is also biadditive (i.e. additive in both arguments), the trace f of D satisfies the relation $f(x+y) = f(x) + f(y) + 2D(x, y)$, for all $x, y \in R$. A biadditive symmetric mapping $D(.,.) : R \times R \longrightarrow R$ is called a symmetric biderivation if $D(xy, z) = D(x, z)y + xD(y, z)$ for all $x, y, z \in R$. Obviously in this case the relation $D(x, yz) = D(x, y)z + yD(x, z)$ is also satisfied for all $x, y, z \in R$.

Gy. Maksa [9] introduced the concept of a symmetric biderivation (see also [10], where an example can be found). It was shown in [9] that symmetric biderivations are related to general solution of some functional equations. Some results on symmetric biderivation in prime and semiprime rings can be found in [12] and [13]. The notion of additive commuting mappings is closely connected with the notion of biderivations. Every commuting additive mapping $f : R \longrightarrow R$ gives rise to a biderivation on R. Namely linearizing $[x, f(x)] = 0$ for all $x \in R$, we get $[f(x), y] = [x, f(y)]$ for all $x, y \in R$ and hence we note that the mapping $(x, y) \mapsto [f(x), y]$ is a biderivation (moreover, all derivations appearing are inner). Deng [5] defined n-centralizing mappings in rings. A mapping $f : R \longrightarrow R$ is said to be n-centralizing (resp. n-commuting) on a non empty subset S of R if $[f(x), x^n] \in Z(R)$ (resp. $[f(x), x^n] = 0$) for all $x \in S$ and n an arbitrary positive integer. There has been ongoing interest concerning the relationship between the commutativity of a ring and the existence of certain specific types of derivations. Recently many authors viz. [1], [2], [3] and [4] have obtained commutativity of prime and semiprime rings with derivations satisfying certain polynomial constraints.

2. Symmetric biadditive maps

In 1992 Daif and Bell [4, Theorem 2] proved that if a semiprime ring R admits a derivation d such that $d([x, y]) - [x, y] \in Z(R)$ for all $x, y \in R$, then R is commutative. Very recently the first author [1] investigated the commutativity of a prime ring R admitting a generalized derivation F satisfying any one of the following properties: (i) $F(xy) \mp xy \in Z(R)$, (ii) $F(xy) \mp yx \in Z(R)$, (iii) $F(x)F(y) \mp xy \in Z(R)$.

Motivated by the aforementioned results in this section, we prove the following: Let R be a semiprime ring of characteristic different from two admitting a biadditive symmetric map D with trace f and L be a nonzero Lie ideal of R. Then $L \subseteq Z(R)$ if for all $x, y \in L$ any one of the following holds: (i) $f(xy) \mp [x, y] \in Z(R)$, (ii) $f(xy) \mp [y, x] \in Z(R)$, (iii) $f(xy) \mp xy \in Z(R)$, (iv) $f(xy) \mp yx \in Z(R)$, (v) $f([x, y]) \mp [x, y] \in Z(R)$, (vi) $f([x, y]) \mp [y, x] \in Z(R)$, (vii) $f([x, y]) \mp xy \in Z(R)$, (viii) $f([x, y]) \mp yx \in Z(R)$, (ix) $f(xy) \mp f(x) \mp [x, y] \in Z(R)$, (x) $f(xy) \mp f(y) \mp [x, y] \in Z(R)$, (xi) $f([x, y]) \mp f(x) \mp [x, y] \in Z(R)$, (xii) $f([x, y]) \mp f(x) \mp [y, x] \in Z(R)$, (xiii) $f([x, y]) \mp f(y) \mp [x, y] \in Z(R)$, (xiv) $f([x, y]) \mp f(y) \mp [y, x] \in Z(R)$, (xv) $f([x, y]) \mp f(xy) \mp [x, y] \in Z(R)$, (xvi) $f([x, y]) \mp f(xy) \mp [y, x] \in Z(R)$, (xvii) $f(x)f(y) \mp [x, y] \in Z(R)$, (xviii) $f(x)f(y) \mp [y, x] \in Z(R)$, (xxi) $f(x)f(y) \mp xy \in Z(R)$, (xx) $f(x)f(y) \mp yx \in Z(R)$.

The following lemmas are essential to prove our theorems.

Lemma 2.1. [2] The centre of a semiprime ring contains no nonzero nilpotent elements.

Lemma 2.2. [8, Lemma 1] Let R be a semiprime ring and L be a nonzero Lie ideal of R. If $[L, L] \subseteq Z(R)$, then $L \subseteq Z(R)$.

Lemma 2.3. Let R be a semiprime ring and L be a nonzero Lie ideal of R. If $L^2 \subseteq Z(R)$, then $L \subseteq Z(R)$.

Proof. Since $xy \in Z(R)$ for all $x, y \in L$, $xy - yx = [x, y] \in Z(R)$ for all $x, y \in L$. Using Lemma 2.2 we get the required result. $\square$

Theorem 2.1. Let R be a semiprime ring of characteristic not two and L be a nonzero Lie ideal of R. Let $D(.,.) : R \times R \longrightarrow R$ be a biadditive symmetric mapping and f be the trace of D. If $f(xy) \mp [x, y] \in Z(R)$ for all $x, y \in L$, then $L \subseteq Z(R)$.

Proof. Suppose

$$f(xy) - [x, y] \in Z(R) \text{ for all } x, y \in L. \tag{2.1}$$

Replacing y by $y + z$ in (2.1) we get

$$f(xy) + f(xz) + 2D(xy, xz) - [x, y] - [x, z] \in Z(R) \text{ for all } x, y, z \in L. \tag{2.2}$$

Since R is of characteristic not two, (2.1) yields that

$$D(xy, xz) \in Z(R) \text{ for all } x, y, z \in L. \tag{2.3}$$

Substituting y for z in (2.3), we get

$$f(xy) = D(xy, xy) \in Z(R) \text{ for all } x, y \in L. \tag{2.4}$$

In view of (2.1), (2.4) yields that

$$[x, y] \in Z(R) \text{ for all } x, y \in L. \tag{2.5}$$

This implies that $[L, L] \subseteq Z(R)$. Hence $L \subseteq Z(R)$ by Lemma 2.2. Similarly, we can prove the result for the case $f(xy) + [x, y] \in Z(R)$ for all $x, y \in L$. □

Theorem 2.2. Let R be a semiprime ring of characteristic not two and L be a nonzero Lie ideal of R. Let $D(., .) : R \times R \longrightarrow R$ be a biadditive symmetric mapping and f be the trace of D. If $f(xy) \mp [y, x] \in Z(R)$ for all $x, y \in L$, then $L \subseteq Z(R)$.

Proof. The proof runs on the same parallel lines as of Theorem 2.1. □

Theorem 2.3. Let R be a semiprime ring of characteristic not two and L be a nonzero Lie ideal of R. Let $D(., .) : R \times R \longrightarrow R$ be a biadditive symmetric mapping and f be the trace of D. If $f(xy) \mp xy \in Z(R)$ for all $x, y \in L$, then $L \subseteq Z(R)$.

Proof. Let

$$f(xy) - xy \in Z(R) \text{ for all } x, y \in L. \tag{2.6}$$

Replacing y by $y + z$ we get

$$f(xy) + f(xz) + 2D(xy, xz) - xy - xz \in Z(R) \text{ for all } x, y, z \in L. \tag{2.7}$$

Using (2.6), we obtain

$$2D(xy, xz) \in Z(R) \text{ for all } x, y, z \in L. \tag{2.8}$$

Since R is of characteristic not two, we have

$$D(xy, xz) \in Z(R) \text{ for all } x, y, z \in L. \tag{2.9}$$

Substituting y for z in (2.9), we get

$$f(xy) = D(xy, xy) \in Z(R) \text{ for all } x, y \in L. \tag{2.10}$$

Using (2.6), we have $xy \in Z(R)$ for all $x, y \in L$. Hence $L^2 \subseteq Z(R)$ and by Lemma 2.3 $L \subseteq Z(R)$. Similarly we can prove the result if $f(xy) + xy \in Z(R)$ for all $x, y \in L$. □

Theorem 2.4. Let R be a semiprime ring of characteristic not two and L be a nonzero Lie ideal of R. Let $D(., .) : R \times R \longrightarrow R$ be a biadditive symmetric mapping and f be the trace of D. If $f(xy) \mp yx \in Z(R)$ for all $x, y \in L$, then $L \subseteq Z(R)$.

Proof. The proof runs on the same parallel lines as of Theorem 2.3. □

Theorem 2.5. Let R be a semiprime ring of characteristic not two and L be a nonzero Lie ideal of R. Let $D(.,.) : R \times R \longrightarrow R$ be a biadditive symmetric mapping and f be the trace of D. If $f([x,y]) \mp [x,y] \in Z(R)$ for all $x, y \in L$, then $L \subseteq Z(R)$.

Proof. Let
$$f([x,y]) - [x,y] \in Z(R) \text{ for all } x, y \in L. \tag{2.11}$$
Replacing y by $y + z$, we have $f([x,y] + [x,z]) - [x,y] - [x,z] \in Z(R)$
i.e. $f([x,y]) + f([x,z]) + 2D([x,y],[x,z]) - [x,y] - [x,z] \in Z(R)$ for all $x, y, z \in L$. Using (2.11), we get
$$2D([x,y],[x,z]) \in Z(R) \text{ for all } x, y, z \in L. \tag{2.12}$$
Substituting y for z in (2.12) and using the fact that R is not of characteristic two, we find
$$f([x,y]) \in Z(R) \text{ for all } x, y \in L. \tag{2.13}$$
In view of (2.11), (2.14) yields that $[x,y] \in Z(R)$ for all $x, y \in L$. Thus we get $[L, L] \subseteq Z(R)$ and by Lemma 2.2 $L \subseteq Z(R)$. Similarly one can prove the result if $f([x,y]) + [x,y] \in Z(R)$ for all $x, y \in L$. $\square$

Using similar arguments as we have done in the proof of Theorem 2.5, we can prove the following:

Theorem 2.6. Let R be a semiprime ring of characteristic not two and L be a nonzero Lie ideal of R. Let $D(.,.) : R \times R \longrightarrow R$ be a biadditive symmetric mapping and f be the trace of D. If $f([x,y]) \mp [y,x] \in Z(R)$ for all $x, y \in L$, then $L \subseteq Z(R)$.

Theorem 2.7. Let R be a semiprime ring of characteristic not two and L be a nonzero Lie ideal of R. Let $D(.,.) : R \times R \longrightarrow R$ be a biadditive symmetric mapping and f be the trace of D. If $f([x,y]) \mp xy \in Z(R)$ for all $x, y \in L$, then $L \subseteq Z(R)$.

Proof. Let
$$f([x,y]) - xy \in Z(R) \text{ for all } x, y \in L. \tag{2.14}$$
Replacing y by $y + z$ in (2.14), we have $f([x,y] + [x,z]) - xy - xz \in Z(R)$
for all $x, y, z \in L$. This implies that
$$f([x,y]) + f([x,z]) + 2D([x,y],[x,z]) - xy - xz \in Z(R) \text{ for all } x, y, z \in L. \tag{2.15}$$
Using (2.14) we obtain
$$2D([x,y],[x,z]) \in Z(R) \text{ for all } x, y, z \in L. \tag{2.16}$$
Since R is of characteristic not two, (2.16) yields that
$$D([x,y],[x,z]) \in Z(R) \text{ for all } x, y, z \in L. \tag{2.17}$$

In particular, if we substitute y for z in (2.17), then we have $f([x,y]) \in Z(R)$ for all $x, y \in L$. Again using (2.14), we get $xy \in Z(R)$ for all $x, y \in L$. Thus $L^2 \in Z(R)$ and application of Lemma 2.3 completes the proof. Similarly we can prove the result if $f([x,y]) + xy \in Z(R)$ for all $x, y \in L$. $\qquad\qquad\square$

Theorem 2.8. Let R be a semiprime ring of characteristic not two and L be a nonzero Lie ideal of R. Let $D(.,.) : R \times R \longrightarrow R$ be a biadditive symmetric mapping and f be the trace of D. If $f([x,y]) \mp yx \in Z(R)$ for all $x, y \in L$, then $L \subseteq Z(R)$.

Proof. The proof runs on the same parallel lines as that of Theorem 2.7. $\qquad\qquad\square$

Theorem 2.9. Let R be a semiprime ring of characteristic not two and L be a nonzero Lie ideal of R. Let $D(.,.) : R \times R \longrightarrow R$ be a biadditive symmetric mapping and f be the trace of D. If $f(xy) \mp f(x) \mp [x,y] \in Z(R)$ for all $x, y \in L$, then $L \subseteq Z(R)$.

Proof. Suppose

$$f(xy) - f(x) - [x,y] \in Z(R) \text{ for all } x, y \in L. \qquad (2.18)$$

Replacing y by $y + z$, we get $f(xy) + f(xz) + 2D(xy, xz) - f(x) - [x,y] - [x,z] \in Z(R)$ for all $x, y, z \in L$. Using (2.18) we obtain

$$f(xz) + 2D(xy, xz) - [x,z] \in Z(R) \text{ for all } x, y, z \in L. \qquad (2.19)$$

Substituting $-z$ for z in (2.19), we get

$$f(xz) - 2D(xy, xz) + [x,z] \in Z(R) \text{ for all } x, y, z \in L. \qquad (2.20)$$

Adding (2.19) and (2.20) we obtain

$$2f(xz) \in Z(R) \text{ for all } x, z \in L. \qquad (2.21)$$

Since R is not of characteristic two, we have $f(xz) \in Z(R)$ for all $x, y \in L$. Using (2.18), we get

$$f(x) - [x,y] \in Z(R) \text{ for all } x, y \in L. \qquad (2.22)$$

Replacing x by $x + z$ in (2.22), we have

$$f(x) + f(z) + 2D(x,z) - [x,y] - [z,y] \in Z(R) \text{ for all } x, z \in L. \qquad (2.23)$$

Again using (2.18) and characteristic of R not two, we find $D(x,z) \in Z(R)$. In particular $f(x) = D(x,x) \in Z(R)$ for all $x \in L$. Since $f(xz) \in Z(R)$ and $f(x) \in Z(R)$, we have $f(xy) - f(x) \in Z(R)$ for all $x, y \in L$. Using (2.18) we get $[x,y] \in Z(R)$ for all $x, y \in L$. Hence Lemma 2.2 completes the proof. The proof is similar if $f(xy) + f(x) + [x,y] \in Z(R)$ for all $x, y \in L$. $\qquad\qquad\square$

Theorem 2.10. Let R be a semiprime ring of characteristic not two and L be a nonzero Lie ideal of R. Let $D(.,.) : R \times R \longrightarrow R$ be a biadditive symmetric mapping and f be the trace of D. If $f(xy) \mp f(y) \mp [x, y] \in Z(R)$ for all $x, y \in L$, then $L \subseteq Z(R)$.

Proof. Let

$$f(xy) - f(y) - [x, y] \in Z(R) \text{ for all } x, y \in L. \quad (2.24)$$

Replacing y by $y+z$, we have $f(xy)+f(xz)+2D(xy, xz)-f(y)-f(z)-2D(y, z)-[x, y]-[x, z] \in Z(R)$ for all $x, y, z \in L$. Using (2.24), we get

$$2(D(xy, xz) - D(y, z)) \in Z(R) \text{ for all } x, y, z \in L. \quad (2.25)$$

Substituting y for z in (2.25) and using the fact that R is not of characteristic two, we find

$$f(xy) - f(y) \in Z(R) \text{ for all } x, y \in L. \quad (2.26)$$

This implies that $[x, y] \in Z(R)$ for all $x, y \in L$. Thus $[L, L] \subseteq Z(R)$. Applying Lemma 2.2, we obtain $L \subseteq Z(R)$. The proof is similar for the case $f(xy) + f(y) + [x, y] \in Z(R)$ for all $x, y \in L$. □

Theorem 2.11. Let R be a semiprime ring of characteristic not two and L be a nonzero Lie ideal of R. Let $D(.,.) : R \times R \longrightarrow R$ be a biadditive symmetric mapping and f be the trace of D. If $f([x, y]) \mp f(x) \mp [x, y] \in Z(R)$ for all $x, y \in L$, then $L \subseteq Z(R)$.

Proof. Suppose

$$f([x, y]) - f(x) - [x, y] \in Z(R) \text{ for all } x, y \in L. \quad (2.27)$$

Replacing x by $x + z$ in (2.27), we obtain

$$f([x, y]) + f([z, y]) + 2D([x, y], [z, y]) - f(x)$$

$$-f(z) - 2D(x, z) - [x, y] - [z, y] \in Z(R) \text{ for all } x, y, z \in L. \quad (2.28)$$

Using (2.27), we have

$$2(D([x, y], [z, y]) - D(x, z)) \in Z(R) \text{ for all } x, y, z \in L. \quad (2.29)$$

Substituting x for z in (2.29) and using the fact that R is not of characteristic two, we obtain

$$f([x, y]) - f(x) \in Z(R) \text{ for all } x, y \in L. \quad (2.30)$$

Again using (2.27), (2.30) yields that $[x, y] \in Z(R)$ for all $x, y \in L$. This implies that $[L, L] \subseteq Z(R)$. Application of Lemma 2.2 completes the proof. Similarly we can prove the theorem if $f([x, y]) + f(x) + [x, y] \in Z(R)$ for all $x, y \in L$. □

Theorem 2.12. Let R be a semiprime ring of characteristic not two and L be a nonzero Lie ideal of R. Let $D(.,.) : R \times R \longrightarrow R$ be a biadditive symmetric mapping and f be the trace of D. If $f([x,y]) \mp f(y) \mp [x,y] \in Z(R)$ for all $x, y \in L$, then $L \subseteq Z(R)$.

Proof. Let

$$f([x,y]) - f(y) - [x,y] \in Z(R) \text{ for all } x, y \in L. \tag{2.31}$$

Replacing y by $y + z$ we get

$$f([x,y]) + f([x,z]) + 2D([x,y],[x,z]) - f(y) - f(z)$$

$$-2D(y,z) - [x,y] - [x,z] \in Z(R) \text{ for all } x, y, z \in L. \tag{2.32}$$

Using (2.31), (2.32) yields that

$$2(D([x,y],[x,z]) - D(y,z)) \in Z(R) \text{ for all } x, y, z \in L. \tag{2.33}$$

Substituting y for z in (2.33) and using the fact that R is of characteristic not two, we get

$$f([x,y]) - f(y) = D([x,y],[x,y]) - D(y,y) \in Z(R) \text{ for all } x, y \in L. \tag{2.34}$$

In view of (2.31), (2.34) yields that $[x,y] \in Z(R)$ for all $x, y \in L$ i.e. $[L,L] \subseteq Z(R)$. Using Lemma 2.2, we have $L \subseteq Z(R)$. Similarly we can prove the theorem if $f([x,y]) + f(y) + [x,y] \in Z(R)$ for all $x, y \in L$. $\square$

Using the similar techniques as we have used in the proof of Theorem 2.11 and Theorem 2.12, we can prove the following:

Theorem 2.13. Let R be a semiprime ring of characteristic not two and L be a nonzero Lie ideal of R. Let $D(.,.) : R \times R \longrightarrow R$ be a biadditive symmetric mapping and f be the trace of D. If $f([x,y]) \mp f(x) \mp [y,x] \in Z(R)$ for all $x, y \in L$, then $L \subseteq Z(R)$.

Theorem 2.14. Let R be a semiprime ring of characteristic not two and L be a nonzero Lie ideal of R. Let $D(.,.) : R \times R \longrightarrow R$ be a biadditive symmetric mapping and f be the trace of D. If $f([x,y]) \mp f(y) \mp [y,x] \in Z(R)$ for all $x, y \in L$, then $L \subseteq Z(R)$.

Theorem 2.15. Let R be a semiprime ring of characteristic not two and L be a nonzero Lie ideal of R. Let $D(.,.) : R \times R \longrightarrow R$ be a biadditive symmetric mapping and f be the trace of D. If $f([x,y]) \mp f(xy) \mp [x,y] \in Z(R)$ for all $x, y \in L$, then $L \subseteq Z(R)$.

Proof. Let

$$f([x,y]) - f(xy) - [x,y] \in Z(R) \text{ for all } x,y \in I. \tag{2.35}$$

Replacing y by $y + z$ in (2.35) we get

$$f([x,y]) + f([x,z]) + 2D([x,y],[x,z]) - f(xy) - f(xz)$$
$$-2D(xy,xz) - [x,y] - [x,z] \in Z(R) \text{ for all } x,y,z \in L. \tag{2.36}$$

Using (2.35) and (2.36), we obtain

$$2(D([x,y],[x,z]) - D(xy,xz)) \in Z(R) \text{ for all } x,y,z \in L. \tag{2.37}$$

Since R is of characteristic not two, we have

$$D([x,y],[x,z]) - D(xy,xz) \in Z(R) \text{ for all } x,y,z \in L. \tag{2.38}$$

Substituting y for z in (2.38), we get

$$f([x,y]) - f(xy) \in Z(R) \text{ for all } x,y \in L. \tag{2.39}$$

Using (2.35), we have $[x,y] \in Z(R)$ for all $x,y \in L$ and Lemma 2.2 completes the proof. The proof is same for the case $f([x,y]) + f(xy) + [x,y] \in Z(R)$ for all $x,y \in L$. $\qquad\square$

Similarly we can prove the following:

Theorem 2.16. Let R be a semiprime ring of characteristic not two and L be a nonzero Lie ideal of R. Let $D(.,.) : R \times R \longrightarrow R$ be a biadditive symmetric mapping and f be the trace of D. If $f([x,y]) \mp f(xy) \mp [y,x] \in Z(R)$ for all $x,y \in L$, then $L \subseteq Z(R)$.

Theorem 2.17. Let R be a semiprime ring of characteristic not two and L be a nonzero Lie ideal of R. Let $D(.,.) : R \times R \longrightarrow R$ be a biadditive symmetric mapping and f be the trace of D. If $f(x)f(y) \mp [x,y] \in Z(R)$ for all $x,y \in L$, then $L \subseteq Z(R)$.

Proof. Suppose

$$f(x)f(y) - [x,y] \in Z(R) \text{ for all } x,y \in L. \tag{2.40}$$

Substituting $y + z$ for y in (2.40), we have

$$f(x)f(y)+f(x)f(z)+2f(x)D(y,z)-[x,y]-[x,z] \in Z(R) \text{ for all } x,y,z \in L. \tag{2.41}$$

Using (2.40) we find

$$2f(x)D(y,z) \in Z(R) \text{ for all } x,y,z \in L. \tag{2.42}$$

Since R is of characteristic not two, we have

$$f(x)D(y,z) \in Z(R) \text{ for all } x,y,z \in L. \tag{2.43}$$

In particular if we replace z by y in (2.43), then

$$f(x)f(y) \in Z(R) \text{ for all } x, y \in L. \tag{2.44}$$

Hence using (2.44) and (2.40) we obtain $[x, y] \in Z(R)$ for all $x, y \in L$ i.e. $[L, L] \subseteq Z(R)$. Application of Lemma 2.2 completes the proof. The proof is same for the case $f(x)f(y) + [x, y] \in Z(R)$ for all $x, y \in L$. $\square$

Theorem 2.18. Let R be a semiprime ring of characteristic not two and L be a nonzero Lie ideal of R. Let $D(.,.) : R \times R \longrightarrow R$ be a biadditive symmetric mapping and f be the trace of D. If $f(x)f(y) \mp [y, x] \in Z(R)$ for all $x, y \in L$, then $L \subseteq Z(R)$.

Proof. The proof runs on the same parallel lines as that of Theorem 2.17. $\square$

Theorem 2.19. Let R be a semiprime ring of characteristic not two and L be a nonzero Lie ideal of R. Let $D(.,.) : R \times R \longrightarrow R$ be a biadditive symmetric mapping and f be the trace of D. If $f(x)f(y) \mp xy \in Z(R)$ for all $x, y \in L$, then $L \subseteq Z(R)$.

Proof. Let

$$f(x)f(y) - xy \in Z(R) \text{ for all } x, y \in L. \tag{2.45}$$

Substituting $y + z$ for y in (2.45), we have

$$f(x)f(y) + f(x)f(z) + 2f(x)D(y, z) - xy - xz \in Z(R) \text{ for all } x, y, z \in L. \tag{2.46}$$

Applying (2.45), we obtain

$$2f(x)D(y, z) \in Z(R) \text{ for all } x, y, z \in L. \tag{2.47}$$

Since R is of characteristic not two, we have

$$f(x)D(y, z) \in Z(R) \text{ for all } x, y, z \in L. \tag{2.48}$$

In particular if we replace z by y in (2.48) and using (2.45), we find

$$f(x)f(y) \in Z(R) \text{ for all } x, y \in L. \tag{2.49}$$

This implies that $xy \in Z(R)$ for all $x, y \in L$ and application of Lemma 2.3 completes the proof. Similarly we can prove the theorem if $f(x)f(y) + xy \in Z(R)$ for all $x, y \in L$. $\square$

Theorem 2.20. Let R be a semiprime ring of characteristic not two and L be a nonzero Lie ideal of R. Let $D(.,.) : R \times R \longrightarrow R$ be a biadditive symmetric mapping and f be the trace of D. If $f(x)f(y) \mp yx \in Z(R)$ for all $x, y \in L$, then $L \subseteq Z(R)$.

Proof. The proof runs on the same parallel lines as that of Theorem 2.19. $\square$

3. Symmetric biderivations

Bell and Martindale [2] proved that if a semiprime ring R admits a derivation d which is nonzero on a nonzero left ideal I of R and centralizing on I, then R must contain a nonzero central ideal. Deng and Bell [6] generalized the result for n-centralizing mappings. In this section we prove the following result concerning symmetric biderivation.

Theorem 3.1. Let R be a semiprime ring and L be a Lie ideal of R such that $u^2 \in L$ for all $u \in L$. Let n be a fixed positive integer. Let R be $n!$-torsion free for $n > 1$ and 2-torsion free for $n = 1$. Suppose there exists a symmetric biderivation $D(.,.) : R \times R \longrightarrow R$ such that the mapping $f : R \longrightarrow R$ is n-centralizing on L, where f stands for the trace of D. Then f is n-commuting on L.

The following lemma due to Deng and Bell [6] is essential to prove our theorem.

Lemma 3.1. Let n be a positive integer and R be an $n!$-torsion free semiprime ring. Let $f : R \longrightarrow R$ be an additive map on R. For $i = 1, 2, ..., n$, let $F_i(x, y)$ be a generalized polynomial which is homogeneous of degree i in the non-commuting indeterminates x and y. Let $a \in R$ and (a) be the additive subgroup generated by a. If $F_n(x, f(x)) + F_{n-1}(x, f(x)) + \dots\dots + F_1(x, f(x)) \in Z(R)$ for all $x \in (a)$, then $F_i(a, f(a)) \in Z(R)$ for $i = 1, 2, ..., n$.

Proof of Theorem 3.1. Assume that $n = 1$. Linearizing the condition $[f(x), x] \in Z(R)$ we have

$$[f(x), y] + [f(y), x] + [2D(x, y), x] + [2D(x, y), y] \in Z(R), \text{ for all } x, y \in I. \tag{3.1}$$

Substituting $-y$ for y in (3.1), we have

$$-[f(x), y] + [f(y), x] - [2D(x, y), x] + [2D(x, y), y] \in Z(R), \text{ for all } x, y \in I. \tag{3.2}$$

Subtracting (3.1) and (3.2), we get $2[f(x), y] + 4[D(x, y), x] \in Z(R)$ for all $x, y \in I$. Replacing y by x^2 in this relation we have $8x[f(x), x] \in Z(R)$ for all $x \in I$. Now commuting this with $f(x)$ and using 2-torsion condition, we obtain $[x[f(x), x], f(x)] = 0$ for all $x \in I$. This implies that $[f(x), x][x, f(x)] = 0$ for all $x \in I$ i.e. $[f(x), x]^2 = 0$ for all $x \in I$. Since the centre of a semiprime ring contains no nonzero nilpotent elements,

$[f(x), x] = 0$ for all $x \in I$.

Now Suppose that $n > 1$. Linearizing the condition $[f(x), x^n] \in Z(R)$ we get

$$[f(x), x^n] + [f(x), x^{n-1}y + \ldots + yx^{n-1}] + [f(y), x^{n-1}y + \ldots + yx^{n-1}]$$

$$+[f(y), x^n] + [2D(x,y), x^n] + [2D(x,y), x^{n-1}y + \ldots + yx^{n-1}] \in Z(R)$$

for all $x, y \in I$. Using Lemma 3.1 and the fact that $[f(x), x^n] \in Z(R)$ for all $x \in I$, we obtain

$$[f(x), x^{n-1}y + \ldots + yx^{n-1}] + [2D(x,y), x^{n-1}y + \ldots + yx^{n-1}]$$

$$+[f(y), x^n] + [2D(x,y), x^n] \in Z(R), \text{ for all } x, y \in I. \tag{3.3}$$

Replacing y by $-y$ in (3.3), we have

$$-[f(x), x^{n-1}y + \ldots + yx^{n-1}] + [2D(x,y), x^{n-1}y + \ldots + yx^{n-1}]$$

$$+[f(y), x^n] - [2D(x,y), x^n] \in Z(R), \text{ for all } x, y \in I. \tag{3.4}$$

Now subtracting (3.3) and (3.4), we get

$$2[f(x), x^{n-1}y + \ldots + yx^{n-1}] + 4[D(x,y), x^n] \in Z(R) \text{ for all } x, y \in I. \tag{3.5}$$

Substituting x^2 for y in (3.5), we find that $2[f(x), nx^{n+1}] + 4[D(x, x^2), x^n] \in Z(R)$ for all $x \in I$. This implies that $2(4 + n)x[f(x), x^n] \in Z(R)$ for all $x \in I$. i.e. $2(4 + n)(x[f(x), x^n])^n \in Z(R)$ for all $x \in I$. Commuting with $f(x)$ and using torsion condition, we get

$$[x^n[f(x), x^n]^n, f(x)] = 0 \text{ for all } x \in I. \tag{3.6}$$

This implies that

$$[f(x), x^n]^{n+1} = 0 \text{ for all } x \in I. \tag{3.7}$$

Lemma 2.1 yields that $[f(x), x^n] = 0$, for all $x \in I$. $\qquad\qquad \square$

Using the similar techniques with slight modifications we can prove the following:

Theorem 3.2. Let R be a semiprime ring and I be a nonzero left ideal of R. Let n be a fixed positive integer. Let R be $n!$-torsion free for $n > 1$ and 2-torsion free for $n = 1$. Suppose there exists a symmetric biderivation $D(.,.) : R \times R \longrightarrow R$ such that the mapping $f : R \longrightarrow R$ is n-centralizing on I, where f stands for the trace of D. Then f is n-commuting on I.

References

[1] M. Ashraf, A. Ali and S. Ali, *Some commutativity theorems for rings with generalized derivations*, Southeast Asian Bull. Math. 31 (2007), 415-421.

[2] H. E. Bell and W. S. Martindale, *Centralizing mappings of semiprime rings*, Canad. Math. Bull. 30 (1987), 92-101.

[3] H. E. Bell and M. N. Daif, *On derivations and commutativity in prime rings*, Acta Math. Hungar. 66 (4) (1995), 337-343.

[4] M. N. Daif and H. E. Bell, *Remarks on derivations on semiprime rings*, Int. J. Math. and Math. Sci. 15 (1) (1992), 205-206.

[5] Q. Deng, *On n-centralizing mappings of prime rings* , Proc. Ir. Acad. 93 (1993), 171-176.

[6] Q. Deng and H. E. Bell, *On derivations and commutativity in semiprime rings*, Comm. Algebra 23 (1995), 3705-3713.

[7] I. N. Herstein, *Rings with involution*, Chicago lectures in mathematics, University of Chicago press, Chicago, III, USA, (1976).

[8] I. N. Herstein, *On Lie structure of an associative ring*, J. Algebra 14 (4) (1970), 561-571.

[9] Gy. Maksa, *A remark on symmetric biadditive functions having non-negative diagonalization*, Glasnik. Mat. 15 (35) (1980), 279-282.

[10] Gy. Maksa, *On the trace of symmetric biderivations*, C. R. Math. Rep. Acad. Sci. Canada 9 (1987), 303-307.

[11] E. C. Posner, *Derivations in prime rings*, Proc. Amer. Math. Soc. 8 (6) (1957), 1093-1100.

[12] J. Vukman, *Symmetric biderivations on prime and semiprime rings*, Aequationes Math. 38 (1989), 245-254.

[13] J. Vukman, *Two results concerning symmetric biderivations on prime rings*, Aequationes Math. 40 (1990), 181-189.

DEPARTMENT OF MATHEMATICS
ALIGARH MUSLIM UNIVERSITY
ALIGARH 202002, INDIA
E-mail address: asma_ali2@rediffmail.com

DEPARTMENT OF MATHEMATICS
ALIGARH MUSLIM UNIVERSITY
ALIGARH 202002, INDIA
E-mail address: faiza.shujat@gmail.com

Proceedings of the Sixth China-Japan-Korea
International Conference on Ring Theory
June 27-July 2, 2011 Suwon, Korea

τ-PROJECTIVE AND STRONGLY
τ-PROJECTIVE MODULES

ISMAIL AMIN, YASSER IBRAHIM, AND MOHAMED YOUSIF*

ABSTRACT. In this paper we introduce the notions of τ-projective
and strongly τ-projective modules relative to a preradical τ. When
$\tau(M) = rad(M)$ we recover all the work that was carried out
on rad-projectivity. New and interesting results are obtained in
the cases where $\tau(M)$ is $soc(M)$, $Z(M)$ or $\delta(M)$, where $soc(M)$,
$Z(M)$ and $\delta(M)$ denotes to the socle, the singular submodule,
and the δ-submodule of M, respectively. New characterizations
of semiperfect and perfect rings in terms of τ-projective covers
are obtained.

1. Introduction

Injective modules, and their dual projective modules play a central
rule in the theory of rings and modules. Both notions are used to characterize semisimple artinian rings, a class of rings that is easily described in
the literature as direct sums of matrix rings over division rings. While,
a result of Eckmann and Schopf [7] asserts that, every right R-module
M can be embedded in an injective envelope (hull) of M, the situation
is different when it comes to projective modules. Indeed, H. Bass in his
remarkable dualization of the notion of injective envelopes [4], has shown
that every right R-module has a projective cover if and only if R is a right
perfect ring, a generalization of artinian rings. Both of these notions have
been utilized to describe some of the most interesting classes of rings,
and a natural question on the subject was to characterize the rings where
both classes of injective and projective modules coincide. This question
was answered by Faith and Walker in their paper [8] where they have
shown that a ring R is quasi-Frobenius if and only if every injective right
R-module is projective; or equivalently every projective right R-module
is injective. Quasi-Frobenius rings are generalizations of semisimple artinian rings and are best known in the literature as those artinian rings
with perfect duality as described by Nakayama in his papers [12], [13]

2010 Mathematics Subject Classification : 16D40, 16D50, 16N80, 16L30, 16L60.
Keywords : Soc-injective and rad-projective modules, injective and projective
modules, τ-projective and strongly τ-projective modules.
*Corresponding author.

and [14]. Indeed, according to Nakayama a ring R is quasi-Frobenius (QF-ring) if R is left (or right) artinian and if $\{e_1, e_2, \cdots, e_n\}$ is a basic set of primitive idempotents of R, then there exists a (Nakayama) permutation σ of $\{1, 2, \cdots, n\}$ such that $soc(Re_k) \cong Re_{\sigma k}/Je_{\sigma k}$ and $soc(e_{\sigma k}R) \cong e_k R/e_k J$, where $J = J(R)$ is the Jacobson radical of R. This remarkable describtion by Nakayama reduces the perfect duality in QF-rings to a duality between the Jacobson radical and the socle of the indecomposable projective components of the basic subring of R. This result was the primary motivation behind the introduction of the concept of *soc-injectivity* [1] and the dual concept of *rad-projectivity* [2], as follows:

Definition 1.1. *If M and N are right R-modules, then M is called socle-N-injective (soc-N-injective) if any R-homomorphism $f : soc(N) \to M$ extends to N. Equivalently, for any semisimple submodule K of N, any R-homomorphism $f : K \to M$ extends to N. A module M, is called soc-quasi-injective if M is soc-M-injective, and is called soc-injective, if M is soc-R-injective. If M is soc-N-injective for all right R-modules N, then M is called strongly soc-injective.*

Definition 1.2. *A right R-module M is called radical N-projective (rad-N-projective) if, for any epimorphism $\sigma : N \to K$ where K is an image of $N/rad(N)$ and any homomorphism $f : M \to K$, there exists a homomorphism $g : M \to N$ such that $f = \sigma \circ g$. The module M is called rad-projective (resp., rad-quasi-projective) if M is rad-R_R-projective (resp., rad-M-projective), and is called strongly rad-projective if M is rad-N-projective for every R-module N.*

In [16] the notions of soc-injective and strongly soc-injective modules were extended to a more general setting in terms of $\sigma [M]$, the full subcategory of the right R-modules whose objects are all right R-modules subgenerated by M. In this paper we generalize and extend the notion of rad-pojectivity by introducing the notions of τ-projective and strongly τ-projective modules relative to any preradical τ. When $\tau(M) = rad(M)$ we recover all the work that was carried out in [2] on rad-projectivity, and obtain new and interesting results in the cases where $\tau(M) = soc(M)$, $\tau(M) = Z(M)$ and $\tau(M) = \delta(M)$, where $soc(M)$, $Z(M)$ and $\delta(M)$ denotes to the socle, the singular submodule and the δ-submodule of M, respectively.

We start by highlighting all the interesting properties of these new notions and show with examples that these notions are non trivial and natural extensions of projective modules. For example, the class of (strongly) τ-projective right R-modules is closed under isomorphisms, direct sums and summands. The notion of τ-projective cover is introduced and new characterizations of semiperfect and perfect rings in terms of τ-projective covers are provided.

Throughout this paper every ring R is associative with unity, all modules are unitary R-modules and we denote by Mod-R for the category of right R-modules. If M is an R-module, we denote by $E(M)$ for the injective hull of M, and if $M = R$, we write $J = J(R) = rad(R)$. We also write $N \subseteq M$ if N is a submodule of M, $N \subseteq^{\oplus} M$ if N is a direct summand of M, $N \subseteq^{ess} M$ if N is an essential submodule of M, and $N \overset{small}{\hookrightarrow} M$ if N is a small submodule of M, i.e. $N + K \neq M$ whenever K is a proper submodule of M. If X is a subset of R, the right annihilator of X in R is denoted by $r_R(X) =: \{r \in R : xr = 0, \text{ for all } x \in X\}$. We refer to [3], [9], and [18] for all the undefined notions in this paper.

2. Preliminaries

A preradical τ of Mod-R assigns to each $M \in Mod$-R a submodule $\tau(M)$ in such a way that for each R-homomorphism $f : M \to N$ we have $f(\tau(M)) \subseteq \tau(N)$. Thus a preradical is a subfunctor of the identity functor of Mod-R. Every preradical τ commutes with direct sums and gives rise to a pretorsion class $T_\tau =: \{M \in Mod\text{-}R : \tau(M) = M\}$ which is closed under under direct sums and factor modules. Clearly $\tau(R)M \subseteq \tau(M)$ for every $M \in Mod$-R. We sometimes call $\tau(M)$ the τ-submodule of M. A preradical is said to be a radical if $\tau(M/\tau(M)) = 0$. Examples of preradicals include:

(1) $rad(M) =: \cap \{N : N \text{ is a maximal submodule of } M\}$
    $= \sum \{L : L \text{ is a small submodule of } M\}$.

(2) $soc(M) =: \sum \{S : S \text{ is a simple submodule of } M\}$
    $= \cap \{N : N \text{ is an essential submodule of } M\}$.

(3) $Z(M) =: \{x \in M : r_R(x) \subseteq^{ess} R_R\}$.

(4) $\delta(M) =: \sum \{L : L \text{ is a } \delta\text{-small submodule of } M\}$
    $= \cap \{N \subset M : M/N \text{ is a simple singular } R\text{-module}\}$.

Where in [19], a submodule N of a right R-module M is called δ-small in M, and denoted by $N \subseteq^{\delta} M$, if $M \neq N + X$ for any proper submodule X of M with M/X singular.

Clearly if M is a right R-module, then $rad(M) \subseteq \delta(M)$ and if M is projective, then $soc(M) \subseteq \delta(M)$.

Definition 2.1. *A right R-module M is called τ-N-projective if, for every diagram:*

$$
\begin{array}{ccc}
 & & M \\
 & \overset{\exists\lambda}{\swarrow} & \downarrow f \\
N & \overset{g}{\longrightarrow} & L \quad \to 0
\end{array}
$$

with L an image of $N/\tau(N)$, equivalentely $\tau(N) \hookrightarrow \ker g$, there exists a homomorphism $\lambda : M \longrightarrow N$ such that $g\lambda = f$. The module M is called τ-projective (resp., τ-quasi-projective) if M is τ-R_R-projective (resp., τ-M-projective), and is called strongly τ-projective if it is τ-N-projective for every R-module N.

If τ is the trivial preradical, $\tau(M) = 0$, for every right R-module M, then the notion of τ-N-projectivity is the usual notion of N-projectivity.

Example 2.2. *(1) Every projective right R-module is strongly τ-projective.*

(2) If M is strongly τ-projective and either $\tau(R) = 0$ or $\tau(M) = 0$, then M is projective. In fact, since every M is a homomorphic image of a free module, there is an exact sequence $R^{(\Lambda)} \xrightarrow{\eta} M \to 0$ for some set Λ. If $\tau(R) = 0$, then $\tau(R^{(\Lambda)}) = (\tau(R))^{(\Lambda)} = 0$ and so $\eta(\tau(R^{(\Lambda)})) = 0$; and if $\tau(M) = 0$, then $\eta(\tau(R^{(\Lambda)})) \subseteq \tau(M) = 0$. In both cases $\tau(R^{(\Lambda)}) \subseteq \ker \eta$ and by the assumption the map η splits. Therefore M is isomorphic to a direct summand of $R^{(\Lambda)}$, and so M is projective.

(3) Since $soc(\mathbb{Z}_\mathbb{Z}) = 0$ and no non-trivial maps from $\mathbb{Q}_\mathbb{Z}$ into $\mathbb{Z}_n$, any diagram:

$$
\begin{array}{ccc}
 & \mathbb{Q}_\mathbb{Z} & \\
 & \downarrow f & \\
\mathbb{Z}_\mathbb{Z} \xrightarrow{\eta} & \mathbb{Z}_n & \to 0
\end{array}
$$

can be completed, and so $\mathbb{Q}_\mathbb{Z}$ is soc-projective, i.e. soc-$\mathbb{Z}$-projective. Since $\mathbb{Q}_\mathbb{Z}$ is not projective, we infer from (2) that $\mathbb{Q}_\mathbb{Z}$ is not strongly soc-projective.

(4) Since $\delta(\mathbb{Z}_\mathbb{Z}) = 0$, it follows from (2) above that every strongly δ-projective $\mathbb{Z}$-module is projective. In particular, if $M = \mathbb{Q}/\mathbb{Z}$, then M as a $\mathbb{Z}$-module is a δ-$\mathbb{Q}_\mathbb{Z}$-projective module with $M = \delta(M)$, which is not strongly δ-projective. Note also that M is not $\mathbb{Q}_\mathbb{Z}$-projective. For, if $\mathbb{Q}/\mathbb{Z}$ were $\mathbb{Q}_\mathbb{Z}$-projective, then the following diagram:

$$
\begin{array}{ccc}
 & \mathbb{Q}/\mathbb{Z} & \\
 & \downarrow id & \\
\mathbb{Q} \xrightarrow{\eta} & \mathbb{Q}/\mathbb{Z} & \to 0
\end{array}
$$

can be completed, and $\mathbb{Z}_\mathbb{Z}$ would be a summand of $\mathbb{Q}_\mathbb{Z}$; a contradiction.

(5) The $\mathbb{Z}$-module $\mathbb{Q} \oplus \mathbb{Q}/\mathbb{Z}$ is an example of a δ-quasi-projective module which is not quasi-projective.

Proposition 2.3. *For right R-modules M and N, the following statements are equivalent:*

(1) M is τ-N-projective

(2) *For every epimorphism* $N \xrightarrow{f} K \to 0$ *of right R-modules with* $\tau(N) \subseteq \ker f$, *the homomorphism*

$$f_* : Hom_R(M, N) \longrightarrow Hom_R(M, K)$$

given by $f_* \alpha = f \alpha$ *is an epimorphism.*

Proof. Let M be a τ-N-projective module, $N \xrightarrow{f} K \to 0$ the given epimorphism and $g \in Hom_R(M, K)$. Since M is τ-N-projective, there exists $\alpha : M \longrightarrow N$ such that $f_*(\alpha) = f\alpha = g$. So f_* is epic. The converse is clear. $\quad\square$

Proposition 2.4.   (1) *If M is τ-N-projective and K is a submodule of N, then M is τ-N/K-projective.*

(2) *A direct sum $\oplus_{i \in I} M_i$ of modules is τ-N-projective iff each M_i is τ-N-projective.*

(3) *A direct summand of a τ-N-projective module is τ-N-projective.*

(4) *If $A \overset{\theta}{\simeq} B$, then M is τ-A-projective if and only if M is τ-B-projective.*

(5) *If M is τ-M_i-projective for all $i = 1, 2, ..., n$, then M is τ-$\oplus_{i=1}^n M_i$-projective.*

(6) *$\oplus_{i=1}^n M_i$ is τ-quasi-projective if and only if each M_i is τ-M_j-projective for all $i, j = 1, 2, 3, ..., n$. In particular, $M \oplus N$ is τ-quasi-projective if and only if both M and N are τ-quasi-projective, M is τ-N-projective and N is τ-M-projective.*

(7) *If M is a τ-projective right R-module and N is a finitely generated right R-module, then M is τ-N-projective.*

(8) *If M is finitely generated and τ-M_i-projective for all $i \in I$, then M is τ-$\oplus_{i \in I} M_i$-projective.*

(9) *If N is a generator, then every finitely generated τ-N-projective module is strongly τ-projective.*

(10) *If M is a finitely generated τ-projective right R-module, then M is strongly τ-projective.*

(11) *If $\tau(R) = 0$, then every finitely generated τ-projective right R-module is projective.*

(12) *If A, B and N are right R-modules with $A \overset{\theta}{\simeq} B$, then A is τ-N-projective if and only if B is τ-N-projective.*

Proof. 1. Assume that M is τ-N-projective, $K \subseteq N$, and consider the following diagram:

$$
\begin{array}{ccccccc}
& & & & M & & \\
& & & & \downarrow f & & \\
N & \xrightarrow{\eta} & N/K & \xrightarrow{\sigma} & L & \to 0 &
\end{array}
$$

with L an image of $(N/K)/(\tau(N/K))$. Since $(K + \tau(N))/K \subseteq \tau(N/K)$ and $N/(K + \tau(N)) \simeq (N/K)/((K + \tau(N))/K)$ we have:

$$N/\tau(N) \overset{onto}{\longrightarrow} (N/K)/((K + \tau(N))/K) \overset{onto}{\longrightarrow} (N/K)/(\tau(N/K)) \overset{onto}{\longrightarrow} L.$$

Therefore L is an image of $N/\tau(N)$. By the τ-N-projectivity of M, f can be lifted to a map $\lambda : M \to N$ with $\lambda \eta \sigma = f$. Now, $\tau = \lambda \eta : M \to N/K$ is the required lifting map, and M is τ-N/K-projective.

2. Let $M = \underset{i \in I}{\oplus} M_i$ be τ-N-projective module, K an image of $N/\tau(N)$, and consider the following diagram:

$$
\begin{array}{c}
M_j \\
\downarrow \mu_j \\
M = \underset{i \in I}{\oplus} M_i \\
\downarrow \pi_j \\
M_j \\
\downarrow f \\
N \overset{\eta}{\longrightarrow} \quad K \quad \longrightarrow 0
\end{array}
$$

where μ_j and π_j are the usual canonical injection and projection maps, respectively, $j \in I$. Since M is τ-N-projective, there exists $\lambda : M \longrightarrow N$ such that $\eta \lambda = f \pi_j$. Now the map $g = \lambda \mu_j : M_j \longrightarrow N$ is the required lifting map, since $\eta g = \eta \lambda \mu_j = f \pi_j \mu_j = f$. Hence M_j is τ-N-projective for all $j \in I$. Conversely, assume that M_j is τ-N-projective for all $j \in I$, K an image of $N/\tau(N)$, and consider the following diagram

$$
\begin{array}{c}
M_j \\
\downarrow \mu_j \\
M = \underset{i \in I}{\oplus} M_i \\
\downarrow f \\
N \overset{\eta}{\longrightarrow} \quad K \quad \longrightarrow 0
\end{array}
$$

Since M_j is τ-N-projective for all $j \in I$, there exists $\lambda_j : M_j \longrightarrow N$ such that $\eta \lambda_j = f \mu_j$. By the universal property of direct sums, there exists a unique $\theta : M \longrightarrow N$ such that $\theta \mu_j = \lambda_j$. Now $\eta \theta \mu_j = \eta \lambda_j = f \mu_j$, and from the uniquness of θ we have $\eta \theta = f$.

3. Follows from (2).

4. Assume that M is τ-A-projective and consider the following diagram:

$$
\begin{array}{c}
M \\
\downarrow f \\
B \overset{\eta}{\longrightarrow} \quad K \quad \longrightarrow 0
\end{array}
$$

with K a homomorphic image of $B/\tau(B)$. Now the isomorphism $A \overset{\theta}{\simeq} B$ induces an isomorphism $A/\tau(A) \simeq B/\tau(B)$, and so K is a homomorphic image of $A/\tau(A)$. Now, consider the following diagram:

$$
\begin{array}{ccc}
 & M & \\
 & \downarrow f & \\
A \overset{\eta\theta}{\longrightarrow} & K & \to 0.
\end{array}
$$

Since M is τ-A-projective, f can be lifted to a homomorphism $g : M \to A$ with $\eta\theta g = f$. Clearly, $\theta g : M \to B$ is the required lifting map of f, and M is τ-B-projective.

5. Without loss of generality, we may assume that $n = 2$. Let $M_1 \oplus M_2 \overset{f}{\longrightarrow} M_1 \oplus M_2/K \to 0$ be the natural epimorphism with $\tau(M_1 \oplus M_2) \subseteq K$, and consider the following commutative diagram:

$$
\begin{array}{ccccccccc}
0 \to & M_1 & \longrightarrow & M_1 \oplus M_2 & \overset{\eta}{\longrightarrow} & M_2 & \to 0 \\
 & \downarrow g & & \downarrow f & & \downarrow \theta & \\
0 \to & M_1 + K/K & \longrightarrow & M_1 \oplus M_2/K & \longrightarrow & (M_1 \oplus M_2)/(M_1 + K) & \to 0 \\
 & \downarrow & & \downarrow & & \downarrow & \\
 & 0 & & 0 & & 0 &
\end{array}
$$

with exact rows and exact columns. Since $\tau(M_1) \subseteq M_1 \cap K = \ker g$, $\tau(M_2) \subseteq (M_1 + K) \cap M_2$ and M is τ-M_i-pojective for $i = 1, 2$, the following diagram:

$$
\begin{array}{ccc}
0 & & 0 \\
\downarrow & & \downarrow \\
Hom_R(M, M_1) & \overset{g_*}{\longrightarrow} & Hom_R(M, M_1 + K/K) \to 0 \\
\downarrow & & \downarrow \\
Hom_R(M, M_1 \oplus M_2) & \overset{f_*}{\longrightarrow} & Hom_R(M, M_1 \oplus M_2/K) \\
\downarrow \eta_* & & \downarrow \\
Hom_R(M, M_2) & \overset{\theta_*}{\longrightarrow} & Hom_R(M, (M_1 \oplus M_2)/(M_1 + K)) \to 0 \\
\downarrow & & \\
0 & &
\end{array}
$$

with exact rows and columns, is commutative. Note that $Hom_R(M, M_1 \oplus M_2) \overset{\eta_*}{\longrightarrow} Hom_R(M, M_2)$ is epic. For, if $h : M \longrightarrow M_2 \in Hom_R(M, M_2)$, define $\alpha : M \longrightarrow M_1 \oplus M_2$ by $\alpha(x) = (0, h(x))$; so α is a well defined homomorphism and $\eta_*(\alpha)(x) = \eta\alpha(x) = \eta(0, h(x)) = h(x)$. Since g_*, θ_* and η_* are all epimorphisms, f_* is epic, and by Proposition 2.3, M is τ-$M_1 \oplus M_2$-pojective.

6. Suppose that $L =: \oplus_{i=1}^n M_i = \oplus_{\substack{i=1 \\ i \neq j}}^n M_i \oplus M_j$ is a τ-quasi-projective i.e., L is τ-L-projective. We have, by (1), L is τ-$L/\oplus_{\substack{i=1 \\ i \neq j}}^n M_i$-projective

and so by (4), L is τ-M_j-projective. Hence as a direct summand of L, M_i is τ-M_j-projective for all $i, j = 1, 2, 3, ..., n$. Now suppose that M_i is τ-M_j-projective for all $i, j = 1, 2, 3, ..., n$. So by part (5), M_i is τ-$\oplus_{i=1}^n M_i$-projective; and hence by (2), $\oplus_{i=1}^n M_i$ is τ-$\oplus_{i=1}^n M_i$-projective, i.e. $\oplus_{i=1}^n M_i$ is τ-quasi-projective. The last statement is clear (take $n = 2$).

7. Since N is finitely generated, N is an epimorphic image of a finitely generated free R-module F. By (3), M is τ-F-projective; and hence τ-N-projective by (1) and (2).

8. Consider the diagram

$$\begin{array}{c} M \\ \downarrow \eta \\ \oplus_{i \in I} M_i \quad \xrightarrow{g} \quad E \quad \rightarrow 0 \end{array}$$

with $\tau(\oplus_{i \in J} M_i) \subseteq \ker g$. Since $\eta(M)$ is finitely generated and g is epic, there exists a finite subset $J \subseteq I$ such that the image of $h := g|_{\oplus_{i \in J} M_i}$ is $\eta(M)$. Now, consider the following diagram:

$$\begin{array}{c} M \\ \downarrow \eta \\ \oplus_{i \in J} M_i \quad \xrightarrow{h} \quad \eta(M) \quad \rightarrow 0. \end{array}$$

Since M is τ-M_i-projective, $i \in J$, it follows from (3) that M is τ-$\oplus_{i \in J} M_i$-projective and there exists $\lambda : M \longrightarrow \oplus_{i \in J} M_i \subseteq \oplus_{i \in I} M_i$ such that $h\lambda = g\lambda = \eta$. Thus M is τ-$\oplus_{i \in I} M_i$-projective.

9. Let A be any right R-module. Since N is a generator, there exists an epimorphism $\theta : \oplus_{i \in I} N \longrightarrow A$. Since M is finitely generated and τ-N-projective, M is τ-$\oplus_{i \in I} N$-projective by (6). Therefore M is τ-$(\oplus_{i \in I} N)/\ker \theta$-projective by (1); and it follows from (2) that M is τ-A-projective. Since the module A was arbitrarily chosen, M is strongly τ-projective.

10. Follows from (9).

11. Let $F \xrightarrow{\eta} M \rightarrow 0$ be the natural epimorphism with F finitely generated and free. By (6), M is τ-F-projective, and since $\tau(F) = 0$, the map η splits and M is projective.

12. Assume that A is τ-N-projective and consider the following diagram

$$
\begin{array}{c}
A \\
\theta \downarrow \uparrow \theta^{-1} \\
B \\
\downarrow f \\
N \xrightarrow{\eta} K \qquad \to 0
\end{array}
$$

with K a homomorphic image of $N/\tau(N)$. Since A is τ-N-projective, there exists $\lambda : A \longrightarrow N$ such that $\eta\lambda = f\theta$. Now $g = \lambda\theta^{-1} : B \longrightarrow N$ is the required lifting map since $\eta g = \eta\lambda\theta^{-1} = f\theta\theta^{-1} = f$, and B is τ-N-projective. $\qquad\square$

Remark 2.5. *Note that if the right R-module M is N-projective, then it is K-projective for every submodule K of N by* [18, 18.2]. *This is not true for τ-N-projective modules. In fact, if $M = \mathbb{Z}_n$, $N = \mathbb{Q}_\mathbb{Z}$ and $K = \mathbb{Z}_\mathbb{Z}$, then $M_\mathbb{Z}$ is rad-N-projective but not rad-K-projective.*

The next corollary is now an immediate consequence of Proposition 2.4.

Corollary 2.6.   *(1) For every family $\{M_i\}_{i\in I}$ of right R-modules, $\oplus_{i\in I} M_i$ is (strongly) τ-projective iff M_i is (strongly) τ-projective for every $i \in I$.*

 (2) A direct summand of a (strongly) τ-projective module is again (strongly) τ-projective.

 (3) If M_R is a finitely generated R-projective module (i.e. projective relative to R_R), then M is projective.

3. τ-Projective Modules

Theorem 3.1. *The following are equivalent for a right R-module N:*

 (1) Every right R-module is τ-N-projective.

 (2) Every homomorphic image of N is τ-N-projective.

 (3) $N = \tau(N) \oplus A$, with A semisimple.

 (4) $N = \tau(N) + soc(N)$.

Proof. 1 $\Rightarrow$ 2. Clear.

2 $\Rightarrow$ 3. Since $N/\tau(N)$ is τ-N-projective, the epimorphism $N \longrightarrow N/\tau(N)$ splits; and so $\tau(N) \subseteq^{\oplus} N$. Let $K/\tau(N)$ be a submodule of $N/\tau(N)$. Since N/K is τ-N-projective, N/K is τ-$N/\tau(N)$-projective by (1) of Proposition 2.4, and so the epimorphism $N/\tau(N) \longrightarrow N/K$ splits. Thus $K/\tau(N) \subseteq^{\oplus} N/\tau(N)$; and consequently $N/\tau(N)$ is semisimple. Hence $N = \tau(N) \oplus A$ with A semisimple.

$3 \Rightarrow 1$. Consider the following diagram:

$$
\begin{array}{ccc}
 & M & \\
 & \downarrow f & \\
N \xrightarrow{\sigma} & K & \to 0
\end{array}
$$

with $\tau(N) \subseteq \ker(\sigma)$. Since σ is an epimorphisim, σ can be factored as $\sigma = \theta\pi$ where $\theta : N/\ker\sigma \longrightarrow K$ is an isomorphism and $\pi : N \longrightarrow N/\ker(\sigma)$ is the natural epimorphism.

Since $\tau(N) \subseteq \ker(\sigma)$, we can define the epimorphism $g : N/\tau(N) \longrightarrow N/\ker\sigma$ by $g(n+\tau(N)) = n+\ker\sigma$; and so we have $\pi = g\eta$ where $N \xrightarrow{\eta} N/\tau(N)$ is the natural epimorphism. Now since $N/\tau(N)$ is semisimple, g splits; and there is a homomorphism $g' : N/\ker\sigma \to N/\tau(N)$ with $gg' = 1_{N/\ker(\sigma)}$. Since $\tau(N) \subseteq^{\oplus} N$, there is a homomorphism $N/\tau(N) \xrightarrow{\eta'} N$ such that $\eta\eta' = 1_{N/\tau(N)}$. Now we have the following diagram:

$$
\begin{array}{ccc}
 & M & \\
 & \downarrow f & \\
N \xrightarrow{\sigma} & K & \to 0 \\
\| & \theta^{-1} \downarrow\uparrow \theta & \\
N \xrightarrow{\pi} & N/\ker(\sigma) & \\
\eta' \nwarrow \searrow \eta & g' \downarrow\uparrow g & \\
 & N/\tau(N) &
\end{array}
$$

So the homomorphism $\alpha =: \eta'g'\theta^{-1}f : M \longrightarrow N$ is the required lifting map since $\sigma\alpha = \sigma\eta'g'\theta^{-1}f = \theta\pi\eta'g'\theta^{-1}f = \theta g\eta\eta'g'\theta^{-1}f = f$.

$3 \Rightarrow 4$. Clear.

$4 \Rightarrow 3$. Suppose $N = \tau(N) + Soc(N)$. If $\tau(N) \cap Soc(N) = 0$, we are done. Otherwise assume that $\tau(N) \cap Soc(N) = B \neq 0$. Since $B \subseteq^{\oplus} Soc(N)$, $N = \tau(N) + Soc(N) = \tau(N) + (B \oplus A) = \tau(N) + A$ for some semisimple submodule A of N. Since $\tau(N) \cap A \subseteq A \cap B = 0$, $N = \tau(N) \oplus A$. $\qquad \square$

Lemma 3.2. [19]. *Let N be a submodule of M. The following conditions are equivalent:*

   (1) $N \subseteq^{\delta} M$.
   (2) If $M = A + N$ for a submodule $A \subseteq M$, then $M = A \oplus B$ for a projective semisimple submodule $B \subseteq N$.

Proposition 3.3. *If N is a right R-module with $\tau(N) \subseteq^{\delta} N$ or $\tau(N)$ is semisimple, then the following conditions are equivalent:*

   (1) Every right R-module is τ-N-projective.
   (2) Every homomorphic image of N is τ-N-projective.

(3) N is semisimple.

Proof. If $\tau(N)$ is semisimple, the result follows directly from Thoreom 3.1. Now suppose that $\tau(N) \subseteq^\delta N$.

$3 \Rightarrow 1 \Rightarrow 2$. Clear.

$2 \Rightarrow 1$. By Theorem 3.1, $N = \tau(N) + soc(N)$ and by Lemma 3.2 $N = Y \oplus soc(N)$ for a projective semisimple submodule Y of N. This means N is semisimple.   □

Corollary 3.4. *The following conditions are equivalent for a finitely generated right R-module N:*

(1) Every right R-module is rad-N-projective.

(2) Every homomorphic image of N is rad-N-projective

(3) N is semisimple.

Corollary 3.5. *The following conditions are equivalent for a right R-module N:*

(1) Every right R-module is soc-N-projective.

(2) Every homomorphic image of N is soc-N-projective.

(3) N is semisimple.

Theorem 3.6. *The following statements are equivalent:*

(1) R is semisimple.

(2) $R/\tau(R)$ is a semisimple τ-projective right R-module with $\tau(R) \subseteq^\delta R_R$.

Proof. $1 \Rightarrow 2$. Clear.

$2 \Rightarrow 1$. Since $R/\tau(R)$ is τ-projective, the epimorphism $R \longrightarrow R/\tau(R)$ splits, and so $\tau(R) \subseteq^\oplus R$. Consequentely $R = \tau(R) \oplus T$ for some right ideal of T of R. Since $\tau(R) \subseteq^\delta R_R$, we infer from Lemma 3.2 that $R = Y \oplus T$ for a semisimple projective right ideal $Y \subseteq \tau(R)$. Since T_R is semisimple, we infer that R is semisimple.   □

Theorem 3.7. *For a right R-module M, the following statements are equivalent:*

(1) Every submodule of a τ-E(M)-projective right R-module is τ-E(M)-projective.

(2) Every submodule of a projective right R-module is τ-E(M)-projective.

(3) Every right ideal of R is τ-E(M)-projective.

(4) Every factor module of $E(M)/\tau(E(M))$ is injective.

Proof. $1 \Rightarrow 2 \Rightarrow 3$. Clear.

$3 \Rightarrow 4$. Let I be a right ideal of R, K a homomorphic image of $E(M)/\tau(E(M))$ and consider the following diagram:

$$
\begin{array}{ccc}
I & \hookrightarrow & R \\
& & \downarrow f \\
E(M) & \xrightarrow{\eta} & K \quad \rightarrow 0.
\end{array}
$$

By the Baer criterion the module K_R is injective if f has an extension from R_R into K_R. Since I is τ-$E(M)$-projective, there is a homomorphism $\theta : I \longrightarrow E(M)$ such that $\eta\theta = f$. Since $E(M)$ is injective, there exists $\lambda : R \longrightarrow E(M)$ such that $\lambda|_I = \theta$. Since $\eta\lambda|_I = \eta\theta = f$, the R-homomorphism $\eta\lambda : R \longrightarrow K$ is the required extension of f.

$4 \Rightarrow 1$. Let B be a τ-$E(M)$-projective module, A a submodule of B, K a homomorphic image of $E(M)/\tau(E(M))$ and consider the following diagram:

$$
\begin{array}{ccc}
A & \overset{id}{\hookrightarrow} & B \\
& & \downarrow f \\
E(M) & \xrightarrow{\eta} & K \quad \rightarrow 0.
\end{array}
$$

To prove that A is τ-$E(M)$-projective, we need to show that f can be lifted to an R-homomorphism from A into $E(M)$. Since K is injective, f extends to an R-homomorphism $\theta : B \longrightarrow K$. Since B is τ-$E(M)$-projective, there exists $\lambda : B \longrightarrow E(M)$ such that $\eta \circ \lambda = \theta$. Now the homomorphism $g = \lambda \circ id : A \longrightarrow E(M)$ is a lifting of f since $\eta \circ g(x) = \eta \circ \lambda \circ id(x) = \eta \circ \lambda(x) = \theta(x) = f(x)$ for all $x \in A$. $\qquad \square$

Corollary 3.8. *For an injective right R-module M, the following statements are equivalent:*

   (1) Every submodule of a τ-M-projective right R-module is τ-M-projective.
   (2) Every submodule of a projective right R-module is τ-M-projective.
   (3) Every right ideal of R is τ-M-projective.
   (4) Every factor module of $M/\tau(M)$ is injective.

Remark 3.9. *Note that if a right R-module M is E-projective for every injective right R-module E, then M is projective (see [11, Exercise 2.16]). This not true for strongly rad-projective modules. In fact, if $R = \mathbb{Z}$ and $M = \mathbb{Q}$, then M is rad-E-projective since $radE = E$ for every injective R-module E. However, M is not strongly rad-projective.*

Lemma 3.10. *If $\tau(N) = \tau(E(N))$ for every right R-module N, then a right R-module M is strongly τ-projective if and only if M is τ-E-projective for each injective right R-module E.*

Proof. Suppose M is τ-E-projective for each injective right R-module E, and let $g : N \to K$ be an R-homomorphism of right R-modules with $\tau(N) \hookrightarrow \ker g$. Consider the following diagram:

$$
\begin{array}{ccc}
 & & M \\
 & & \downarrow f \\
N & \xrightarrow{g} & K & \to 0.
\end{array}
$$

We need to show that f can be lifted to a map $\lambda : M \to N$ with $g\lambda = f$. Let $id : N \to E(N)$ be the inclusion map and $< \alpha, \beta >$ the pushout of $< g, id >$ with $D = (E(N) \oplus K)/U$ where $U =: \{(id(n), -g(n)) : n \in N\}$, $\alpha(x) = (x, 0) + U$ and $\beta(k) = (0, k) + U$ for each $x \in E(N)$ and $k \in K$.

$$
\begin{array}{ccc}
 & & M \\
 & & \downarrow f \\
N & \xrightarrow{g} & K & \to 0 \\
\curvearrowright\downarrow id & & \downarrow \beta \\
E(N) & \xrightarrow{\alpha} & D
\end{array}
$$

Inasmuch as g is epic, we infer that α is epic. We claim that $\tau(E(N)) \hookrightarrow \ker \alpha$. To see this, let $s \in \tau(E(N)) = \tau(N)$. Since $\tau(N) \hookrightarrow \ker g$, $g(s) = 0$ and $(s, 0) = (id(s), -g(s)) \in U$. Thus $\alpha(s) = (s, 0) + U = U = \bar{0}$ and so $\tau(E(N)) \hookrightarrow \ker \alpha$, proving the claim. Now since M is τ-$E(N)$-projective, there exists $\lambda : M \longrightarrow E(N)$ with $\alpha\lambda = \beta f$.

If $x \in M$, then $(\lambda(x), 0) + U = \alpha\lambda(x) = \beta f(x) = (0, f(x)) + U$, and so $(\lambda(x), -f(x)) \in U$. Consequently there exists $y \in N$ such that $(\lambda(x), -f(x)) = (id(y), -g(y)) = (y, -g(y)$. Thus $\lambda(x) = y$ and $f(x) = g(y)$, and so $g\lambda(x) = g(y) = f(x)$. Since x was an arbitrary element of M, $g\lambda = f$ as required. The converse is clear. $\square$

Corollary 3.11. *A right R-module M is strongly soc-projective if and only if M is soc-E-projective for each injective right R-module E.*

Theorem 3.12. *If $\tau(N) = \tau(E(N))$ for every right R-module N, then the following conditions are equivalent:*

(1) *Every submodule of a strongly τ-projective right R-module is strongly τ-projective.*

(2) *Every submodule of a projective right R-module is strongly τ-projective.*

(3) *Every right ideal of R is strongly τ-projective.*

(4) *If M is an injective module, then every homomorphic image of $M/\tau(M)$ is injective.*

Proof. $1 \Rightarrow 2 \Rightarrow 3$. Clear.

$3 \Rightarrow 4$. Le M_R be an injective module, K a submodule of M, I a right ideal of R and consider the following diagram:

$$
\begin{array}{ccc}
I & \hookrightarrow R \\
\downarrow f \\
M & \xrightarrow{\eta} M/K & \to 0
\end{array}
$$

To prove that M/K is injective, it is enough by the Baer criterion to show that f has an extension from R into M/K. Since I_R is srongly τ-projective, there is a homomorphism $\theta : I \longrightarrow M$ such that $\eta\theta = f$. Since M is injective, there exists $\lambda : R \longrightarrow M$ such that $\lambda|_I = \theta$. Since $\eta\lambda|_I = \eta\theta = f$, $\eta\lambda : R \longrightarrow M/K$ is the required extension of f.

$4 \Rightarrow 1$. Let M be a strongly τ-projective module and L a submodule of M. To prove that L is strongly τ-projective it is enough by Lemma 3.10 to prove that L is τ-E-projective for every injective right R-module E. Now consider the following diagram:

$$
\begin{array}{ccc}
L & \overset{id}{\hookrightarrow} M \\
\downarrow f \\
E & \xrightarrow{\eta} K & \to 0
\end{array}
$$

with K a homomorphic image of $E/\tau(E)$. We need to prove that f can be lifted to an R-homomorphism $g : L \longrightarrow E$. By the hypothesis, K is injective and so f can be extended to an R-homomorphism $\theta : M \longrightarrow K$ with $\theta|_L = f$. Since M is τ-E-projective, there exists $\lambda : M \longrightarrow E$ such that $\eta\lambda = \theta$. Now the homomorphism $g = \lambda \circ id : L \longrightarrow E$ is the required lifting of f since $\eta \circ g(x) = \eta \circ \lambda \circ i(x) = \eta \circ \lambda(x) = \theta(x) = f(x)$ for all $x \in L$. $\qquad\square$

Corollary 3.13. *The following conditions are equivalent:*

(1) Every submodule of a strongly soc-projective right R-module is strongly soc-projective.

(2) Every submodule of a projective right R-module is strongly soc-projective.

(3) Every right ideal of R is strongly soc-projective.

(4) If M is an injective module, then every homomorphic image of $M/socM$ is injective.

Proposition 3.14. *Let M and N be right R-modules. If $N/\tau(N)$ is semisimple, then the following are equivalent:*

(1) M is τ-N-projective.

(2) Every R-homomorphism $f : M \to N/\tau(N)$ can be lifted to an R-homomorphism $g : M \to N$ such that $f = \eta \circ g$, where $\eta : N \to N/\tau(N)$ is the natural R-epimorphism.

Proof. $1 \Rightarrow 2$. Clear.

$2 \Rightarrow 1$. Consider the the following diagram:

$$
\begin{array}{ccc}
 & M & \\
 & \downarrow f & \\
N \stackrel{\pi}{\longrightarrow} & K & \to 0
\end{array}
$$

with $\tau(N) \hookrightarrow \ker \pi$. We need to show that f can be lifted to an R-homomorphism $\lambda : M \to N$ with $\pi\lambda = f$. If $\eta : N \to N/\tau(N)$ is the natural R-epimorphism, then the map π induces an R-epimorphism $\phi : N/\tau(N) \to K$ such that the following diagram is commutative:

$$
\begin{array}{ccc}
 & K & \\
\pi \nearrow & \uparrow \varphi & \\
N \stackrel{\eta}{\longrightarrow} & N/\tau(N) & \to 0.
\end{array}
$$

Since $N/\tau(N)$ is semisimple, φ splits and there exists an R-homomorphism $g : K \to N/\tau(N)$ with $\varphi g = 1_K$. Now consider the following diagram:

$$
\begin{array}{ccc}
 & M & \\
\exists\lambda \swarrow & \downarrow f & \\
N \stackrel{\pi}{\longrightarrow} & K & \to 0 \; . \\
\eta \searrow & \varphi \uparrow\downarrow g & \\
 & N/\tau(N) &
\end{array}
$$

By (2), there exists an R-homomorphism $\lambda : M \to N$ such that $\eta\lambda = gf$. Now λ is the required map since $\pi\lambda = \varphi\eta\lambda = \varphi g f = 1_K f = f$. $\square$

Corollary 3.15. *If $R/\tau(R)$ is semisimple and M is a right R-module, then the following conditions are equivalent:*

(1) M is strongly τ-projective.

(2) For every right R-module N, every R-homomorphism $f : M \to N/\tau(N)$ can be lifted to an R-homomorphism $g : M \to N$ such that $f = \eta \circ g$, where $\eta : N \to N/\tau(N)$ is the natural R-epimorphism.

Corollary 3.16. *If $R/\tau(R)$ is semisimple and M is a right R-module with $M = radM$, then M is strongly τ-projective.*

Proof. Let N be a right R-module and consider the following diagram:

$$
\begin{array}{ccc}
 & M & \\
 & \downarrow f & \\
N \stackrel{\eta}{\longrightarrow} & N/\tau(N) & \to 0
\end{array}
$$

224 ISMAIL AMIN, YASSER IBRAHIM, AND MOHAMED YOUSIF

where η is the natural R-epimorphism. Since $R/\tau(R)$ is semisimple, $N/\tau(N)$ is semisimple, and so $f(M) = f(radM) \subseteq rad(N/\tau(N)) = 0$. Clearly f can be lifted to an R-homomorphism from M into N. By Corollary 3.15 M is strongly τ-projective. $\qquad\square$

Example 3.17. *If $R = \mathbb{Z}_{(p)}$ is the localization of $\mathbb{Z}$ at the prime ideal generated by p, then the field of fractions of R is the field of rational numbers $\mathbb{Q}$. By Corollary 3.16, since $R/\delta(R)$ is a division ring and $\mathbb{Q}_R$ has no maximal submodules, $\mathbb{Q}_R$ is strongly δ-projective module which is not projective (since projective modules have maximal submodules).*

Example 3.18. *In general, if R is a discrete valuation ring, i.e. a principal ideal domain with exactly one non-zero maximal ideal, and K is its quotient field (field of fractions), then K_R as a right R-module has no maximal submodules. For, if M is the unique maximal right ideal of R, write $M = xR$ for some $x \in R$. It can be shown that the R-submodules of K are 0, K and $x^i R$, $i \in \mathbb{Z}$, from which we can easily infer that $rad(K_R) = K$. Since R is a local ring, it follows from Corollary 3.16 that K_R is strongly δ-projective which is not projective. More specifically, if k is a field and $R = k\,[[x]]$ is the formal power series with one indeterminate variable x, and K is its quotient field, then K is strongly δ-projective that is not projective.*

Example 3.19. *If R is a discrete valuation ring and K is its quotient field, then both R_R and K_R are strongly δ-projective right R-modules. By Corollary 2.6, $M =: R \oplus K$ is a strongly δ-projective right R-module that is not projective with $M \neq radM$.*

Recall that the right ideal $T_R(M) = \sum\{\mathrm{Im}(f) : f \in Hom_R(M, R)\}$ is called the trace ideal of M in R, and a right R-module M is a generator if $T_R(M) = R$. The next propostion is an extension of a well know result on projective modules.

Proposition 3.20. *If $\tau(R) \subseteq \delta(R)$ and M_R is a τ-projective module, then the following conditions are equivalent:*

(1) M is a generator.
(2) M generates every simple right R-module.
(3) $Hom_R(M, S) \neq 0$ for all simple right R-modules S.

Proof. $1 \Rightarrow 2 \Rightarrow 3$. Clear.

$3 \Rightarrow 1$. Assume to the contrary that $T_R(M) \neq R$, and let A be a maximal right ideal containing $T_R(M)$. By the hypothesis, let $f : M \longrightarrow R/A$.

be a non-zero R-homomorphism and consider the following diagram:

$$M$$
$$\downarrow f$$
$$R \xrightarrow{\eta} R/A \to 0$$

where $R \xrightarrow{\eta} R/A \to 0$ is the canonical quotient map. If R/A is a singular module, then $\delta(R/A) = 0$. Therefore $\eta(\tau(R)) \subseteq \eta(\delta(R)) \subseteq \delta(R/A) = 0$, and $\tau(R) \subseteq \ker(\eta)$. By the τ-projectivity of M, the above diagram can be completed. If R/A is a nonsingular module, then R/A is projective and $A \subseteq^{\oplus} R$. In this case also the above diagram can be completed. Let $\lambda : M \longrightarrow R$ be an R-homomorphism with $\eta\lambda = f$. Since $Im\lambda \subseteq T_R(M) \subseteq A = \ker(\eta)$; $f = \eta\lambda = 0$, a contradiction. Thus $T_R(M) = R$, and M is a generator. $\square$

A right R-module M is called a small module if $M \overset{small}{\hookrightarrow} E(M)$.

Proposition 3.21. *If $\tau(R) \subseteq \delta(R)$, then the following conditions are equivalent:*

(1) Every simple singular right R-module is injective.

(2) Every small submodule of a right R-module is projective.

(3) Every small submodule of a right R-module is τ-projective.

(4) Every small right R-module is τ-projective.

Proof. $1 \Rightarrow 2$. Let K be a small submodule of a right R-module M. If $K = 0$, there is nothing to prove. Otherwise, suppose $0 \neq x \in K$ and let A be a maximal submodule of xR. We claim that xR/A is non-singular, otherwise if xR/A is singular, then by the hypothesis xR/A is a direct summand of M/A. But, since K is a small submodule of M, xR/A is a small submodule of M/A, and so $xR/A = 0$, a contradiction. Therefore xR/A is non-singular and hence projective. Thus every maximal submodule A of xR is a summand of xR, and hence xR, and consequently K is semisimple. Write $K = \oplus_{i \in I} K_i$ with each K_i a simple submodule of K. If, for some $i \in I$, K_i is singular, then by the hypothesis, K_i is injective and hence a summand of M. This implies that $K_i = 0$, since K is a small submodule of M. Thus each K_i, $i \in I$, and consequently K, is non-singular and hence projective.

$2 \Rightarrow 3 \Rightarrow 4$. Clear.

$4 \Rightarrow 1$. Let M be a simple singular right R-module. If M is small in its injective hull $E(M)$, then by the hypothesis M is τ-projective. We claim that M is projective. To see this, consider the following diagram:

$$M$$
$$\exists h \swarrow \quad \downarrow id$$
$$R \xrightarrow{g} M \to 0.$$

Since $g(\tau(R)) \subseteq g(\delta(R)) \subseteq \delta(M) = 0$, $\tau(R) \subseteq \ker(g)$. By the τ-projectivity of M, there exists an R-homomorphism $h : M \longrightarrow R$ such that $gh = id$, and so g splits and M is projective. This is a clear contradiction since M is singular. Thus M is not small in $E(M)$, and $E(M) = K + M$ for a proper submodule K of $E(M)$. Clearly, $K \cap M = 0$, and so M is a direct summand of $E(M)$, and hence injective. $\square$

4. τ-Projective Covers

Lemma 4.1. *If $\tau(R) \subseteq \delta(R)$, then every simple τ-projective right R-module is projective.*

Proof. Let S be a simple right R-module. If S is nonsingular, then S is projective. If S is singular and $\eta : R \to S$ is the canonical quotient map, then $\tau(R) \subseteq \delta(R) \subseteq \ker \eta$. Since S is τ-projective, η splits and so S is projective. $\square$

Lemma 4.2. *Let $\bar{R} = R/J(R)$.*

   (1) If M_R is τ-projective , then $(M/radM)_{\bar{R}}$ is τ-projective.
   (2) If $\tau = \delta$, soc or rad, M_R is τ-projective and $(M/radM)_R$ is simple, then $(M/radM)_{\bar{R}}$ is projective.

Proof. 1. Let $g : \bar{R}_{\bar{R}} \to K_{\bar{R}}$ be an $\bar{R}$-epimorphism with $\tau(\bar{R}_{\bar{R}}) \subseteq \ker(g)$ and $f : (M/radM)_{\bar{R}} \to K_{\bar{R}}$ an $\bar{R}$-homomorphism. We need to show that f can be lifted to an $\bar{R}$-homomorphism from $(M/radM)_{\bar{R}}$ into $\bar{R}_{\bar{R}}$. Consider the following diagram in mod-R:

$$
\begin{array}{ccc}
 & M & \\
 & \downarrow \eta & \\
 & M/radM & \\
 & \downarrow f & \\
\bar{R}_{\bar{R}} & \overset{g}{\longrightarrow} & K_{\bar{R}} \quad \to 0
\end{array}
$$

with $\eta : M \longrightarrow M/radM$ the canonical quotient map. By (1) of Proposition 2.4, M_R is τ-$\bar{R}$-projective and so there exists an R-homomrphism $\lambda : M_R \longrightarrow \bar{R}_R$ such that $g\lambda = f\eta$. Since $\lambda(radM) \subseteq rad(\bar{R}) = 0$, there exists an R-homomorphism $\gamma : (M/radM)_R \longrightarrow \bar{R}_R$ such that $\gamma\eta = \lambda$. Since $g\gamma\eta = f\eta$ and η is an epimorphism, $g\gamma = f$. Since γ can be viewed as an $\bar{R}$-homomorphism with $g\gamma = f$, we infer that $(M/radM)_{\bar{R}}$ is τ-$\bar{R}$-projective.

2. Clearly if $\tau = \delta$, soc or rad, then $\tau(\bar{R}) \subseteq \delta(\bar{R})$. Now the result follows from (1) and Lemma 4.1. $\square$

Definition 4.3. *Let R be a ring and Ω a class of right R-modules which is closed under isomorphisms. An R-homomorphism $\phi : P \to M$ is called an Ω-cover of the right R-module M, if $P \in \Omega$ and ϕ is an epimorphism* with ker $\phi \overset{small}{\hookrightarrow} P$. In particular, if Ω is the class of (strongly) τ-projective right R-modules, the R-homomorphism $\phi : P \to M$ is called (strongly) τ-projective cover of M.

It is well-known that a ring R is semiperfect if and only if every simple R-module has a projective cover. In the next theorem we show that if R is a ring with $\tau = \delta$, soc or rad, then R is semiperfect if and only if every simple R-module has a τ-projective cover.

Theorem 4.4. *If $\tau = \delta$, soc or rad, then the following statements are equivalent:*

(1) R is semiperferct.

(2) Every finitely generated right R-module has a strongly τ-projective cover.

(3) Every finitely generated right R-module has a τ-projective cover.

(4) Every finitely generated right R-module has a τ-quasi-projective cover.

(5) Every 2-generated right R-module has a τ-quasi-projective cover.

(6) Every simple right R-module has a τ-projective cover.

Proof. $1 \Rightarrow 2 \Rightarrow 3 \Rightarrow 4 \Rightarrow 5$. Clear.

$5 \Rightarrow 6$. Let M be a simple right R-module and $g : R \to M$ the natural R-epimorphism. By the hypothesis, let $f : P \longrightarrow R \oplus M$ be a τ-quasi-projective cover of the right R-module $R \oplus M$ and set $K =: \ker \pi f$, where $\pi : R \oplus M \longrightarrow R$ is the canonical projection map. By the projectivity of R, the following diagram:

$$
\begin{array}{ccc}
 & & R \\
\exists \lambda \swarrow & & \downarrow id \\
P & \overset{\pi f}{\longrightarrow} & R \quad \to 0
\end{array}
$$

can be completed by an R-homomorphism $\lambda : R \longrightarrow P$ with $\pi f \lambda = id$. Since $P = \operatorname{Im} \lambda \oplus K$ and λ is a monomorphism, $P \simeq R \oplus K$ and $R \oplus K$ is τ-quasi-projective. Therefore by (6) of Proposition 2.4, K is τ-R-projective, i.e. K is τ-projective. If $\theta = f|_K : K \longrightarrow R \oplus M$ is the restriction of f to K, then θ is an R-homomorphism from K onto M. Clearly θ is an R-homomorphism from K into M. To see that it is onto, let $m \in M$. Now $m = f(p)$ for some $p \in P$, and if we write $p = \lambda(r) + k$ where $r \in R$ and $k \in K$, then $m = f(p) = f\lambda(r) + f(k)$, and so $0 = \pi(m) = \pi f \lambda(r) + \pi f(k) = r$. Therefore $m = f(k) = \theta(k)$. Next, we show that $\ker \theta$ is small in K. To see this, let $N + \ker \theta = K$ for some submodule $N \subseteq K$. Since $P = \operatorname{Im} \lambda \oplus K = (\operatorname{Im} \lambda \oplus N) + \ker \theta = (\operatorname{Im} \lambda \oplus N) + \ker f$ and $\ker f$ small in P, $P = \operatorname{Im} \lambda \oplus N$. If $x \in K \subseteq P = \operatorname{Im} \lambda \oplus N$, write $x = y + n$ where $y \in \operatorname{Im} \lambda$ and $n \in N \subseteq K$.

Thus $x - n = y \in K \cap \operatorname{Im}\lambda = 0$, and so $x = n \in N$, $N = K$ and $\theta : K \longrightarrow M$ is a τ-projective cover of M.

$6 \Rightarrow 1$. We first show that $\bar{R} = R/J(R)$ is a semisimple ring. Let S be a simple right $\bar{R}$-module. By the hypothesis, S_R has a τ-projective cover P_R, say $f : P \longrightarrow S$ with $radP = \ker f =: K \overset{small}{\hookrightarrow} P$. Since $P/radP \cong S$ is simple, we infer from Lemma 4.2 that $(P/radP)_{\bar{R}}$ is projective. Consequently $S_{\bar{R}}$ is projective, and so $\bar{R}$ is semisimple. Since $J(R) \subseteq \delta(R)$, there is an R-epimorphism $R/J(R) \longrightarrow R/\delta(R)$ and so $R/\delta(R)$ is semisimple. Write $R/\delta(R) = \oplus_{i=1}^{n} K_i$ with each K_i is simple as a right R-module. Let L_i be a τ-projective cover of K_i, $1 \leq i \leq n$, as right R-modules.

Now, let K be an arbitrary simple right R-module. We need to show that K_R has a projective cover. If K is non-singular, then K is projective. Suppose that K is singular. Since $K\delta(R) = 0$, K is simple as an $R/\delta(R)$-module. Therefore $K \cong K_i$ for some i, $1 \leq i \leq n$, and so in order to prove that R is a semiperfect ring, we only need to show that each K_i has a projective cover. It is enough to show that each $L_i, 1 \leq i \leq n$, is projective as a right R-module. Clearly $L = \oplus_{i=1}^{n} L_i$, as a right R-module, is a τ-projective cover of $R/\delta(R)$. Consider the following diagram:

$$\begin{array}{ccc} & & L_R \\ \exists g \swarrow & & \downarrow f \\ R_R \overset{\eta}{\longrightarrow} & R/\delta(R) & \to 0 \end{array}$$

with f being the τ-projective cover of $R/\delta(R)$, and η the canonical R-epimorphism. Since L is τ-projective and $\tau(R) \subseteq \delta(R) = \ker\eta$, f can be lifted to a map $g : L \longrightarrow R$ such that $\eta g = f$. Since $R = \operatorname{Im} g + \ker\eta = \operatorname{Im} g + \delta(R)$ and $\delta(R) \subseteq^{\delta} R$, we infer from Lemma 3.2 that $R = \operatorname{Im} g \oplus Y$, where Y is a semisimple projective submodule of $\delta(R)$. Therefore $\operatorname{Im} g$ is projective, and so $L = \ker g \oplus B$ for a submodule B of L with $B \cong \operatorname{Im}(g)$. Since $\ker g \subseteq \ker f \overset{small}{\hookrightarrow} L$, we infer that $L_R = B \cong \operatorname{Im} g$ is projective. Therefore each L_i is a projective right R-module as required. $\qquad \square$

A right R-module M is R-projective, if it is projective relative to the right R-module R_R. An R-homomorphism $\phi : P \to M$ is called an R-*projective cover* of the right R-module M, if P is R-projective, ϕ is an epimorphism, and $\ker(\phi)$ is small in P. Since R-projective modules are τ-projective, as an immediate consequence of Theorem 4.4, the next corollary provides new characterizations of semiperfect rings.

Corollary 4.5. *The following statements are equivalent:*

(1) R is semiperfect.

(2) Every 2-generated right R-module has a quasi-projective cover.

(3) Every 2-generated right R-module has a rad-quasi-projective cover.

(4) Every 2-generated right R-module has a soc-quasi-projective cover.

(5) Every 2-generated right R-module has a δ-quasi-projective cover.

(6) Every simple right R-module has an R-projective cover.

(7) Every simple right R-module has a rad-projective cover.

(8) Every simple right R-module has a soc-projective cover.

(9) Every simple right R-module has a δ-projective cover.

Lemma 4.6. *Let $\tau(R) \subseteq \delta(R)$ and M be a τ-projective right R-module. If $M = A + B$ where A is a maximal submodule of M and M/A has a projective cover $f : P \longrightarrow M/A$, then $M = Q \oplus T$ where $Q \subseteq B$ and $Q \cong P$ is cyclic indecomposable projective module.*

Proof. Clearly P is cyclic and indecomposable and $\ker f = radP$ is maximal and small in P. Consider the following diagram:

$$
\begin{array}{c}
M \\
\downarrow g \\
P \xrightarrow{\ f\ } M/A \quad \to 0
\end{array}
$$

with g the natural R-epimorphism. Since $\tau(R) \subseteq \delta(R)$, P is projective and any preradical commutes with direct sums, we may assume without loss of generality that $\tau(P) \subseteq \delta(P)$. We claim that the map g can be lifted to an R-homomorphism from M into P. If M/A is singular, then $f(\tau(P)) \subseteq f(\delta(P)) \subseteq \delta(M/A) = 0$. Therefore $\tau(P) \subseteq \ker f$. Since M is τ-projective, the map g can be lifted as claimed. If M/A is nonsingular, then M/A is projective and the epimorphism $P \xrightarrow{\ f\ } M/A \to 0$ splits. Since $\ker f \overset{small}{\hookrightarrow} P$, $\ker f = 0$ and f is an isomorphism, proving the claim. Now, let $\lambda : M \longrightarrow P$ be an R-homomorphism such that $f\lambda = g$, and consider the following commutative diagram:

$$
\begin{array}{c}
B \\
\lambda|_B \swarrow \qquad \downarrow g|_B \\
P \xrightarrow{\ f\ } \quad M/A \quad \to 0.
\end{array}
$$

Since $\mathrm{Im}(\lambda|_B) + \ker f = P$ and $\ker f \overset{small}{\hookrightarrow} P$, we infer that $\lambda|_B : B \longrightarrow P$ is epic. By the projectivity of P, $\lambda|_B$ splits, and so there exists a monomorphism $\theta : P \longrightarrow B$ such that $\lambda\theta = 1$. Therefore $M = Q \oplus \ker \lambda$ with $Q =: \theta(P)$. Clearly $Q \cong P$ is cyclic indecomposable projective module. □

With the help of Lemma 4.6 above and an argument similar to the one provided by Ketkar and Vanaja in [10, Theorem 1], we can establish the next theorem.

Theorem 4.7. *Let R be a semiperfect ring with $\tau(R) \subseteq \delta(R)$. If M_R is a τ-projective module with small radical, then M_R is projective.*

Proof. Let M be a τ-projective right R-module with $radM \overset{small}{\hookrightarrow} M$. Let $x \in M$ with $x \notin radM$. If A is a maximal submodule of M with $x \notin A$, then $M = xR + A$. By Lemma 4.6, M/A has a cyclic indecomposable projective cover $f : P \longrightarrow M/A$ with $M = P \oplus N_1$ and $P \subseteq xR$. Write $xR = y_1R \oplus x_1R$ where $x = y_1 + x_1$, $P = y_1R$ and $x_1R = xR \cap N_1$. By (2) of Corollary 2.6, N_1 is a τ-projective module. Clearly $radN_1 \overset{small}{\hookrightarrow} N_1$. If $x_1 \notin radN_1$, then by dublicating the preceding argument we can write $N_1 = y_2R \oplus N_2$ with y_2R cyclic indecomposable projective direct summand of M contained in x_1R and $x_1 = y_2 + x_2$ such that $x_1R = y_2R \oplus x_2R$ where $x_2R = x_1R \cap N_2$. We claim that this process can be repeated only for finitely many times. Otherwise, we obtain an infinite direct sum $y_1R \oplus y_2R \oplus \cdots \oplus y_nR \oplus \cdots$ inside xR such that for each n, $y_1R + y_2R + \cdots + y_nR$ is a cyclic projective module generated by $y_1 + y_2 + \cdots + y_n$.

For each n, define $g_n : R \longrightarrow (y_1 + \cdots + y_n)R$ by $g_n(1) = y_1 + \cdots + y_n$. Each of these maps is a split epimorphism with $\ker g_n = ann_R(y_1 + \cdots + y_n)$. Clearly $\ker g_1 \supseteq \ker g_2 \supseteq \cdots \supseteq \ker g_n \supseteq \cdots$ is a descending chain of direct summands of R. Since R is semiperfect, we infer that $\ker g_n = \ker g_{n+1}$ for some $n \geq 1$. Therefore $y_1R \oplus \cdots \oplus y_nR \cong y_1R \oplus \cdots \oplus y_nR \oplus y_{n+1}R$, a contradiction since each y_iR is a non-zero indecomposable module. Now, let

$$F = : \{y|\ 0 \neq y \in M \text{ and } yR \text{ is cyclic indecomposable}$$
$$\text{projective direct summand of } M\}.$$

Then the preceding argument together with the fact that $radM \overset{small}{\hookrightarrow} M$ shows that $M = \sum_{y \in F} yR$. Let Ω be the family of subsets B of F satisfying the conditions:

(a) $\sum_{y \in B} yR = \bigoplus_{y \in B} yR$ and

(b) For $y_1, \cdots, y_n \in B$, $y_1R + \cdots + y_nR$ is a direct summand of M.

Clearly Ω is a non-empty set which can be partially ordered by the usual inclusion. By Zorn's lemma, let B_0 be a maximal element in Ω. Then $T =: \sum_{y \in B_0} yR = \bigoplus_{y \in B_0} yR$ is a projective right R-module. We claim that $T = M$. Since $M = \sum_{y \in F} yR$ and $radM \overset{small}{\hookrightarrow} M$, it suffices to prove that $F \subseteq T + rad(M)$. Let $y \in F$. We consider two cases:

Case (1). $T \cap yR = 0$. Then $B_0 \subsetneq B_0 \cup \{y\} \subseteq F$. By the maximality of B_0 we can find $y_1, \cdots, y_n$ in B_0 such that $(U =: y_1 R \oplus \cdots \oplus y_n R) \oplus yR$ is not a direct summand of M. By condition (b) on B_0, we can write $M = U \oplus E$ for a submodule $E \subseteq M$. Then $U \oplus yR = U \oplus ((U \oplus yR) \cap E)$, and so $(U \oplus yR) \cap E \cong yR$. Let $(U \oplus yR) \cap E = zR$. Then zR is a cyclic indecomposable projective submodule of E and zR cannot be a direct summand of E. Now the above argument coupled with the fact that E is τ-projective, being a direct summand of M, shows that $z \in rad(E) \subseteq rad(M)$. Therefore $y \in T + rad(M)$, proving our claim in this case.

Case (2). $T \cap yR \neq 0$. If $y \in T$, we are done. Assume that $y \notin T$. Let $0 \neq yr = t \in T \cap yR$. Since yR is a non-zero projective module, the right annihilator of y in R, denoted by $r_R(y)$, is a direct summand of R. Let $r_R(y) = eR$, for some $e^2 = e \in R$. Choose a finite subset $B \subseteq B_0$ such that $t \in \sum_{z \in B} zR$. Then $\sum_{z \in B} zR$ is a direct summand of M. Let $\lambda : M \longrightarrow \sum_{z \in B} zR$ be the natural projection. If $\acute{y} = \lambda(y)$, then $(y - \acute{y})e = 0$. We also have $(y - \acute{y})r = 0$ since $\acute{y}r = \lambda(y)r = \lambda(yr) = \lambda(t) = t = yr$. Thus $r_R(y) \subsetneq r_R(y - \acute{y})$. We claim that $(y - \acute{y})R$ does not contain any non-zero projective summand. Assume to the contrary N is a non-zero projective summand of $(y - \acute{y})R$. Since $r_R(y) \subsetneq r_R(y - \acute{y})$ we can define an R-epimorphism $f : yR \longrightarrow (y - \acute{y})R$ by $f(y) = (y - \acute{y})$. Let $g : (y - \acute{y})R \longrightarrow N$ be the natural projection map. Then $gf : yR \longrightarrow N$ is a split R-epimorphism. Since yR is indecomposable, gf is an isomorphism. Therefore $r_R(y - \acute{y}) \subseteq r_R(y)$, a contradiction. This proves our claim. It follows that $y - \acute{y} \in radM$, and so $y \in T + rad(M)$. Thus $M = T$ is projective. $\qquad\square$

Since every right R-module over a right perfect ring has a small radical, the next result is an immediate consequence of Theorem 4.7.

Corollary 4.8. *Over a right perfect ring R with $\tau(R) \subseteq \delta(R)$, every τ-projective right R-module is projective.*

A ring R is called *quasi-Frobenius* (*QF-ring*) if R is right (or left) artinian right (or left) self-injective. Equivalently [8], R is QF-ring if every projective right R-module is injective. Since every QF-ring is right (and left) perfect ring, the next result is now an immediate consequence of Corollary 4.8.

Corollary 4.9. *If R is a ring with $\tau(R) \subseteq \delta(R)$, then R is quasi-Frobenius if and only if every τ-projective right R-module is injective.*

Remark 4.10. *It is also well-known that R is QF iff every injective right R-module is projective, see [8]. Such a result can not be extended to strongly δ-projective modules. In fact, if $p_i \in \mathbb{Z}, 1 \leq i \leq 2$, are two* *distinct prime numbers, and $R =: \left\{ \frac{m}{n} : m, n \in \mathbb{Z} \text{ and } p_i \nmid n \right\}$, then R is a commutative, semilocal domain such that $M = radM = \delta(M)$, where M is any injective R-module. Now, since R is semilocal, it follows that every injective R-module is strongly δ-projective. To see this, consider the following diagram:*

$$M$$
$$\downarrow f$$
$$L \xrightarrow{\eta} K \rightarrow 0$$

with K a homomorphic image of $L/\delta(L)$. Since $radL \subseteq \delta(L)$ and R is semilocal, $rad(K) = 0$ and the only map from M into K is the trivial map. This means every such diagram can be completed and so M is strongly δ-projective. However R is not a perfect ring, and hence not quasi-Frobenius.

Theorem 4.11. *If $\tau = \delta$, soc or rad, then the following statements are equivalent:*

(1) R is right perfect.

(2) Every right R-module has a strongly τ-projective cover.

(3) Every right R-module has a τ-projective cover.

(4) Every semisimple right R-module has a strongly τ-projective cover.

(5) Every semisimple right R-module has a τ-projective cover.

Proof. $1 \Rightarrow 2 \Rightarrow 4 \Rightarrow 5$ and $1 \Rightarrow 3 \Rightarrow 5$. Clear.

$5 \Rightarrow 1$. By Theorem 4.4, R is a semiperfect ring. Now, let M be a semisimple right R-module and $f : P \longrightarrow M$ a τ-projective cover of M. Clearly $rad(P) = ker(f) \overset{small}{\hookrightarrow} P$, and so by Theorem 4.7, P is projective. Thus every semisimple right R-module has a projective cover, and so R is right perfect. $\square$

As an immediate consequence of Theorem 4.11, the next corollary provides new characterizations of right perfect rings.

Corollary 4.12. *the following statements are equivalent:*

(1) R is right perfect.

(2) Every right R-module has an R-projective cover.

(3) Every semisimple right R-module has an R-projective cover.

(4) Every semisimple right R-module has a rad-projective cover.

(5) Every semisimple right R-module has a soc-projective cover.

(6) Every semisimple right R-module has a δ-projective cover.

It is well-known that a ring R is right perfect if and only if every flat right R-module is projective. In the next theorem we show that if R is a ring with $\tau(R) \subseteq \delta(R)$, then R is right perfect if and only if every flat right R-module is τ-quasi-projective, but first we need the following definition.

Definition 4.13. [6] *Let M be an R-module. A homomorphism g : $F \longrightarrow M$ is called a flat cover of M if the following conditions are satisfied:*

(1) F is a flat right R-module.

(2) For every homomorphism θ : $H \longrightarrow M$ with H a flat right R-module, there exists a homomorphism α : $H \longrightarrow F$ such that $g\alpha = \theta$, i.e.

$$\begin{array}{ccc} & & F \\ \exists\alpha \nearrow & & \downarrow g \\ H & \xrightarrow{\theta} & M \end{array}$$

(3) Every endomorphism λ : $F \longrightarrow F$ with $g\lambda = g$ is an automorphism.

If we choose H in (2) to be a free right R-module and $\theta : H \to M$ to be an R-epimorphism, then it follows easily from condition (2) above that the map g is an epimorphism. Thus every flat cover is an epimorphism. According to [6, Theorem 3], every R-module has a flat cover.

Theorem 4.14. *If R is a ring with $\tau(R) \subseteq \delta(R)$, then the following statements are equivalent:*

(1) R is right perfect.

(2) Every flat right R-module is strongly τ-projective.

(3) Every flat right R-module is τ-quasi-projective.

Proof. $1 \Rightarrow 2 \Rightarrow 3$. Clear.

$3 \Rightarrow 1$. Let A be a semisimple right R-module. Write $A = D \oplus M$ where $D = \oplus_{j \in J} K_j$ and $M = \oplus_{i \in I} K_i$ with each K_j simple nonsingular right R-module and each K_i simple singular right R-module. Clearly D is projective. We need to show that M has a projective cover. Let $f : N \longrightarrow M$ be a flat cover of M, F a free right R-module and g an R-epimorphism from F into M. Since $\tau(R) \subseteq \delta(R)$, we infer that $f(\tau(F) \subseteq f(\delta(F) \subseteq \delta(M) = \delta(\oplus_{i \in I} K_i) = \oplus_{i \in I} \delta(K_i) = 0$, and so $\tau(F) \subseteq \ker g$. Since $T =: N \oplus F$ is flat, it follows from the hypothesis that T is τ-quasi-projective. By (6) of Proposition 2.4, N is τ-F-projective. Now, consider the following diagram:

$$\begin{array}{ccc} & N & \\ \exists\lambda \swarrow \nearrow \exists\theta & \downarrow f & \\ F \xrightarrow{g} & M & \to 0. \end{array}$$

There exists an R-homomorphism $\lambda : N \longrightarrow F$ such that $g\lambda = f$, since $\tau(F) \subseteq \ker g$ and N is τ-F-projective. By (2) of Definition 4.13, since f :

$N \longrightarrow M$ is a flat cover of M, there exists a homomorphism $\theta : F \longrightarrow N$ with $f\theta = g$, and so $f = g\lambda = f\theta\lambda$. By (3) of Definition 4.13, $\theta\lambda$ is an automorphism of N, and so $\theta : F \to N$ is a split epimorphism. Therefore N is projective. Finally, we only need to show that $\ker f \overset{small}{\hookrightarrow} N$; for if this is the case, then $f : N \longrightarrow M$ will be a projective cover of M, and by [15, Theorem B.38 page 272] R will be right perfect. Now, let $L + ker f = N$ for a submodule L of N. Since flat covers are epic, $f|_L : L \longrightarrow M$ is an epimorphism, and by the projectivity of N, there exists an R-homomorphism $h : N \longrightarrow L \subseteq N$ with $fh = f$. By (3) of Definition 4.13, h is an automorphism of N, and so $N = \text{Im}(h) \subseteq L$. Therefore $N = L$ and $\ker f \overset{small}{\hookrightarrow} N$, as required. $\square$

Corollary 4.15. *the following statements are equivalent:*

(1) R is right perfect.

(2) Every flat right R-module is quasi-projective.

(3) Every flat right R-module is rad-quasi-projective.

(4) Every flat right R-module is soc-quasi-projective.

(5) Every flat right R-module is δ-quasi-projective.

Example 4.16. *Strongly τ-projective right R-module need not be flat. For, if p_1 & p_2 are two distinct prime numbers and*

$$R =: \left\{ \frac{m}{n} : m, n \in \mathbb{Z}, \; p_i \nmid n \right\},$$

then R is a commutative semilocal domain such that $E(R/p_iR), 1 \leq i \leq 2$, is a strongly δ-projective R-module which is not flat.

Acknowledgement: *Part of this work was carried out while the second author was visiting the mathematics department of the Ohio State University during the Autumn of 2011. The second author would like to express his gratitude to the members of the mathematics department at OSU for the hospitality and the financial support.*

References

[1] I. Amin, M.F. Yousif and N. Zeyada, *Soc-injective rings and modules*, Comm. Algebra 33 (2005), 4229-4250.

[2] I. Amin, Y. Ibrahim and M.F. Yousif, *Rad-projective and strongly rad-projective Modules*, Comm. Algebra, to appear.

[3] F.W. Anderson and K.R. Fuller, *Rings and Categories of Modules*, Springer-Verlag, Berlin and New York, 1974.

[4] H. Bass, *Finitistic dimension and a homological generalization of semiprimary rings*, Trans. Amer. Math. Soc. 95 (1960), 466-488.

[5] R. Baer, *Abelian groups that are direct summands of every containing abelian group*, Bull. Amer. Math. Soc. 46, (1940). 800–806.

[6] L. Bican, R. El Bashir and E. Enochs, *All modules have flat covers*, Bull. London Math. Soc. 33 (2001), 385-390.

[7] B. Eckmann and A. Schopf, *Über injektive moduln*, Arch. Math. (Basel) 4, (1953), 75–78.

[8] C. Faith and E.A. Walker, *Direct sum representations of injective modules*, J. Algebra 5 (1967), 203-221.

[9] F. Kasch, *Modules and Rings*, L.M.S. Monograph No. 17, Academic Press, New York, 1982.

[10] R.D. Ketkar and N. Vanaja, *R-projective modules over a semiperfect ring*, *Canad. Math. Bull.* 24 (1981), 365-367.

[11] T.Y. Lam, *Exercises in Modules and Rings*, Problem Books in Mathematics, Springer, New York, 2007.

[12] T.Nakayama, *On Frobeniusean algebras I*, Ann. Math. 40 (1939), 611-633.

[13] T. Nakayama, *On Frobeniusean algebras II*, Ann. Math. 42 (1941), 1-21.

[14] T. Nakayama, *On Frobeniusean algebras III*, Japan J. Math. 18 (1942), 49-65.

[15] W.K. Nicholson and M.F. Yousif, *Quasi-Frobenius Rings*, Cambridge Tracts in Math. 158, Cambridge Univ. Press, Cambridge, UK, 2003.

[16] A.Ç. Özcan, D.K. Tütüncü and M.F. Yousif, *On some injective modules in $\sigma[M]$*. Modules and comodules, 313–329, Trends Math., Birkhäuser Verlag, Basel, 2008.

[17] F.L. Sandomierski, *On semiperfect and perfect rings*, Proc. Amer. Math. Soc. 21 (1969), 205-207.

[18] R. Wisbauer, *Foundations of Module and Ring Theory*, Gordon and Breach, Philadelphia, 1991.

[19] Y. Zhou, *Generalization of perfect, semiperfect, and semiregular rings*, Algebra Colloquium 7 (2000), 305-318.

DEPARTMENT OF MATHEMATICS
FACULTY OF SCIENCE
CAIRO UNIVERSITY
GIZA, EGYPT
E-mail address: isamin@live.com

DEPARTMENT OF MATHEMATICS
FACULTY OF SCIENCE
CAIRO UNIVERSITY
GIZA, EGYPT
E-mail address: yfibrahim@aucegypt.edu

DEPARTMENT OF MATHEMATICS
THE OHIO STATE UNIVERSITY
LIMA, OHIO 45804, U.S.A.
E-mail address: yousif.1@osu.edu